The Nature of Nebraska

The Nature
of Nebraska

Ecology and Biodiversity

Paul A. Johnsgard

UNIVERSITY OF NEBRASKA PRESS • LINCOLN AND LONDON

Publication of this volume was assisted by
The Virginia Faulkner Fund,
established in memory of Virginia Faulkner,
editor-in-chief of
the University of Nebraska Press.

Library of Congress
Cataloging-in-Publication Data
Johnsgard, Paul A.
The nature of Nebraska : ecology and biodiversity
/ Paul A. Johnsgard.
p.   cm.
Includes bibliographical references (p.) and index.
ISBN 0-8032-2596-2 (cloth : alk. paper)
1. Ecology—Nebraska
2. Biological diversity—Nebraska
I. Title
QH105.N2 J629 2001
577'.09782—DC21   2001027065

Dedicated to the children of Nebraska,
and especially the fourth-grade class
at Elwood Public School

Contents

Illustrations

Maps

Tables

Preface

If you plan for one year, plant rice. If you plan for 10 years, plant trees. If you plan for 100 years, educate mankind.

Kuan-Tzu

There is a place in America where East and West merge together as smoothly as one river flows into another. That place is called the Great Plains. There is a river in America that gave sustenance to perhaps a hundred thousand migrants who trudged westward in the mid-nineteenth century along the Mormon and Oregon Trails. That river is called the Platte. There is a vast region of sandy grasslands in America that represents the largest area of dunes and the grandest and least disturbed region of tallgrass prairies in all of the Western Hemisphere. That region is called the Sandhills. There is an underground reservoir in America that at maximum may be close to 1,000 feet deep and provides the largest known potential source of unpolluted water to be found anywhere. That reservoir is called the Ogallala aquifer. There is a state in America that offers unhindered vistas of the West, contains stores of vast fossil deposits that shed light on our collective past, and boasts an enlightened citizenry that has built an enviable human history and looks confidently toward the future. That state is called Nebraska.

Nebraska has now been a state for more than 130 years, and I have lived within its borders for almost a third of that entire period. During this time I have learned to love it and its wild places even more than I have loved my native North Dakota. Yet it has taken many years to discover its beauties, which are often little known. I have increasingly realized that, even with my access as a university professor to a variety of books, journals, and other resources that are available to few others living in the state, it is sometimes difficult or nearly impossible to extract even rather simple information on a species or a habitat, to say nothing of trying to understand an entire ecosystem. Furthermore, many of these species and their habitats have literally disappeared before our eyes in my lifetime, and others are in the process of vanishing. Once they are gone, they are gone forever, and tomorrow's Nebraskans will be the poorer for it, even if they may be unaware of their loss. It is hard to mourn something or some creature that one has never been exposed to, but it is even harder to refrain from mourning those whom we have come to know personally and will see no more. There are no longer any wild elk to announce the coming of each prairie autumn in eastern Nebraska with their melancholic bugling calls. The great flocks of Eskimo curlews that arrived each March after a long flight from the Argentine pampas have been replaced by clouds of blackbirds and grackles. The sounds of uncountable bison herds thundering over the Nebraska plains can now only be heard in our imagination, as can the wail of prairie wolves on a star-drenched night in the Sandhills. But the calls of sandhill cranes still spill forth from our skies every spring, and the rattling dances of sharp-tailed grouse on a frosty March sunrise somehow bring to mind the lost cultures and nearly forgotten traditions of the Pawnee, the Otoe, and the Dakota, who knew these birds much better than most of us ever can hope to.

In talking with young schoolchildren and even with university students, it is apparent to me that, rather than filling our newspapers and magazines with unending news of scheduled athletic events and their outcomes, we would do better to take note of when the pasque flowers bloom or when the snow geese materialize each spring and fall. Even more important, we should hope to gain some sense of what the role of the pasque flower and the snow goose might be in the greater scheme of things, whether it may be source of pollen for some obscure insect or as a source of inspiration for an equally obscure poet. We can never know the purpose of all things in nature; it is often impossible in our short lifetimes to understand our own individual roles and possible purpose for living. Probably nonhumans don't even ask themselves these questions, so it is our human responsibility to try to fathom their secrets, and, failing that, to act to preserve these natural treasures long enough for future generations to ask these same questions if they should care to do so.

With respect to the admonition of Kuan-Tzu, I have never planted rice and have planted only a few trees. But I hope that in writing this book I will have planted some lasting ideas in the minds of readers.

Acknowledgments

There is an African saying to the effect that while you are planning a journey you are in control, but once you begin that journey the journey is in control. So too with writing a book; I usually think I know what I am going to need to do when I plan a book, but soon after I start the writing, the book begins to take command and tells me what I must do. Thus, I need only follow its directions. So it was with this enterprise.

I first got the idea for writing this book some years ago, when speaking to JoD Blessing's fourth-grade class in Elwood, Nebraska. It was apparent that, although the students were keenly interested in nature, they had virtually no access to Nebraska-oriented references that would tell them what kinds of animals and plants might occur right around them or what sorts of behavior and habitats might characterize these organisms. I remembered my own small-town North Dakota childhood and how I ached for a book that would at least give me an entry into understanding a little bit of the natural world that was immediately around me.

I have also increasingly realized that many university-level students arrive at college with only the most rudimentary knowledge of ecology and know virtually nothing of the Earth's geologic history, of our own human evolutionary heritage, or even of our own state's geography, ecology, and natural history. This situation reflects our entire educational system, which in our scientifically oriented age will increasingly need to be concerned with providing children a firm grounding in the basic physical and biological sciences. In a recent nationwide survey, Nebraska's schools, as well as those of most other states, received only a mediocre grade in terms of providing training in the principles of biological evolution, our basic tool for understanding life on earth. Partly for this reason, I have included a rather lengthy glossary that includes definitions of many terms that biologists or geologists might consider unnecessary. I have also included in the references not only hundreds of fairly technical journal citations but also a comparable amount of the "popular" literature relating to Nebraska, such as articles from the Game and Park Commission's magazine *Nebraskaland* and similar nontechnical publications of the Nebraska State Museum. I hope that even small-town libraries might thus have access to the latter sources of information.

In preparing this book, I have drawn on half a lifetime of personal experiences across Nebraska and at times also rely on the expertise of many other people. Among these are several employees of the Nebraska Game and Parks Commission. They include Barbara Voeltz, staff librarian, who has invariably provided me with wonderful computerized literature searches, often at very short notice. Several biologists with the Nebraska Game and Parks Commission provided me with useful reports or advice, including Gerry Steinauer. Gerry supplied me with a copy of a then still-unpublished report he and Steve Rolfsmeier produced for the Game and Parks Commission, "Terrestrial Natural Communities of Nebraska," which gave me a method for organizing most of my chapter contents and which has proven so useful that I have extracted and condensed much of its information. I am much in their debt. Another person from Game and Parks deserving thanks is John Farrar. His many articles on Nebraska wildlife and natural communities, and especially his field guide to wildflowers, have been invaluable.

Colleagues in the School of Biological Sciences, especially Robert Kaul and Jim Rosowski, offered their respective expertise on various plant and animal groups. Several Nebraska State Museum personnel, especially Brett Ratcliffe and Tom Labedz, were of similar valuable assistance with questions on invertebrate and vertebrate groups, respectively. Michael Voorhies kindly reviewed my account of Nebraska's geologic history, and Patricia Freeman and Hugh Genoways similarly reviewed and edited my mammal list. Marion Ellis helped out with bee queries. Librarians in the University of Nebraska Biological Sciences library must have tired of locating obscure books and again hauling them back to their shelves after I had returned them, but they never complained. Jackie Canterbury shared with me some of the hard-to-locate literature that she had assembled for her graduate research on Nebraska ecosystems. Photos were offered by Scott Johnsgard, Bill Scharf, Ken Fink, Ron Marquardt, and others; some of these served as a basis for drawings. Amy Richert provided me with base maps of Nebraska (courtesy University of Nebraska Department of Geosciences). Scott Johnsgard, Richard Voeltz, Barbara Voeltz, Linda Brown, Jackie Canterbury, and many others read various sections of the manuscript critically or for readability, and the entire manuscript was reviewed by Charles Brown and an unnamed reviewer. To all these people I offer my sincere thanks.

I also wish to thank the Nature Conservancy for helping to preserve and protect some of Nebraska's most important biological environments and for raising the consciousness level of all Nebraskans to our precious natural heritage. All the royalties from this book have been assigned to the Conservancy. As a final point: in consideration of my intended audience I have used the generally more familiar English system of weights and measurements rather than the metric ones normally used by scientists. All of the drawings in the book are my own, except as indicated.

The Nature of Nebraska

PART ONE

Nebraska and Its Place in the Great Plains

CHAPTER I

The Geology and Landforms of Nebraska

A Legacy of Wind, Water, and Ice

The History of Life in Nebraska: A Time-Travel Adventure

It is difficult for humans, whose lives are measured in years and decades, to fathom the age of the Earth, whose history is patiently but inexorably written on thin pages of landscape, each lasting millions of years or more. As an exercise in Earth-time, let a single mile represent a million years. Thus .5 mile would represent 500,000 years, .10 mile equals 100,000 years, .01 mile (52 feet) equals 10,000 years; .001 mile (5.2 feet) equals 1,000 years, and about 6 inches equal 100 years. A decade would equal about half an inch. It is 450 miles from the 60th Street on-ramp on Interstate 80 (I-80) in downtown Omaha to the westernmost I-80 exit at the Wyoming border, which may be thought of as equal to 450 million years. This period of nearly half a billion years encompasses most of the time that evidence of animal life has been found on Earth, but Earth itself is more than 4 billion years old, or ten times older than the time scale described here, and the earliest known algal fossils are 3.8 billion years old. To start at the approximate age of the planet Earth, we would have to travel about 4,500 miles, rather than 450 miles, to get a sense of the time involved (see table 1 and figure 1). The age of the universe is perhaps three or four times greater, or more than half the distance around the globe at Nebraska's latitude, assuming the rate of 1 mile per 1 million years.

As we join I-80 in Omaha, some 450 million years ago, we are in the middle of the Paleozoic era, the "era of ancient life." Then we would need

Era	Period	Epoch	Years before Present	Major Events (*particularly in Nebraska*)
CENOZOIC	Quarternary	*Holocene*	0–11,000	Postglacial period to present
		Pleistocene	0.01–1.6 MYA*	Several glaciations; Sandhills form; also widespread loess, till, and alluvial deposits
	Tertiary	*Pliocene*	1.6–5.2 MYA	Grasslands spreading; cooler and drier
		Miocene	5.2–23.3 MYA	Ashfall Beds (10 MYA); Agate Fossil Beds (20 MYA)
		Oligocene	23.3–35 MYA	Mountain building in the West; early grasslands and grazing mammals on plains
		Eocene	35–60 MYA	Major modern mammal and bird groups appear
		Paleocene	60–65 MYA	Modern plants appear; Nebraska emerging
MESOZOIC	Cretaceous		65–135 MYA	Last of dinosaurs; Nebraska mostly submerged
	Jurassic		135–197 MYA	Peak of dinosaurs; early birds and mammals
	Triassic		197–225 MYA	Early dinosaurs appear
PALEOZOIC	Permian		225–280 MYA	Cooler and drier; many extinctions worldwide
	Carboniferous		280–345 MYA	Early reptiles appear
	Devonian		345–405 MYA	Seed plants appear
	Silurian		405–425 MYA	First land plants and early amphibians appear
	Ordovician		425–500 MYA	Early fishes appear
	Cambrian		500–570 MYA	Abundant marine life; many invertebrates

*MYA = millions of years ago

Table 1. Geologic Timetable

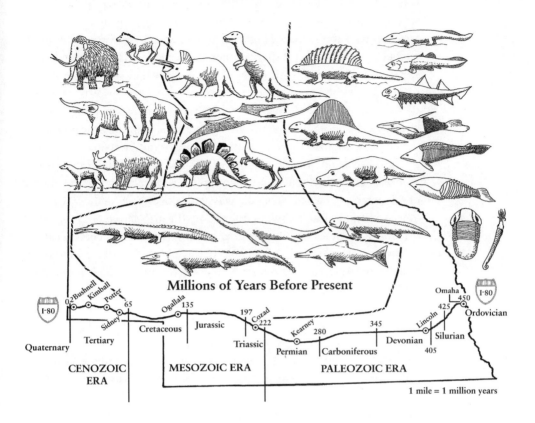

Millions of Years Before Present

I-80 — 0.2 Bushnell — Kimball — Potter — 65 — Sidney — Ogallala — 135 — 197 — Cozad — 222 — Kearney — 280 — 345 — Lincoln — Omaha — 450 — I-80 — 425 — Ordovician

Quaternary — Tertiary — Cretaceous — Jurassic — Triassic — Permian — Carboniferous — Devonian — 405 — Silurian

CENOZOIC ERA — MESOZOIC ERA — PALEOZOIC ERA

1 mile = 1 million years

a boat, as Nebraska was actually submerged under a great inland sea, and the animals present would be mostly corals, sponges, and mollusks, plus a few primitive fishes, such as sharks. Evidence of animal life comparable to these early types can be found in limestone outcrops such as those near Weeping Water, not far south of Omaha in southeastern Nebraska.

In traveling the 50 miles to Lincoln, we cover 50 million years of late Ordovician and Silurian times, to 400 million years ago, and are entering the Devonian period, and the warm waters of the Paleozoic sea still surround us. We must drive (or float) all the way to Kearney, some 270 million years ago, before the last remnants of this great sea have retreated and left us on dry land. To reach the start of the age of dinosaurs, the Mesozoic era (the era of "middle animal life"), which began some 220 million years ago, we must get to Cozad, traversing both the Carboniferous and Permian periods in the process. From there to Ogallala, a distance that includes the Triassic and Jurassic periods and encompasses the peak of the dinosaurian reign, there is no clear geologic record in Nebraska. However, abundant dinosaur fossils occur in the Mesozoic deposits of eastern Wyoming and western South Dakota and tell us much of what

Fig 1. The geological history of Nebraska, with the time scale represented as a million years before present for each mile along I-80. Representative extinct animals of the three major geologic eras are depicted in east-to-west (older to newer) sequence. Animals of a single geologic era are generally arranged vertically upward from older to newer forms, with a few exceptions. Not all of these North American fossil types are known from Nebraska.

is now known of these beasts. These regions were probably then mostly warm, moist bottomlands and evergreen forests of primitive trees and ferns, with an abundance of plant life and other animals. Crouching among all these large beasts were the first mammals, small and timid creatures that perhaps lived on insects and such small reptiles as they could easily capture or on the eggs and young of larger species. Also, early birds were starting to share the skies with flying reptiles. Even these rather primitive birds had larger brains and more advanced wing structures than the pterosaurs, which were doomed to extinction at the end of the Mesozoic.

The third and last segment of the Mesozoic, the Cretaceous period, is approximately represented by the 65-mile distance from Ogallala to Sidney. It was a time when Nebraska was again covered by shallow seas, with long-necked plesiosaurs swimming swiftly through the waters in search of fish and sometimes rivaling the largest dinosaurs in size. There were also carnivorous mosasaurs, sea-going lizards with large crushing jaws, killing and eating large sharks and bony fishes. A few large but flightless loonlike birds swam along the ocean's shorelines, also in search of fish, and some smaller ternlike ones hovered about its surface. Even then the mammals and birds were still relatively inconspicuous and of little ecological significance in the overall ecology of Mesozoic life.

All of this life came to a crashing halt about 65 million years ago. The scenario is still being debated, but many scientists believe that the curtain was brought down on the Mesozoic era with the arrival of a huge asteroid that entered the Earth's atmosphere and struck the Caribbean Sea, immediately or eventually killing much of the Earth's terrestrial and aquatic life and altering the climate permanently. Even such giant marine animals as the plesiosaurs, mosasaurs, ichthyosaurs, and the largest marine turtles met their end, as did many marine invertebrates such as mollusks and mollusklike brachiopods. Warm-water marine phytoplankton that depended on calcium supplies for forming their shells also disappeared, as did many other calcium-dependent marine groups. It was not only the end of an era but also the end of the entire Mesozoic lifestyle, both on land and at sea.

Thus, near Sidney, about 65 miles from the Wyoming border, we have just entered the Cenozoic era, the age of "modern animal life," also called the Age of Mammals. Over this 65-million-year period what is now Nebraska was covered sequentially (over millions of years) by tropical forests, then savanna-like mixtures of trees and grasses, and finally true grasslands, an interval called the Tertiary period. This progressive drying of Nebraska's climate was largely the result of the rising of the Rocky Mountains to the west, which shut down the supply of moisture-rich air masses from the Pacific Ocean and produced a much drier climate with greater seasonal temperature variations. Great elephant-sized titanotheres lumbered across the uplands of Nebraska nearly 40 million years ago, or near Potter on our I-80 time scale. Many other browsing and grazing mammals were present, with grassland-adapted forms progressively replacing the forest-adapted browsers as the woods and forests slowly gave way to prairies. These grazing mammals include camels, rhino, horses, and many others.

It isn't until we reach Kimball, about 20 million years ago, that Nebraska began to resemble the grassy plains we know today. At that time grazing animals such as horses, rhino, camels, grass-eating rodents, and wolflike dogs were common, and by 17 million years ago the earliest true cats had appeared. The first elephants arrived from Asia 14 million years ago; these were largely browsing mastodons and four-tuskers. These species would be replaced later by more modern varieties of elephants that were better adapted for grazing.

About 10 million years ago, or near Bushnell on our I-80 time scale, vast volcanic clouds of dust settled on Nebraska, gradually choking herds of horses, camels, and rhino and providing the skeletal framework for what is now the Ashfall Fossil Beds State Historical Park in northeastern Nebraska. Birds such as cranes similar to Africa's present-day crowned cranes were also caught in this cataclysmic event. Yet elsewhere in western Nebraska, ancestral sandhill cranes were already gathering and clamoring each spring, probably along the edges of the similarly ancestral river that would ultimately become Nebraska's lifeblood, the Platte.

A little more than 2 million years ago, or just 2 miles from the end of our road, an enormous, continentwide glacier swept slowly southward out of Canada, bringing with it a variety of arctic-adapted mammals. This was the first of many long glacial periods, which were separated by interglacial periods of similar lengths. Only two of these glaciers actually brought ice as far south as eastern Nebraska, but one glacial lobe even reached northeastern Kansas. These glaciers deposited great beds of potential soils in the form of glacial till, scraped off from lands much farther north. They also occasionally and randomly left enormous boulders, called glacial "erratics," that they had carried south from Minnesota or the Dakotas. During the rather dry interglacial periods, strong winds stirred up the vast beds of sand, silt, and clay that had eroded off the highlands of Colorado and Wyoming and that had previously been carried to western Nebraska by meltwater rivers from the High Plains and Rocky Mountains. Thus were formed the Nebraska Sandhills. The finer, silty materials were similarly worked by the winds and were carried varying distances to the south and east, mostly accumulating south of the Platte and reaching eastward to Iowa and perhaps even beyond.

The first humans, the antecedents of Native Americans, probably reached North America about 12,000 years ago, or about 75 feet from the end of our trip. This was near the end of the last glaciation, when there was still a relatively dry Bering Sea causeway connecting Alaska with the great Asian landmass. However, remnants of the glaciers in the central

Great Plains were then melting and forming enormous but temporary lakes, such as Lake Agassiz in the Minnesota–North Dakota border region, and in other locations farther north in Canada. Probably caribou, musk-oxen, and other arctic-adapted mammals followed the slowly retreating ice sheets northward, feeding on the tundralike grassy vegetation at their edges. By then there were already two kinds of bison on the plains of Nebraska. One was a very large species with massive horns that became extinct about 11,000 years ago, when the last mammoths also disappeared from America. The other bison was smaller and the direct ancestor of the modern bison, which would become the keystone species for survival of Native Americans on the High Plains for nearly a millennium. We have now reached the last major milepost of time, the start of the Holocene epoch, during which humankind would come to control the destiny of the planet.

America wasn't discovered by Europeans until about five centuries ago, or about 3 feet from the end of our road. Nebraska was mostly settled by Europeans little more than a century ago, or about 6 inches from the last turnoff. Also, by about then the passenger pigeon became extinct, as did the Carolina parakeet, both of which were well known to the Native Americans who had lived in eastern Nebraska. The last Eskimo curlew to be shot in America, and one of the last ever to be seen alive anywhere, was killed in Nebraska about 90 years ago. The gray wolf, elk, pronghorn, bighorn sheep, and free-ranging herds of bison were also gone from Nebraska by then, as were the last free-living Native Americans. The people who are alive today, and the interstate highway on which we have mentally traveled, represent only a few inches of the Earth's total history, based on our 450-million-year scale between Omaha and the Wyoming border. A single year, which sometimes seems endless to us humans, corresponds with about the thickness of the steel on the last Nebraska I-80 exit sign as we end our imaginary trip.

The Present-day Landscapes of Nebraska

Lying near the center of the Great Plains of North America, Nebraska shows a variety of geographic and ecologic influences on its bird fauna. Prior to Euroamerican settlement, probably close to 90 percent of the land area of Nebraska was covered by native grasslands. Of the total land area of 77,510 square miles, almost a fourth, or 19,000 square miles, is comprised of the Sandhills grasslands. Other major grassland components are the tallgrass bluestem prairie of the eastern third of the state, the mixed-grass prairie generally lying to the west of the tallgrass prairies in the loess plains, the sandsage prairie of southwestern Nebraska, the mixed-grass prairie of the southernmost counties, the drier mixed- and shortgrass

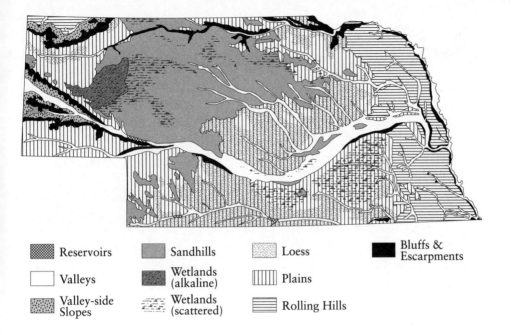

▨ Reservoirs	▨ Sandhills	▨ Loess	■ Bluffs & Escarpments
□ Valleys	▨ Wetlands (alkaline)	▥ Plains	
▨ Valley-side Slopes	≋ Wetlands (scattered)	▤ Rolling Hills	

prairie of the Pine Ridge area, and the shortgrass prairie of the High Plains in western Nebraska.

The two largest river valleys in the state are the Missouri Valley and the Platte Valley, which tend to be quite broad and fairly shallow, while the Niobrara Valley in the northern part of the state tends to be deeper and narrower, the shorelines often lined with steep bluffs. Bluffs and escarpments are also typical of the Pine Ridge area in the northern Panhandle, the Wildcat Hills in Scotts Bluff and Banner Counties, and the upper portions of the North Platte Valley. These features are shown on map 1. Also shown are the areas of the state that are loess covered, areas of wetlands in the Sandhills, and the Rainwater Basin wetlands south of the Platte.

In driving across or, even better, flying over Nebraska today, we are presented with a snapshot in time, the land surface carrying the visible wrinkles of its past and the usually temporary if conspicuous signatures of the present. Nebraska may be thought of as resembling a billiard table that has been designed for use by a masochist; one having had its southwestern corner cut out and its northeastern corner irregularly rounded, and with shorter legs on the eastern, especially the southeastern, corners. Thus the lowest point in the state is about 850 feet above sea level. This occurs at its very southeastern corner, namely the Missouri River at the Kansas border, near Rulo. The state's highest point is near its southwestern corner in Kimball County, south of Bushnell. This inconspicuous hilltop location, Panorama Point, is situated less than 1 mile north of the Colorado border and about 2 miles east of the Wyoming border. Its highest point is at 5,424

Map 1. Landforms of Nebraska, based on a map by the University of Nebraska Division of Conservation and Survey, but with major wetland areas also shown, and regions of surface loess deposition indicated by stippling.

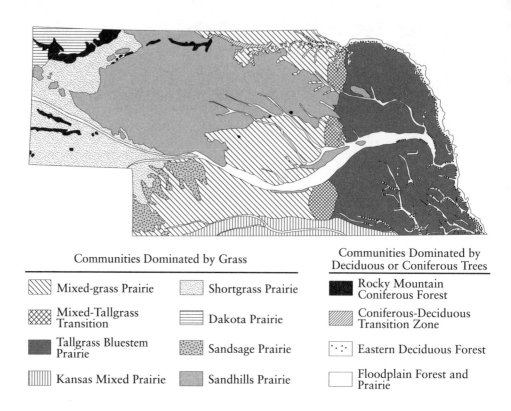

Communities Dominated by Grass

▨ Mixed-grass Prairie   ▦ Shortgrass Prairie

▨ Mixed-Tallgrass
Transition   ▤ Dakota Prairie

■ Tallgrass Bluestem
Prairie   ▦ Sandsage Prairie

▥ Kansas Mixed Prairie   ▦ Sandhills Prairie

Communities Dominated by
Deciduous or Coniferous Trees

■ Rocky Mountain
Coniferous Forest

▨ Coniferous-Deciduous
Transition Zone

⋅⋅⋅ Eastern Deciduous Forest

□ Floodplain Forest and
Prairie

Map 2. Pre–Euroamerican settlement native vegetation types in Nebraska, based on a map by Kaul (1975).

feet elevation. From there the state tilts eastwardly toward the Missouri River at a gradient of roughly 10 feet per mile. Additionally, average annual precipitation in the state increases from less than 14 to 15 inches in the central Panhandle to more than 36 inches in the southeastern corner of the state, or an average northwest-to-southeast precipitation gradient of about 1 inch per 25 miles. The length of the frost-free growing season also increases from the northwest to the southeast, ranging from less than 120 days in the northwestern Panhandle to approximately 180 days in the lower Missouri Valley.

Because of this eastward tilt, Nebraska rivers invariably flow to the east and southeast at a consistently leisurely pace of only a few miles per hour. This is true even for the still unpolluted and beautiful streams originating in the Sandhills, which are spring-fed by the vast Ogallala aquifer that underlies most of the state. Since the average annual precipitation in Nebraska also increases from west to east, so Nebraska's rivers tend to increase in size and in number of tributaries as they flow eastwardly. All the state's other large rivers have their primary origins in the highlands and mountains of Wyoming and Colorado. As a result, they are variably laden with dissolved and suspended clay and silt. This has been the case for millions of years, during which time immeasurable quantities of materials have eroded off these elevated lands and have flowed into Nebraska, filling lowlands and helping to fill in the shallow seas that once inundated nearly the entire state. They have thus effectively covered over all the sediments laid down in Paleozoic times as well as the later Mesozoic ones. Newer Cenozoic deposits have in turn buried earlier ones. However, wind and water erosion has since worn these more recent sediments away in some areas, thus exposing earlier and often fossil-rich strata in cliffsides, bluff faces, and irregularly eroded badlands.

The best examples of these exposed pages of geological history are to be found in western Nebraska, where such well-known geographic features as Scotts Bluff, Chimney Rock, Castle Rock, Courthouse Rock, and Jail Rock bear mute testimony to history writ large in the state's landscape. These locations were not only important landmarks for early immigrants measuring their progress westward on the Mormon and Oregon Trails, but they also provide us with a sense of the vastness of the time that surrounds us.

Scotts Bluff rises 800 feet above the North Platte Valley. At its floor, the river slowly makes its way through sedimentary deposits (the Brule Formation of the White River Group) laid down in the early Oligocene about 35 million years ago. Hidden below ground and river level are even earlier Eocene deposits. These are the Chadron Formation layers of the same geologic group, which are readily visible farther north in the eroded badlands exposures in the Chadron area as well as in the South Dakota Badlands still farther north. Above this, at least 10 million years of history are piled on top of one another in the form of pink-tinted silt and clay. These sediments are interrupted twice by layers of pale whitish volcanic ash that were laid down when volcanoes to the west spewed out clouds of fine dust that settled over much or all of western Nebraska. The top of Scotts Bluff is capped by a layer of ashy marl, marking the boundary between the Oligocene and Miocene and dating from about 24 million years ago. Late Miocene deposits can also be seen in the rocky exposures visible on the south side of Lake McConaughy, where the Ogallala Formation occupies much of the 200-foot cliffside exposures and marks the period from about 15 to 7 million years ago. Beds of Miocene materials similar in age and composition to those just mentioned can likewise be seen on the steep and exposed slopes of the Niobrara Canyon in north-central Nebraska. Above these deposits is a mantle of sand and gravel of Pliocene age (4–2 million years old), overlain in turn by finer and loose silty materials, namely wind-carried loess of late Pleistocene age, far less than a million years old. Collectively, the layers of soft rock, clay, and silt that are exposed in these several localities represent over 40 million years of Earth history, encompassing most of the entire time that mammals and birds have dominated life on Earth.

Unlike the rock-solid mountains of Colorado and Wyoming, the some-times mountainlike peaks of the Pine Ridge and Wildcat Hills escarpments are not primarily the result of faulting and folding of the geological strata below. Instead, they are the ancient resistant remains of uplands, around which the rest of the land has disappeared through erosion. It is a mind-bending exercise to stand on the top of Scotts Bluff or Courthouse Rock and to try to imagine the vast amounts of land that have evaporated from sight, only to reappear as new and wholly re-formed landscapes some-where out of sight to the east.

In the very middle of the state, the Sandhills region lies like a now-caged but still-rebellious ocean, its stabilized dunes resembling waves that have been frozen in time and its sandy cellar actually dripping with water from the Ogallala aquifer below. Its long dune crests sometimes rise as much as 300 feet above the interdune troughs and may be 20 miles or more in length. They ripple out over the state's interior in a generally southwest to northeast orientation, having captured and frozen the strong northwest winds of the late Pleistocene in place as effectively as ancient human footprints have been immortalized in the sun-baked clays of East Africa. Even today, the Sandhills also offer the adventurous tourist an unobstructed view of a grassy, softly undulating Eden, still appearing much as it did when Native Americans first saw it. However, the bison are now missing as indeed are most of the Native Americans.

To the south and east of the Sandhills, great areas of the state are cov-ered by a usually rather thin mantle of loess, its transport and deposition further testimony to the awesome powers of long-forgotten winds. The Sandhills themselves were certainly the regional source of some of these finer materials, but other origins may also have been involved. There were several periods of loess deposition associated with each major glaciation. The most widespread loess deposits, which extend east to the Missis-sippi and Ohio River valleys, are the Loveland loesses, dating from about 140,000 years ago. The most recent and some of the thickest deposits are the Peorian loesses, which were deposited as recently as 23,000 to 13,000 years ago. These deposits often form slightly undulating plains, depending in part upon their thickness and the shapes of the landforms lying immediately below. Rather tall hills were formed where the loess is especially thick, as is the case along the Iowa side of the Missouri River valley and near the middle Platte River. In these locations, some of the layers of loess may be more than 100 feet thick. Over time this loess mantle has been extensively eroded into ridges, fairly steep slopes, and nearly flat valleys. Because of the uniform sizes of their silt particles, loess deposits can erode or be cut into nearly vertical slope faces without collapsing.

The Niobrara River is, at least geologically speaking, Nebraska's most interesting river, as it is the only river that flows over bedrock across much of the state. This fact makes it the only river with real rapids and thus

Nebraska's most attractive river for canoeists. Its exposed horizontal strata are also clearly laid out on the valley slopes for geologists to read as they drift downstream. In a similar manner, ecologists can read the changes in plant and animal life as western species gradually merge with eastern biological elements, especially as the Niobrara approaches and flows into the Missouri River.

The Missouri River valley probably gained much of its modern-day appearance by being the escape channel for the vast quantities of water that were produced as the melting glaciers slowly retreated in Montana and the Dakotas. The Missouri River and its northern tributaries must have been much larger then than they are today. Even now it exerts a powerful geological influence when it floods periodically, and if that part of its lower valley that is shared with the Mississippi River is included, it is easily the longest and most important river in North America. Its geological age is also attested to by the fact that it still supports such ancient fish as lampreys, paddlefish, sturgeons, and gars, species that belong to some of the oldest of all surviving fish families.

The state's other larger rivers, the Platte and Republican, have broad floodplains of little geologic interest; their restrained beauty and the easy access to their surface and subsurface waters have made them fair game for exploitation by irrigators and other water-hungry interests for nearly a century. They also have sandy bottoms, the sand and gravels in their streambeds having traveled far from their origins in the mountains of Colorado and Wyoming.

Thus, through soil erosion and redeposition, most of present-day Nebraska came originally from someplace else. Erosion by Nebraska's rivers continues to remove parts of the state's surface layers and pass them on downstream, only to be deposited again somewhere else, continuing an endless cycle of erosion and redeposition.

If wind and flowing waters had such tremendous effects on the landscapes of Nebraska, glacial ice also did its part. The entire eastern fifth of Nebraska, from Boyd County in the northeast to Jefferson County in the southeast, bears the clear evidence of the glaciers' tracks in the form of till deposits, hilly moraines, and occasional erratics. Farther west along the major river systems are beds of Pleistocene sands and gravels that were dropped by glacier-fed streams, especially east of the Sandhills. Sometimes later covered with layers of wind-borne loess, these materials provided the basis for the deep, calcium-rich, loamy soils that typify eastern Nebraska. These soils also once supported the lush tallgrass prairies that quite literally took root in the wake of the last glacier's passing. Like the bison, the tallgrass prairies have virtually disappeared, too.

In 1676 Isaac Newton wrote in a letter to Robert Hooke to the effect that that if he had seen farther than other scientists, it was by standing on the shoulders of giants. When one stands on the summit of Scotts

Bluff, or indeed any promontory, one should not only look to the farthest horizon but also look below, for he or she is standing on the shoulders of unimaginable numbers of biological generations that may lie unknown and forever embedded below and on incalculable layers of dust and sand that are ultimately as old as the Earth itself. It should be a humbling yet also exalting experience to be able to interpret the land and its life, of which we are but a tiny and very temporary part.

CHAPTER 2

An Entangled Bank

Nebraska's Biomes, Natural Communities, and Ecoregions

Imagine asking one of the mid–nineteenth century immigrants on the Oregon or Mormon Trails how he or she would characterize Nebraska in a single word or short phrase and the answer would probably have been the Platte River. Ask the modern person on the street to describe the state's most famous attribute and the answer would most likely be the University of Nebraska and its associated Cornhusker football team. Ask a field ecologist and the answer might well be the tallgrass prairie.

All of these statements are partially true, and yet all are variously wanting. The Platte River is now a mere emaciated and sad remnant of its former self, as compared to a time when it provided security but sometimes also danger for the undaunted immigrants destined for California or Oregon, as they traversed nearly its entire length. Identifying the state's university with its quasi-professional football team does not do justice to either. Nebraska's greatest botanist, Charles Bessey, once noted that football occupies the same relationship to education as bullfighting does to farming.

Finally and most sadly, the once horizon-bounded tallgrass prairies of Nebraska are now mostly a memory. The realities are confined to tiny relicts in cemeteries, railroad rights-of-way, and other purposefully or accidentally protected sites. We must not be remorseful about changing times, and there will never again be a time when the West beckons an entire population to begin a new life in an essentially unknown land. But we can work to save what is left of the Platte, the prairies, and the other still-surviving parts of Nebraska's original natural heritage. These include the conspicuous and inconspicuous, the beautiful, the plain, and sometimes even the ugly. In order to love all these parts, we must first try to understand them. In order to understand them, we must first identify and describe them. Perhaps only then we will know what is most important to try to save and why.

Over the twentieth century many attempts were made to map the distribution of species and more comprehensive natural collective entities of North America. Some of these maps have relied largely or entirely on weather or climatic information, including the now-abandoned nineteenth-century "Life Zones" concept of C. H. Merriam. Others have relied more on physical factors such as landforms and common regional geologic histories. Still others have utilized the dominant plants of native, fully stabilized climax communities that have coevolved in regions of similar climate and soil substrates. This approach to classification represents the Plant Formation concept of the University of Nebraska's famous botanists Frederick Clements and John Weaver. This idea holds that, within single broad climatic regions, interactive plant and biotic communities evolve over time. Through their ecological relationships, these communities acquire interdependent characteristics that are greater and more complex than are represented by the sum of their individual species. This book is broadly organized along ecological principles.

Several very useful maps of Nebraska's landforms and their related natural communities have appeared in recent years, which provide a basis for discussing the state's biological geography (see maps 1 and 2). Perhaps the most useful from a botanical perspective is that of Robert Kaul, whose original (1975) map of the state's historic vegetational geography as it probably existed about 1850, or prior to agrarian settlement, was updated in 1993 with the help of Steven Rolfsmeier. This map relied almost entirely on historical data and current evidence in the form of remnant plant communities for establishing boundaries of the predevelopment plant communities of Nebraska. It identified 13 major types of terrestrial plant communities plus several additional aquatic or wetland types. There are four major forest types (ponderosa pine forests and savannas, red cedar forests and savanna, deciduous riparian forests, and upland deciduous forests). The major subtypes of grassland are upland tallgrass prairie, lowland tallgrass prairie, loess mixed-grass prairie, Sandhills mixed-grass prairie, and sandsage mixed-grass prairie. There are also four additional intermediate, substrate-dependent, or mosaic types of mixed-grass prairies recognized in the more recent version, including the transition of mixed-grass prairie to shortgrass prairie.

Map 3 is based mainly on Kaul's earlier map but has some updating, especially of wetland distribution. It shows all of Kaul's original major vegetation types but not the more recently identified red cedar forests

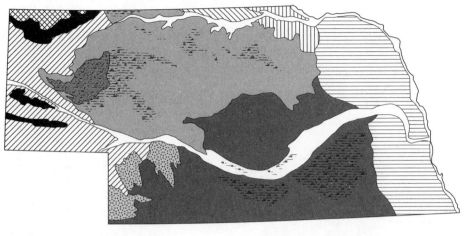

Northwestern Great Plains

⊠ Shortgrass & Sagesteppe Prairies

⧄ Mixed-grass Prairie

Western High Plains

■ Ponderosa Pine Escarpments

⧄ Shortgrass–Mixed-grass Mosaic

Central Great Plains (Loess Hills)

Loess Mixed-grass Prairies

Northwestern Glaciated Plains

⦀ Sandhills–Mixed-grass Ecotone

Nebraska Sandhills

Sandhills & Border Prairies

Sandsage-Sandhills Ecotone

Western Cornbelt (Glaciated) Plains

≡ Tallgrass Prairies

Major River Valleys

☐ Riparian Forests

Wetlands (alkaline)

Wetlands (scattered)

and savannas, which are relatively local around the southern edge of the Sandhills and probably were of little significance before the suppression of fires on the prairies became commonplace. The map also excludes several of his more recently recognized mixed-grass subtypes and defines as "shortgrass prairie" a large region that Kaul and Rolfsmeier have since described as consisting of a mosaic of shortgrass and mixed-grass types. Map 3 additionally shows wetland regions, which are not shown in map 2 but instead have been transferred to the map of landforms (map 1).

Another mapping approach, utilizing that of variably scaled ecoregions rather than historic native vegetational patterns, is still undergoing refinement by governmental agencies but in broad outline has been described in technical papers by scientists such as J. M. Omernik (1987) and in more extended book form by R. G. Bailey (1995). This system uses biotic information (native vegetation, wildlife) as well as nonbiological data (geology, physiography, soils, hydrology, and prior land use) to identify regions of broad ecological or conservation significance for

Map 3. Ecoregions and major plant communities of Nebraska, based on a map by Omernik (1987).

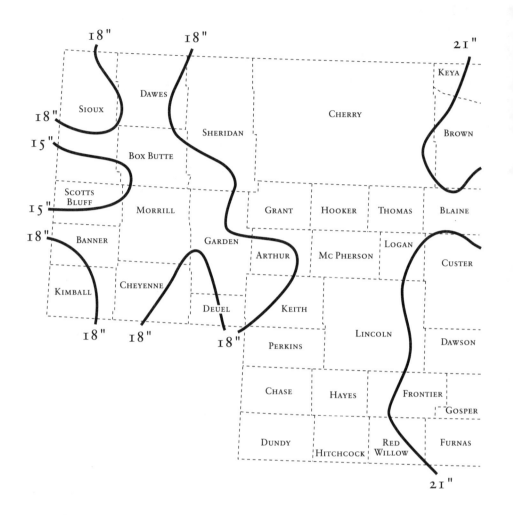

18"   18"   21"

Map 4. Nebraska counties,
plus zones of annual
precipitation (in inches;
40-year average) (Lawson et
al. 1977).

recognition and consideration by private and governmental resource management agencies.

These regions are identified and classified hierarchically, beginning
with the largest category (Level I, totaling 15 North American regions),
followed by Level II (a total of 51 North American regions). There are
98 Level III ecoregions in North America, of which Nebraska contains
6 (map 3). They include the Northwestern Great Plains (the dry uplands
north of the Pine Ridge), the Western High Plains (the uplands west of
the Sandhills and central loess plains), the Central Great Plains (the loess
plains south of the Sandhills plus the Platte River valley), the Northwestern Glaciated Plains (the Niobrara Valley and the glaciated uplands to

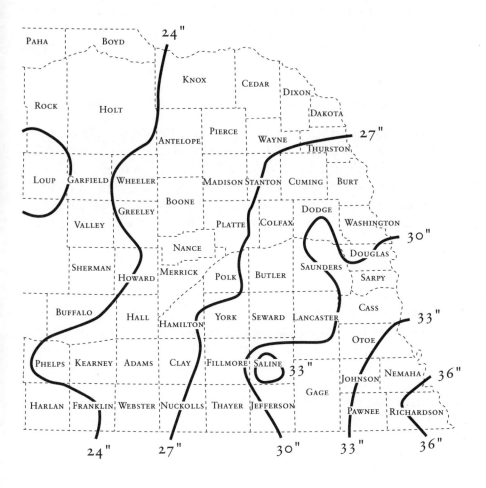

PAHA   BOYD   24"

ROCK   HOLT

LOUP GARFIELD WHEELER

VALLEY

SHERMAN

BUFFALO

PHELPS KEARNEY ADAMS

HARLAN FRANKLIN WEBSTER NUCKOLLS THAYER JEFFERSON

24"   27"   30"   33"   36"

KNOX   CEDAR   DIXON

DAKOTA

PIERCE   WAYNE   27"

ANTELOPE   THURSTON

MADISON STANTON CUMING BURT

BOONE   DODGE

GREELEY   WASHINGTON

NANCE   PLATTE COLFAX   30"

DOUGLAS

MERRICK   POLK BUTLER SAUNDERS   SARPY

HOWARD

HALL   YORK SEWARD LANCASTER   33"

HAMILTON   OTOE

CLAY FILLMORE SALINE 33"

JOHNSON NEMAHA   36"

GAGE

PAWNEE RICHARDSON

CASS

the north), the Nebraska Sandhills, and the Western Cornbelt Plains (the eastern glaciated till and loess plains plus the Missouri Valley).

The fourth level of ecoregion classification provides for further subdivisions: Nebraska contains 28 currently identified fourth-level ecoregions. Map 3 shows the boundaries of all the Level III regions but merges or ignores some of the Level IV types for simplicity and clarity. Additionally, some of the Level IV boundaries (mainly of wetlands) have been modified to try to conform more closely with the comparable boundaries of wetland regions identified by Kaul and Rolfsmeier. A map of Nebraska's counties has been included to help readers correlate these natural regions with current political boundaries, to which the state's average annual precipitation patterns have been added (map 4).

Still lower levels of ecological classification of the state's natural plant communities have been provided by Gerry Steinauer and Steven Rolfsmeier (2000). They have classified the state's upland plant communities into 35 subtypes, and the floodplain, shoreline, and aquatic communities into 28 subtypes. Their classification has been largely adopted here, and a summary of the upland and terrestrial wetland community types appears in table 2.

Distribution, Fragmentation, and Degradation of Nebraska's Ecosystems

Using an estimate by John Weaver, it seems likely that under pre–Euroamerican settlement conditions perhaps only 2.5 to 3 percent of Nebraska's total surface area was originally covered by forest. A somewhat higher forest coverage occurs today, owing to the repression of prairie fires and subsequent woody plant growth, supplemented by human tree-planting activities. These forests may be divided between coniferous and deciduous forests. The coniferous forests of western Nebraska are almost entirely dominated by ponderosa pine, plus some red cedar or mixed forests, whereas the deciduous forests are mostly eastern riverine, or gallery, forests that extend westward varying distances into grasslands along rivers and smaller streams. A survey of Nebraska's forest lands in the mid-1990s indicated that there were about 225,000 acres (350 square miles) of coniferous timberlands in the state and about 625,000 acres (1,000 square miles) of mixed forests and hardwoods, largely gallery forests of elm, cottonwood, hackberry, and other floodplain species. Together these forests comprised 848,000 acres.

Although these forests now comprise only 2 percent of the state's land area, they support almost half (48 percent) of the state's overall breeding avifauna (of 202 species), based on studies by the author. The majority of these bird species are eastern or northern in their geographic affinities; only a few are western or southern. The biodiversity of Nebraska's breeding birds is highest in the deciduous forests of the lower Missouri River valley (with an estimated 138 breeding species, based on a 1999 analysis by the author) but diminishes northward and westward along the Platte and Niobrara Valleys (to about 110 species along the upper Platte and upper Niobrara). The coniferous forests of the Pine Ridge provide a secondary peak in biodiversity, with an estimated 128 breeding species in Sioux County.

Nearly all of the rest of Nebraska's surface, or more than 95 percent, was probably dominated by perennial grasslands prior to Euroamerican settlement. Nearly 30 percent of these grasslands were comprised of Sandhills prairie, including a small area of sandsage prairie. The tallgrass prairie may have constituted another third and the mixed-grass and shortgrass prairies the rest. All told, the native grasslands of Nebraska supported

Floodplain Communities	Sparsely Vegetated Communities
FORESTS	DRY CLIFF
Cottonwood-Dogwood Floodplain Forest	ROCK OUTCROP
Eastern Lowland Forest	BADLANDS
WOODLANDS	
Eastern Floodplain Woodland	
Western Floodplain Woodland	
SHRUBLANDS	
Willow Shrubland and Perennial Sandbar	
Dogwood Floodplain Shrubland	

Upland Communities

FORESTS	GRASSLANDS
Bur Oak Forest	**Shortgrass types**
Southeastern Upland Forest	Shortgrass Prairie
Northeastern Upland Forest	Western (Alkaline)
Northern Springbranch Canyon Forest	Floodplain Meadow
Western Coniferous Forest	**Mixed-grass types**
	Loess Bluff Prairie
WOODLANDS	Loess Mixed-grass Prairie
Oak Woodland	Western Mixed-grass Prairie
Northwestern Canyonbottom Deciduous Woodland	Northwestern Mixed-grass Prairie
	Northern Mixed-grass Prairie
Western Coniferous Woodland	Wheatgrass Basin Prairie
Mixed Coniferous Woodland	Western Sandy-slope Prairie
Juniper Woodland	Pine Ridge Sandy-slope Prairie
SHRUBLANDS	**Tallgrass types**
Sumac-Dogwood Shrubland	Tallgrass Prairie
Mountain Mahogany Shrubland	Wet Mesic Prairie
Buckbrush Shrubland	Sand/Gravel Prairie
Silver Sagebrush Shrubland	
Greasewood Shrub Prairie	**Sandhills and Sandsage types**
	Sandsage Prairie
	Sandhills Dune Prairie
	Sandhills Needle-and-thread Prairie
	Sandhills Dry Valley Prairie
	Sandhills Wet Mesic Prairie

about 15 percent of the state's breeding avifauna, or about 30 species. This would seem to represent a disproportionately small proportion of Nebraska's avifauna, considering the vast area that grasslands once occupied in the state, but the grasslands of the Great Plains were rather late in evolving as compared with the North American forests. The grasslands also

Table 2.
Classification of Nebraska's Terrestrial Plant Communities

have far less three-dimensional structural complexity and offer a less varied diversity of potential plant and animal food sources than do the forests. Furthermore, perhaps not enough time passed to permit the evolution of a more diverse avifauna within the grassland ecosystem. There are, for example, far fewer grass-dependent species of *Ammodramus* sparrows in the Great Plains grasslands (5) than there are species of *Dendroica* warblers in the coniferous and hardwood forests to the east and west (21). The list of grassland-adapted birds also includes many of the species that are most rapidly declining in population on a national level, probably because of the high rate of prairie destruction that has occurred during the past century.

Wetland habitats probably historically comprised far less than 1 percent of the state's surface area, and even today, with the many large reservoirs, the collective wetlands area still represents only about 1 percent of the state's surface area. Although reservoirs have increased in number, drainage of native wetlands, as in the Rainwater Basin, has tended to offset increases from reservoirs. However, in spite of their rarity, these diverse wetland habitats (such as lakes, marshes, ephemeral playa wetlands, fens, and wet meadows) support at least 30 percent of the state's breeding avifauna. Thus, from the standpoint of providing habitats for breeding birds, the wetlands are the most important habitat type relative to their abundance. The Sandhills region supports about 110 breeding bird species, of which nearly half are actually wetland- or meadow-adapted rather than true grassland birds, and many of the rest are shrub- or tree-dependent species.

Sage scrub, consisting of various species of sagebrush and other similarly arid-adapted shrubs in a shortgrass matrix, occurs in a rather small part of the northern Panhandle, representing only about two percent of the state's surface area. It also supports only a few unique breeding birds, likewise representing about 2 percent of Nebraska's total avifauna. These species only rarely consume sage directly but often nest within its protective vegetative cover. They may also exploit the shade it casts for cooling or use its tallest branches for songposts or lookout points.

The remaining few bird species, approximately 5 percent, cannot be easily tied to a particular and vegetation type for their breeding. They instead may have specific substrate needs (for example, rock ledges, burrows, human-made structures) or have foraging requirements that are more important than other requirements that may be provided by any single identifiable vegetation type.

Vegetation maps such as Kaul's typically represent often unrealized or hypothetical, if not "perfect," conditions and thus may be vastly different from the present situation. These maps not only exclude human-altered environments but also usually ignore transitional stages of natural plant succession that occur following various kinds of disturbance (secondary succession). They also exclude similar temporal changes occurring on newly formed or massively disrupted landscapes (primary succession), as might occur following landslides, lake drainage, and the like. Of all Nebraska's native landscapes, probably the Sandhills are least modified from their original state, simply because it has long been recognized that the Sandhills communities are so fragile that almost any disturbance to their vegetation would have an immediate and dire impact on the entire environment, including the very stability of the substrate itself. However, the tallgrass prairies have been so effectively exploited and fragmented for agricultural purposes that probably less than 3 percent of the North American tallgrass prairies can be found in anything approaching their original condition and diversity, and the figure is sometimes believed to be closer to 1 percent.

Kaul and Rolfsmeier point out that, in addition to the approximately 1,400 species of native vascular plants in Nebraska, about 350 more have been accidentally or purposefully introduced and now survive here without benefit of cultivation. Most of these are Eurasian in origin and include all of the state's most serious weeds. At the same time many of Nebraska's native species have become relatively rare, although none are known to have completely disappeared from the state's floral list. Many counties are known to have had from fewer than 400 to as many as almost 700 native plant species, but now many counties have more than 100 introduced ones, and in some highly altered counties practically no native vegetational communities still survive. The destruction of any community structure, whether human or natural, eventually destroys its integrity and productivity.

Just as it is possible for a person to travel on I-80 from the Missouri River bridge to the Wyoming border without pausing to think of the state's geological history, it is also possible to be oblivious to the many ecological zones and habitats that are being traversed at 75 miles per hour. One of the purposes of this book will be to try to point out that the story of Nebraska is not to be discovered at the monstrous historical steel arch spanning I-80 near Kearney or even solely in the yellowing pages stored in files at the State Historical Society in Lincoln.

Instead, it is also to be found by meditating under an ancient cottonwood growing along the Mormon and Oregon Trails that once may have given succor and shade to the westward-bound pioneers. It hides among the head-high prairie grasses and wildflowers that bison once tramped through and slept in as they rested from their long migrations to and from the Platte and Republican River valleys. It can be clearly heard in the song of a meadowlark perched on a roadside fencepost, should one only take time to stop and listen. It is present in the gently waving grasses of the green-capped loess hills and the yucca-peppered Sandhills that are easily visible from the noisy interstate, beckoning one like quiet oases in a cacophonous bedlam. It silently calls out in the smell of ponderosa pines among the Wildcat Hills, along the Pine Ridge, and on the crest of Scotts Bluff. The story is there and always has been. How long the story may last and what lessons we may learn from it are up to us.

PART TWO

Nebraska's Ecological Regions and Biological Communities

Ponderosa Pin
and Rimrock

The Western Escarpments and Their Coniferous Forests

The Nebraska Panhandle is the portion of the state that, at least on the basis of its geology, probably should have become part of Wyoming but somehow by chance and luck became attached to Nebraska. The region includes the state's most scenically attractive sites as well as the scenes of perhaps its saddest and most tragic human histories. Together with the shortgrass-dominated High Plains and Black Hills of South Dakota to the north, this is the land of the Oglala Sioux and the martyred Crazy Horse. His bones still probably lie buried in some hidden place near Fort Robinson, secreted there by friends and relatives who feared that even his remains would be desecrated by the soldiers who murdered him. One can only hope that golden eagles still occasionally fly over his grave site, that prairie dogs still burrow nearby, and that it has not already been converted to an irrigated crop field.

The Panhandle is the driest part of Nebraska, with only about 14 inches of precipitation annually in its northwestern corner. Here wind-eroded sandstone formations comparable to those of the South Dakota Badlands can be found in places such as Toadstool Geologic Park, and steep rimrock-edged escarpments interrupt the horizon. It is also a land that is darkly painted with stands of ponderosa pine, especially on ridge-tops and on the north- and east-facing canyon slopes of the Pine Ridge and Wildcat Hills, reminding one of South Dakota's Black Hills. It is likewise a land rich in Cenozoic mammal fossils, particularly those of Miocene age, where paleontologists from eastern universities and from the University of

Nebraska at Lincoln once competed with one another for the best fossil discoveries.

Agate Fossil Beds National Monument in Sioux County was the site of many of these spectacular finds, but almost any eroding hillside in this region might be a place for encountering fossil remains such as teeth, jaw fragments, or turtle shells. Some of the most wonderful fossil finds have occurred simply by accident while people were engaged in other activities, such as farming, excavating, or road building. But one must always remember to tread softly on this land. It is easily defiled, for it houses the memories of millions of years of mammalian history and hundreds if not thousands of years of our own more immediate relatives' footprints.

Ecology of the Pine Ridge Escarpments and Forests

The western escarpments of the Nebraska Panhandle (map 5) consist of two irregular wrinkles in the landscape, both of which gradually diminish in height and extent as they extend east from the Wyoming border. The larger of these is the Pine Ridge, which separates the western High Plains ecoregion to the south from the northwestern Great Plains and Missouri Plateau to the north. It covers about 2,700 square miles in Nebraska and varies in topography from being merely undulating to locally quite rugged, at an average elevation of about 4,000 feet.

Much of the land is covered by ponderosa pines, ranging from open, parklike stands to closed forest. Beginning as early as the 1880s, nearly all this region was logged, and forest fires have occasionally also affected the entire region. However, trees located on the steeper escarpments and deeper valleys may have escaped most of the lightning-caused grassland fires that frequently ravaged the plains. A few trees more than a century old may still be found in some deeper canyons. Much of the original forest is now at least partly protected from logging by being part of state park or national forest lands.

Plants typical of the Rocky Mountain flora that occur in the Pine Ridge include not only the ponderosa pine but also mountain mahogany, serviceberry, sego and mariposa lilies, and several montane grasses and sedges. Representatives from the Great Basin flora include Great Plains yucca and various cacti and sagebrush species. Southern or southwestern grasses that occur in the Pine Ridge region include buffalograss and hairy grama, whereas little and big bluestem have penetrated to the Pine Ridge from the east. There are also many mixed-grass prairie species, especially on south-facing slopes and open ridges, some of which originated in the Sandhills region to the east.

Southward, just south of the North Platte River, the Wildcat Hills escarpment similarly extends eastward, but this region is only about 70 miles long and occupies considerably less than 1,000 square miles. Like the

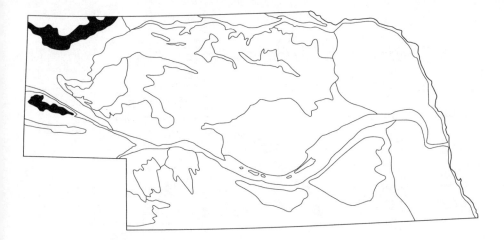

Pine Ridge, its topography is irregular and sometimes rather steep-sided, the higher ridges and buttes occasionally exceeding 4,600 feet of elevation. Among the higher landmarks within this general region are (from west to east) Scotts Bluff, Castle Rock, Chimney Rock, Courthouse Rock, and Jail Rock. Ponderosa pine often occurs in sparse to dense stands on the crests and north-facing slopes, whereas the valleys are often covered by hardwoods.

Both of these regions receive up to about 20 inches of annual precipitation, of which the majority occurs during the roughly 120-to-150-day (April to September) growing season. Much of the annual precipitation in the Panhandle is dependent on moisture originating in the west, via fronts coming over the Rockies from the Pacific Coast. By comparison, farther east in the state there is an increasing dependence upon Gulf Coast moisture that is drawn northward through the Mississippi and lower Missouri Valleys and released when cooler arctic fronts collide with warm subtropical air masses, often producing violent summer thunderstorms. Additionally, the average relative humidity in the Panhandle is considerably below that of eastern Nebraska, producing more comfortable summer temperatures and also more rapid short-term and seasonal changes in temperature.

As part of a biotically diverse region, the Pine Ridge and its adjoining High Plains areas support a high diversity of breeding birds. Sioux County alone supports an estimated 128 breeding species of birds, slightly fewer than occur in the southeastern corner of Nebraska but more than in any other area in the state except for the lower Niobrara and upper Missouri Valleys. There are actually a larger number of avian endemics in this northwestern corner of the state than in the opposite southeastern corner. These include many pine-adapted or general forest-related western

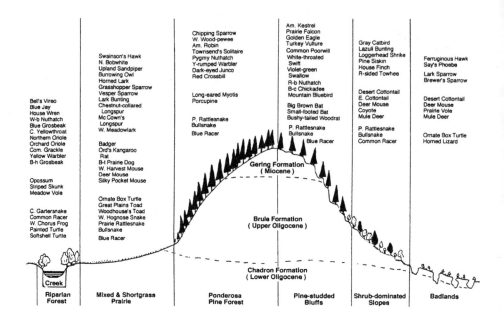

Fig. 2. Profile view of typical Pine Ridge and High Plains habitats and some associated vertebrates. After Johnsgard (1995).

species (including Lewis' woodpecker, pygmy nuthatch, mountain blue-bird, Townsend's solitaire, western tanager, yellow-rumped warbler, red crossbill, and dark-eyed junco). Some of these, such as Townsend's soli-taire, are known to breed regularly in the Black Hills not far to the north but may not breed in the Pine Ridge, at least on a regular basis. There is also a mixture of shortgrass species (for example, McCown's longspur and chestnut-collared longspur), woodland- or shrub-dependent species (such as sage thrasher, Brewer's sparrow, cordilleran flycatcher, Cassin's kingbird, plumbeous vireo, and pinyon jay), and some cliff-dependent forms (especially the crevice-nesting white-throated swift and some cliff-nesting raptors).

Sioux County supports only a few species of fish (8), but has at least 19 species of herps, including the horned lizard, native in Nebraska to the northern Panhandle. There are also 49 mammal species, including the endemic least chipmunk, long-legged myotis, and Townsend's big-eared bat. The greatest number of vascular plant species reported for any county in the region is 760, in Dawes County, as compared with only 290 in Deuel County in the southern Panhandle.

Major Terrestrial Community Types

Western Coniferous Forest

Of all the coniferous trees of North America's western forests, the pon-derosa pine is the most widespread, ranging north to British Columbia and south into Mexico. It extends farther east into Nebraska than anywhere else in its historic North American range, namely to the central Niobrara Valley, but it does best in deep and well-drained soils. The ponderosa pine forest of western Nebraska is rather similar to that of the Black Hills of South Dakota and those of the Rocky Mountain outlier ranges of eastern Wyoming, but it lacks lodgepole pine or any other pines. Indeed, the only other native pine in western Nebraska is the limber pine, which occurs in scattered locations in Kimball County near the Wyoming line, usually on dry ridges. Otherwise, unlike the multispecies deciduous forests of eastern Nebraska, this western forest is dominated completely by ponderosa pine. Some smaller deciduous trees, such as American elm and Rocky Mountain juniper, may be present. In addition, there are taller shrubs, including wolfberry and chokecherry, and other rather tall shrubs such as gooseberry and western wild rose. Shorter shrubs, including Oregon grape and poison ivy, also occur, but grasses and forbs are not numerous. In part the lack of grasses and forbs is a result of the accumulated litter of pine needles, which upon decay produce an acidic, less fertile substrate. However, a few grasses and sedges and sometimes ferns are present. The pines can grow on quite steep slopes, and their rather deep roots allow them to survive the hot and often dry summer months.

Western and Mixed Coniferous Woodland

Two broad types of woodlands are associated with the escarpments in western Nebraska and to varying degrees also appear farther east along the Niobrara Valley in Brown, Cherry, and Keya Paha Counties in the transition area to deciduous forests and woodlands. One of these is a savanna community type. Dominated by open stands of ponderosa pines, savanna communities alternate with and grade into stands of little bluestem. These communities are best developed over loamy soils on gentle to steep slopes of the Pine Ridge escarpment. The second major type is geographically associated with the Wildcat Hills and also consists of scattered ponderosa pines but lacks little bluestem. Instead, grama grasses, introduced Kentucky bluegrass (in disturbed areas), and other grasses or sedges may be found. In both, a subcanopy or lower understory of Rocky Mountain juniper may be present, and various shrubs such as aromatic (skunkbrush) sumac, wolfberry, chokecherry and prairie rose are often also present. Mixed coniferous woodlands can be found farther east along the Niobrara River.

Northwestern Canyonbottom Woodland

Woodlands differ from true forests in that there is less overall canopy cover in the tree stratum (up to 60 percent canopy cover), although it may be clumped and thus quite discontinuous. In the deeper canyons and steep ravines of the Pine Ridge, and to a lesser degree in those of the

Wildcat Hills, a deciduous woodland community is developed that consist of many eastern-affiliated and mesophytic, or riparian, tree species such as box elder, hackberry, and especially green ash and American elm. At some sites quaking aspen may also be present, and rarely the mountain birch may even occur. Connections to the eastern Nebraska deciduous flora are evident via comparable ravine communities that are present in the Niobrara Valley. The usual assortment of western shrubs, including those mentioned earlier for the woodland and forest communities, are also present. Additional typical shrubs include wild plum, smooth sumac, red osier, and hawthorn. Several attractive flowering forbs occur, including two orchids, the spotted coral-root and northern green orchid. The blue larkspur is also often present.

Juniper Woodland

The juniper woodland community type occurs on steep slopes in the Panhandle, where it usually exists in association with ponderosa pine woodlands. It is best developed in west-central Nebraska and is therefore discussed in chapter 5.

Mountain Mahogany Shrubland

Shrublands are communities in which the dominant plants consist of a shrub layer less than about 18 feet tall (in Nebraska much lower), with a canopy coverage ranging from complete to as little as 10 percent. Such communities are fairly common in western Nebraska. In the Wildcat Hills, a variant type up to about 6 feet tall and dominated by an open canopy of mountain mahogany is sometimes present on ridges and steeper slopes. These communities are often situated close to coniferous woodlands, and thus scattered ponderosa pines or Rocky Mountain junipers may be present. This shrubland type may also develop within very open, savanna-like stands of ponderosa pine. Side-oats grama is typically present, as are introduced brome grasses and fringed sage.

Buckbrush Shrubland

The buckbrush shrubland, which is widespread in western Nebraska, differs from the mountain mahogany shrubland in that it is dominated by wolfberry ("buckbrush") rather than mountain mahogany and is of lower stature, up to about 3 feet tall. Yet it often exhibits denser canopy coverage, which sometimes approaches an impenetrable 100 percent. A few taller shrubs, such as chokecherry, skunkbrush sumac, and buffalo currant, may also be scattered among the wolfberry. Poison ivy often forms a dense ground cover, and vines such as woodbine and western clematis may be found. There is little herbaceous diversity, and what is present is likely to consist of exotic, weedy species. Where the canopy cover is less, producing scattered openings, native grasses of the mixed-grass assemblage tend to occupy the available spaces.

Wheatgrass Basin Prairie

In the immediate vicinity of Scotts Bluff National Monument and perhaps elsewhere in the Panhandle, a mixed-grass upland prairie type dominated by western wheatgrass is present, supplemented by other mixed-grass species including side-oats grama, needle-and-thread, little bluestem, and other rarer native grass species such as green needlegrass. Weedy exotic grasses, such as annual bromes, and a variety of annual native or exotic forbs may also occur. Soils supporting this community type tend to be fairly deep but poorly drained. The community type is seemingly highly localized and remains almost completely unstudied.

Western Sandy-Slope Prairie

On the gentle slopes of the Panhandle, especially on colluvial soils that develop below sandstone escarpments and rock outcrops, a mixed- to tallgrass prairie type develops that is dominated by blue grama, prairie sandreed, and needle-and-thread, particularly the latter two species. Western wheatgrass may be locally present as well, especially on level sites. Where the soil is finer needle-and-thread is more evident, and sand bluestem and various sandsage prairie species appear on steeper sandy slopes. Shrubs are generally rare or absent. The community type is well developed at Agate Fossil Beds National Monument. It locally intergrades with western mixed-grass prairie and may also intergrade with western sandy slope prairie. It extends eastward along the Niobrara and Platte Rivers to Keith County in the case of the Platte Valley drainage.

Pine Ridge Sandy-Slope Prairie

The Pine Ridge sandy-slope prairie community type is similar to the western sandy-slope prairie but is limited to the Pine Ridge region, with possible extensions eastward along the Niobrara River. Like the western sandy-slope prairie type, it is associated with gentle slopes and colluvial soils located below sandy outcrops and escarpments. It differs slightly in that the dominant grasses are prairie sandreed, needle-and-thread, and little bluestem, as well as several other mixed-grass or tallgrass species. It also often contains sand bluestem and other plants typical of Sandhills prairie. Scattered small ponderosa pines may be present, but shrubs are rare or absent. A good example of this community type occurs in the Soldier Creek Wilderness Area of Sioux County.

Dry Cliff

Steep cliffs, such as at Scotts Bluff and Chimney Rock, result from erosion of the soft sandstone or siltstone layers laying underneath a capping of more resistant materials, producing sharp drop-offs that sometimes measure hundreds of feet. Among the few plants that sometimes can survive on such steep slopes are ten-petal mentzelia and sawsepal penstemon.

This unstable substrate is clearly not a desirable habitat for plants but makes a fine refuge for hole-nesting birds, and the occasional level ledges are perfect perching sites or nest sites for raptors. Birds of cliffsides and butte faces include ledge-nesters such as the prairie falcon, golden eagle, ferruginous hawk, and occasional red-tailed hawks. Cliff swallows nest below overhanging ledges, and Say's phoebes and white-throated swifts search for crevices and holes. Rock wrens and poorwills also find these areas to be perfect nesting habitats.

Rock Outcrop

In association with dry cliffs, rock outcrops occur where more resistant layers of sandstone or siltstone resist weathering more effectively than does the land around and underneath them, producing ridges, capstone rocks on bluffs, and the like. A mixture of shrubs and herbaceous plants often is found at such sites, the shrubs most often being skunkbrush sumac, with rabbitbrush and mountain mahogany less frequently encountered. Depending on the degree of soil development, grasses such as various mulhy and grama grasses may be common, at least in better soils. On poorer sites, grasses are less common than are the forbs. Legumes able to manufacture their own nitrogen through their nitrogen-fixing root nodules, such as scurfpeas and milk vetches, are typical. The draba milk vetch has longer leaves and more pronounced flowering parts in wetter years or conditions and a more matlike appearance in drier ones. Also, cushion-shaped plants such as Hood's phlox are present that likewise often seem to be barber-trimmed, their aboveground parts with an earth-hugging profile that keeps their leaves and flowering parts away from winds that would otherwise suck the plant dry.

Profiles of Keystone and Typical Species

Ponderosa Pine

Few pines, or indeed any other tree, have been more important in the history of the American West than has the ponderosa pine. It has the broadest distribution of any North American pine and is one of the most drought tolerant. The ponderosa pine is the first tree to be encountered at the forest-grassland transition in the western High Plains, where it often occurs in savanna-like stands similar to the stands of scattered bur oaks in the tallgrass prairies of eastern Nebraska. It is well adapted to a climate of cold winters and warm, dry summers and to situations where recurring but low-intensity forest fires are a part of the ecological pattern. Like other conifers, resistance to fires among pines is affected by the thickness of their bark, the rate of seedling growth, their sprouting abilities, and, in some species such as lodgepole pine, the presence of cones that will open and drop their seeds only after being subjected to substantial heat (serotinous cones). Although the ponderosa pine does not have serotinous cones, the bark of mature trees is quite thick and thus quite fire resistant. The ponderosa pine also grows rapidly, helping to place the young saplings out of danger from low-intensity ground fires. It is also self-pruning, at least in denser stands, so that its lower branches tend to die and fall away, placing the upper needles and branches out of the reach of substrate-limited fires. Thus the ponderosa pine tends to do well in areas where limited fires occur at fairly frequent intervals of about 5 to 25 years. Such fires do not kill older trees but prevent young growth from making the accumulated dead needles and ground litter so thick and the sapling layer so dense as to promote a rapid spread of fires that may become hot enough to kill even the mature trees. They also open up the forest enough to allow some new ponderosa seedlings to find sufficient light to thrive, eventually maturing and replacing the oldest trees.

The moisture typically provided by winter snow and spring rains allows these trees to survive and shade out most competitors, and the acidic soil environment produced by the tree's resins is also not favorable for the growth of most other tree species. But these forests tend to build up flammable forest litter in the form of dead needles and pruned branches, facilitating the fairly frequent return of substrate fires.

Very long-lived, the ponderosa pine may survive several hundred years, with the first cones usually being produced after about 20 years. Grouse (none in Nebraska), jays (such as the pinyon jay), and Clark's nutcrackers all consume the seeds, as do red crossbills, whose bills are especially well adapted for prying seeds out of their otherwise relatively safe recesses. Pine cone scales are armed with only rather small spines, which do not seem to represent a major deterrent to such species as crossbills. Chipmunks (the least chipmunk in Nebraska) and squirrels (the fox squirrel in Nebraska) eat the seeds and, through their caching activities, help disseminate them. Each seed also bears a small propeller-like "wing" that helps disseminate the seed crop by aerially carrying the seeds well away from their parent trees.

The ponderosa pine is one of the most important timber trees in the West, and its wood, which ranges from fairly soft to hard, is used in construction, crates, furniture, mill products, and the like. Its rather yellowish cast is the basis for the name "yellow pine," and it is sometimes also used in rustic furniture and may then be known as "knotty pine." In Nebraska, the trees usually are no more than 60 feet tall but may be up to about 2 feet in diameter near the base.

Bark beetles are a common source of damage to these pines, and a fungus (*Elytroderma*) may also cause the tree to die. Woodpeckers such as the Lewis' woodpecker and the three-toed woodpecker are attracted to damaged pines, especially fire-damaged trees that support good beetle populations. Pygmy nuthatches and violet-green swallows frequently nest in the trees' woodpecker holes as well. As of the early 1990s there were about 175,000 acres of pine forests in Nebraska.

Prairie Falcon

One of the fastest-flying birds in Nebraska is the prairie falcon. It is a near relative of the peregrine and also of the gyrfalcon, both of which are only seasonally present in the state and are much rarer. Like the winter-visiting gyrfalcon, the prairie falcon relies on blinding speed in level or nearly level flight, rather than the peregrine's even swifter high-altitude dives, to capture its prey. It can outfly small songbirds such as meadowlarks and horned larks quite easily, and both of these species are fairly regular prey items. But it also is a master at taking prairie dogs and ground squirrels in low-altitude sneak attacks, like a radar-avoiding stealth fighter, by coming in silently, nearly at ground level, and taking its prey before it is even detected by the usually alert rodents. Rodents are preferred prey, especially during the breeding season.

Although the prairie falcon may be well named in terms of its foraging habitat, nesting is always done on elevated sites, preferably steep cliffs with inaccessible ledges protected from above by an overhang and from below by 30 feet or more of nearly vertical rock surface. After returning from its winter quarters, the smaller male establishes a territory, which includes a potential nest site and a surrounding territory extending hundreds of yards in all directions. The birds are monogamous, and pair bonds are formed or renewed shortly after both sexes return. There are several so-called ledge-displays performed by both members on the nesting ledge, and mutual soaring behavior is also common.

Once the female begins incubation, the male must provide food for himself, his mate, and any resulting young, at least until they are old enough to allow the female to take short hunting forays on her own. Peregrines are mortal enemies of nesting prairie falcons and are sometimes killed by them; golden eagles and great horned owls present real dangers to both species, especially to immatures. Owls, hawks, and eagles are all threatened or attacked when seen near the nest.

Fig. 3. Prairie falcon adult and young with green-winged teal.

Bobcat

Bobcats are widespread in the state but probably are most common in the broken country of the Pine Ridge and Wildcat Hills, where shallow rocky recesses, crevices, and steep-sided slopes offer denning and hiding places. Bobcats are slightly larger and much more widespread than the Canada lynx, which may soon be federally listed as threatened and which has only very rarely been seen in the state. The most recent case was in 1999, when a radio-tagged lynx that had been released by wildlife biologists in Colorado was illegally killed in western Nebraska. Unlike lynxes, bobcats do not depend upon deep snow for effective hunting and caching of food. Instead, the bobcat likes rocky outcrops with ledges for dens but can also use hollow logs, pile or brush, or similar sites for their activities. They are primarily nocturnal, stalking and killing prey that range in size from

Fig 4. Bobcat, immature.

small rodents to deer. In turn, they are at times killed by mountain lions and coyotes.

Like other wildcats, bobcats have highly variable but sometimes enormous home ranges, extending from less than 1 to about 70 square miles, with the slight larger males having somewhat greater home ranges than females. In ideal habitat their density may reach nearly one animal per square mile. Adults of the same sex usually have nonoverlapping home ranges or territories; these areas are often marked by exposed feces deposits or by urine-marked scent-posts. Bobcats mature rapidly, at one or two years, and immature animals often disperse until they can locate unoccupied areas in which to establish themselves. Only the mother bob-cat looks after the young, with the period to independence about seven months. Except for breeding, little contact between the sexes occurs among adults.

Mule Deer

Of Nebraska's two species of deer, the mule deer is more common in the west and the white-tailed in the east. The zone of a 50:50 ratio extends from Cherry and Sheridan Counties in the north to Gosper and Furnas Counties in the south. Whitetails have been able to thrive in the intensively farmed and wooded river valley areas of the east, but mule deer prefer open, more arid grassland and brushy areas. When alarmed, mule deer are likely to head for heavy brush and rough terrain, fleeing in a distinctive bouncy manner that easily identifies them, as do their larger ears and smaller, black-tipped tail. The branching pattern of the males' antlers also differentiates the two, with the mule deer having repeated, elklike forks on their tines, whereas the whitetail has a series of tines coming off a main, forwardly curving antler on each side.

At one time both mule deer and white-tailed deer had been nearly extirpated from Nebraska, but decades of protection gradually brought the herds back. Enemies such as the prairie wolf and mountain lion are now gone, and smaller predators such as coyotes and bobcats are threats only to fawns and sick individuals.

Mule deer are both grazers and browsers and can live off a wide variety of foods, with grasses and other herbaceous materials more important in spring and summer and woody species more vital in fall and winter. Although in mountainous areas mule deer may be quite migratory, this is not so in Nebraska. Females maintain home ranges of about 1 square mile, and males have ranges of about twice that size. During fall the males' necks begin to swell and rutting begins. Like other deer, males try to attract and control harems of several females through fights with other males and coercion of females. Females typically have their first offspring when between two and three years old and in later years usually have twins. About four months are required for weaning. When about a year old, males disperse to form new home ranges, whereas females are likely to remain near their mothers.

Vignettes of Endangered and Declining Species

Mountain Lion

The mountain lion, or cougar, the largest North American cat, has an enormous geographic range extending from southern Alaska to southern South America over a wide variety of habitats. In Nebraska, it apparently is a rare transient in the western part of the state, but sightings here are becoming increasingly common, perhaps as a result of young animals dispersing out of the Black Hills or the mountains of eastern Wyoming. In 1999 alone there were three definite reports, making a total of at least five in the state during the 1990s. One was shot near Crawford in 1991, and another was found dead near Crawford in 1996. In 1999 one was killed hear Harrison, and another was shot near Alliance. A third was seen near Gering. The species is now fully protected by state law, and there are substantial fines for killing one.

Female mountain lions weigh an average of about 90 pounds, and males average about 140 pounds. Their primary prey are ungulates, especially deer. They are largely nocturnal (most if not all television scenes of mountain lion attacks made during the daytime are set up), often traveling up to 5 miles or more per night. Females have smaller home ranges (averaging about 50 square miles) than males (averaging about 100 square miles). There is little overlap between the ranges of adult males, but those of females and males may overlap somewhat. Except during a brief breeding period, the adults are solitary. However, cubs remain with their mother and are dependent upon her for more than a year.

Attacks on humans used to be extremely rare, but decreases in wilderness areas and increased human intrusion into isolated mountain lion habitats have changed that. There is now an average of about four attacks on humans, and one fatality, annually in North America. The victims usually are people, especially children, who are alone.

Elk

Elk were a common sight for early explorers of the North American prairie, as were bison, deer, and antelope. However, elk were soon eliminated from that habitat for their meat and the males' trophy antlers, which are perhaps the most majestic of all North American ungulates. Wild elk were last seen in eastern Nebraska in 1877, or about 70 years after Lewis and Clark first observed them near the Missouri River of northeastern Nebraska and adjoining South Dakota. By the 1950s and 1960s a few were again observed in western Nebraska, probably as a result of relocation efforts in eastern Wyoming. The herds slowly multiplied, and by 1986 limited sport hunting was allowed for the first time in the twentieth century. The elk are almost entirely confined to the Pine Ridge region, although strays may sometimes occur elsewhere. By the year 2000 about 200 to 250 were present in the state.

Compared with mule deer and white-tailed deer, elk are enormous, with females averaging about 450 to 600 pounds and males 500 to 800 pounds. Elk are more much dedicated grazers than browsers, although in winter shrubby foods may become important for survival.

Fig 5. Elk, adult male.

Although yearling elk may be sexually developed, only older males have any chance of attracting and controlling females. During the fall rut the strange bugling whistle of males can be heard daily, and the sound may carry a mile or more under good conditions. It serves as a challenge to other males, and fights often erupt as a result. One photographer at Nebraska's Safari Park near Omaha used the recording of a bugling elk to attract a male close to his car, which worked better than expected and resulted in a few antler holes in the car's exterior plus a very shaken photographer.

After rutting the male loses interest in the females, which produce one, or rarely two, young after a gestation period of about 240 days. The young remain with their mothers and other herd members indefinitely. Under protected conditions elk may live for 20 years or more.

Lewis' Woodpecker

The Lewis' woodpecker (named after Meriwether Lewis) is Nebraska's rarest woodpecker and one for which actual recent breeding records seem to be rare, although both Lawrence Bruner and Myron Swenk reported them to be breeders in the Pine Ridge earlier in the twentieth century. They certainly nest in the open pine and cottonwood stands of the Black Hills of South Dakota. Although not specifically associated with ponderosa pines, open stands of ponderosa pines, followed by logged or burned forests and riparian woodlands, are the species' most typical habitats. The Lewis' woodpecker is an opportunistic species, responding to insect outbreaks. It does not excavate trees as is typical of other woodpeckers but instead relies on emergent adults, which it typically catches in flight like a flycatcher. It is also not a predictably migratory species in some areas but is most likely to be found in oak woodlands, riparian woodlands, or orchards during migration. Fruits, berries and especially acorns may be eaten then. Being a poor excavator, these woodpeckers are attracted to old burns where the wood is well decayed, allowing nest excavation.

Males are larger than females and have longer bills, but there seems to be no measurable differences in food intake and foraging behavior between the sexes. Flycatching by short hawking flights is the typical foraging mode, with the bird quickly returning to a perch to deal with its captured prey. The Lewis' woodpecker also seeks out acorns during the winter rather than excavating individual crevices for each item. It stores supplies by shoving acorn meat into natural crevices. The acorn woodpecker may recover and eat such stored supplies at any time of the year, but the Lewis' woodpecker only uses acorns as a winter food. The Lewis' woodpecker closely resembles the red-headed woodpecker, which is also a flycatching species to some degree, and occasionally makes caches of insect prey for later consumption.

During the breeding season the Lewis' woodpecker male utters harsh churring calls and drums on wood to advertise territories. A visual display is the "wing-out" posture, performed while perching with the wings extended, the head depressed, and the silvery throat and breast feathers expanded. The male also engages in circular flights around the nest tree. At least in some areas the birds may pair bond permanently, but in migratory populations this seems unlikely.

CHAPTER 4

In the Shadow of the Badlands

The High Plains and Shortgrass Prairies

Walking on the High Plains of northwestern Nebraska is like taking a time-capsule trip back a century or more. The horizon is limitless, the air is sweet, and there are no traffic noises or billboards to assault the senses and sensibilities. The land and its life are simpler here; climatic adversity has winnowed the species list down to those few that are particularly adapted to withstanding frigid winters and sometimes murderously hot summers and are able wisely to use or conserve the limited amounts of available fresh water. Plants huddle in the protective shade of large rocks or become cushionlike as searing breezes desiccate their exposed tips. Many mammals become more nocturnal, and songbirds increasingly confine their activities to the hours around sunrise and sunset, when it is light enough to hunt but not too hot to move about energetically.

In some places winds of the past eons have sculpted the land into grotesque shapes, as at Toadstool Geologic Park, and both hawks and eagles have learned to ride the daily thermal winds effectively in their constant searches for prey. Reptiles such as horned lizards are common here, and scorpions are also to be encountered. Where it does occur, surface water is often too bitter in dissolved salts to be drunk or to be used by most plants; cacti and some other spiny plants have protected their valuable tissues and stored water by evolving adaptations that reduce or prevent grazing.

Birds of the shortgrass prairies include the McCown's longspur and chestnut-collared longspur, the former adapted to lower grasses and drier

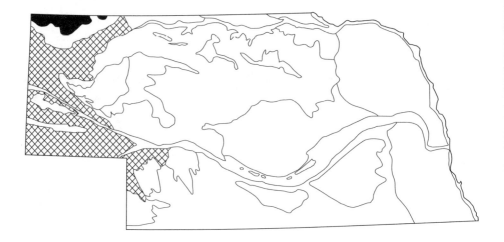

Map 6. Shortgrass and Sagesteppe Prairies subregion (inked) of the Northwestern Great Plains ecoregion, plus the Shortgrass–Mixed-grass Mosaic subregion (cross-hatched) of the Western High Plains ecoregion.

conditions than the latter. Horned larks are common, and western meadowlarks are often fairly abundant. Rock wrens nest in the crevices of eroding cliffs, and the poorly named mountain plover is usually found in nearly barren but level landscapes.

Where shrubs such as sagebrush, rabbitbrush, or greasewood occur and provide both song perches and potential nesting sites, there are likely to be Brewer's sparrows. Sage thrashers are sometimes also seen, but their nesting status in the state needs to be confirmed. The green-tailed towhee is another sage-adapted bird whose breeding status in Nebraska is in doubt, and the sage grouse has long been extirpated from the state. A dryland shrew, the Merriam shrew, has been reported only once or twice from Nebraska in recent years in similar dryland habitats. Several bats also occur only in this part of the state but are more associated with the rocky topography of this region and the nearby Pine Ridge escarpment than with any specific vegetation type.

Ecology of the Shortgrass Prairies

Perhaps the best examples of the shortgrass prairies in Nebraska occur just north of the Pine Ridge in an area called the Pierre Hills in Sioux and Dawes Counties, especially within the Oglala National Grasslands. In Nebraska, only about 600 square miles of this ecoregion (shortgrass and shrubsteppe prairies component; northwestern Great Plains ecoregion) exists; it is much more widespread in South Dakota and eastern Wyoming. The soils have developed from clayey shales, and the topography is mostly gently undulating but with some low buttes and local areas of badlands, for example, at Toadstool Geologic Park, at Agate Fossil Beds National Monument, and along the White River near Chadron. This is one of the driest parts of the entire state, with annual precipitation averaging 14 to

17 inches, much of it occurring as thunderstorms between May and July. Trees are mostly limited to the edges of permanent streams, and cactus such as prickly pear is often common among the perennial grasses, especially where grazing is prevalent.

This is classic pronghorn and sage country, and together with buffalo grass, prairie dogs, and burrowing owls these species might be regarded as indicators of shortgrass prairie. The pronghorn has remarkably large eyes (1.5 inches in diameter), long eyelashes that operate as sunshades, and acute hearing. Lightning speeds (up to about 60 miles per hour in short bursts and up to 45 miles per hour over a prolonged route) were once needed to escape its classic but now extirpated enemy, the prairie wolf. But evolution failed to prepare it for jumping over even fairly low obstacles such as wire fences; it would much rather go miles out of its way in trying to detour around a woven-wire fence than try jump over one, a fact that sometimes leads to its demise. Big sagebrush is the most common shrub associate of the pronghorn and is a major dietary item, even in summer when green grasses are available. During the summer this sage species produces large, juicy leaves that are relished by the pronghorn. In the winter these large leaves are shed and are replaced by smaller and drier ones that nevertheless allow for some photosynthetic activity to occur even in subfreezing weather and provide above-snow-level food resources for the pronghorn.

During the drought years of the 1930s, shortgrass species moved eastward, occupying much of the area that today is usually classified as mixed-grass prairie in the Panhandle and in southwestern Nebraska. The hatched areas on map 1 indicate this region, which Kaul and Rolfsmeier have recently classified as a mosaic of shortgrass and mixed-grass prairies and which might be thought of as a dynamic ecotone easily shifted in one direction or another under varied conditions.

Major Terrestrial Community Types

Shortgrass Prairie

Of all the native prairie types of North America, the shortgrass prairie still persists over the greatest area, for it historically occurred on land too dry for most dryland farming and usually has been located in regions where no reliable source of water for irrigation is present. This is traditional "cowboy country," which is used primarily for grazing cattle (or sheep in even drier areas). Many of the national grasslands in the western states are the legacies of unsuccessful efforts to farm this land, an endeavor that proved to be the result of too much optimism and too little logic. The droughts of the 1930s caused massive bankruptcies in the western states and a return of the land to the federal government. Thus, the Oglala National Grasslands and a dozen or more similarly preserved grasslands were established and have gradually led to a slow rehabilitation of the land and its biota. Two of the biotic community's most characteristic, if not controlling, mammals, the bison and the prairie dog, had historic distributions that encompassed the shortgrass prairie. Recurrent seasonal grazing by bison may have contributed to keeping the low-stature species such as buffalo grass dominant over the taller ones that are typically commonly found in mixed-grass prairie.

Although best developed in the Panhandle, shortgrass prairie occurs locally over much of the western half of the state, especially where cattle grazing shifts the balance of dominance toward shorter species, just as bison grazing did earlier. The remaining native grasslands of western Nebraska are therefore largely a mosaic of mixed-grass and shortgrass types. Typical shortgrass species are buffalo grass and blue grama, both only about a foot or so in maximum height. Included among these shorter grasses are some somewhat taller ones, including side-oats grama, prairie three-awn, western wheatgrass, needle-and-thread, and little bluestem. There are few shrubs but numerous forbs, including nitrogen-fixing legumes such as milk vetches, locoweeds, and wild alfalfa. Other attractive forbs include scarlet guara, cutleaf ironplant, prairie coneflower and scarlet globemallow.

Northwestern Mixed-Grass Prairie

On upland soils of the White River and Hat Creek basins of Dawes, Sheridan, and Sioux Counties, a mixed community of shortgrass and mixed-grass species occur, with blue grama and western wheatgrass the usual co-dominants. Buffalo grass and green needlegrass are abundant. There are few shrubs, but big sagebrush is sometimes present, and where the soils are more claylike silver sagebrush, fringed sage, rabbitbrush, and greasewood may also appear. Native forbs include several legumes such as two-grooved milk vetch and Drummond's milk vetch, and scarlet globe mallow is also often present.

Western (Alkali) Floodplain Meadow

From Sioux and Dawes Counties south to Kimball County, a distinctive meadow community occurs on level and somewhat alkaline soils near rivers and streams. These soils may be flooded occasionally and support a community of variously alkali-tolerant grasses and sedges, including saltgrass, green needlegrass, and western wheatgrass. The taller grasses such as western wheatgrass may attain heights of up to about 3 feet, whereas saltgrass is of lower stature. The wheatgrass-dominated type is more common on the less alkaline sites, whereas saltgrass is more prevalent on the distinctly alkaline sites, which also often support other alkali-tolerant grasses and forbs. Shrubs are uncommon in the latter environment, but under less extreme soil conditions western snowberry, silver sagebrush, and occasional cottonwood trees may be present. Many of these sites are now badly degraded, owing to overgrazing by cattle.

Silver Sagebrush Shrub Prairie

On the floodplain terraces of Hat Creek basin in Sioux County as well as in various locations within the Oglala National Grasslands, a community dominated by mid- to short-stature grasses is present, with silver sagebrush usually also occurring in rather low densities. Other plants are mostly those of shortgrass and mixed-grass prairies. The silver sagebrush shrub prairie is too little studied to know its relation to other community types in the region.

Greasewood (Arvada) Shrub Prairie

The greasewood (arvada) shrub prairie community is known from the Hat Creek basin and occurs on upland floodplain terraces and especially on poorly drained clay soils. The alkaline-tolerant shrub greasewood may occasionally be present, as may rabbitbrush. Alternatively, prickly pear cactus may be common, particularly on clay pans. There are few forbs, but the attractive two-grooved milk vetch is typical. Like some others of its genus, this species accumulates poisonous selenium, but it is rarely eaten by grazing animals. Often prairie dogs find this soil type to their liking, and ferruginous hawks and prairie falcons likewise find good hunting.

Badlands

Although South Dakota and North Dakota can lay claim to the most famous areas of badlands, Nebraska also has a few such areas in Dawes, Sioux, and Scotts Bluff Counties. Of these, Toadstool Geologic Park in the Oglala National Grasslands is best known. The eroded slopes and irregularly shaped landforms are visually attractive but support relatively few plants and animals. Scattered shrubs, primarily silver sagebrush, shadscale, and rabbitbrush, are most typical, but greasewood may occur in more alkaline soils. Many annual forbs, mostly exotic weeds, occur here, along with some native perennials. The perennials include such legumes as the selenium-accumulating alkali milk vetch, the similarly poisonous white locoweed, and pulse milk vetch.

Profiles of Keystone and Typical Species

Buffalo Grass

Buffalo grass is as important to the shortgrass prairies as the bluestems are to the taller prairies. It is a sod-forming grass of low stature that is able to spread by relatively shallow roots and especially by aboveground stolens.

Fig 6. Buffalo grass: *above,* seed-bearing plant, *below,* stamen-bearing plant. Illustration from Hitchcock (1935).

Thus it forms a matlike turf at the ground surface, beginning its growth in late spring and continuing throughout the summer. To a greater degree than any of the other short grasses, it tolerates repeated grazing, and it also withstands a very wide variety of soil and climatic conditions. Highly palatable to both bison and cattle, it helps reduce erosion on overgrazed pastures. Buffalo grass is unusual in that the sexes are on separate plants, their frequency being about equal. Also unlike the other prairie grasses, its seeds are enclosed in a hard bur that persists on the plant. Without treatment, such as chilling, soaking, or abrasion, few of these seeds will germinate. In natural stands up to 100 pounds of seed-containing burs may be produced per acre on natural stands of buffalo grass.

Because of the low germination rate of buffalo grass seeds without treatment, it has been suggested that this grass has evolved a special mechanism favoring herbivore dispersal. That is, ingestion of its ungulate-preferred foliage causes an associated passage of the less tasty burs through the animals' digestive tracts, favoring both the seeds' dispersal and their improved germination rates. This is the "foliage is the fruit" hypothesis, and it has been proposed that this array of traits has favored the spread of buffalo grass from its geographic origins in Mexico north throughout the drier Great Plains region. Additionally, the sex-separate trait of buffalo grass allows for some sex-specific differences in foliage traits. Male plants tend to allocate more of their biomass to roots and vegetative parts, whereas female plants use more of their energy for reproduction. Separation of the sexes may also allow for ecological segregation to occur under varied environmental conditions.

Black-Tailed Prairie Dog

Prairie dogs, pronghorns, bison, and burrowing owls are all as American as apple pie and ice cream and are as closely associated with one another as the presidential heads on Mount Rushmore. The prairie dog is certainly the glue that holds the burrowing owl to this assemblage; the pronghorn can perhaps get along one its own, but where there are pronghorns or bison there are likely also to be prairie dogs and owls. Unlike bison, pronghorns do not seem to seek out prairie dog towns for grazing or use their loosened soil for initiating wallows, but all four of these species seem to prefer open country with unlimited visibility and rather low vegetation. As with the American bison, we can never know and scarcely imagine the one-time abundance of prairie dogs on the Great Plains. Perhaps 700 million acres of the plains were occupied by prairie dogs only a century or so ago. "Dog towns" of seemingly endless size were commonplace. One such site in Texas once covered 25,000 square miles (an area larger than the Nebraska Sandhills) and may have contained 400 million animals. Now the species is so rare it is being considered for national listing as a threatened species.

Of the several species of prairie dogs, only the black-tailed occurs in Nebraska. The social structure of all of the species seems to be similar and is unusually complex for a rodent. That of the black-tailed prairie dog is the most complex, and its colonies can attain the largest sizes. Large colonies are divided, like city blocks, into geographically definable "wards" that tend to be separated vegetationally or topographically. In each ward are a number of social units called "coteries" that consist of related animals. A typical coterie contains one or more adult males, varying numbers of adult and immature females, and juveniles of both sexes. The population of a coterie may consist of up to 30 adults and yearlings plus an indefinite number of juveniles. These groups cooperatively defend their coterie against other groups. Each coterie may encompass an acre or more of area, and represents the home range of any of its members, with its boundaries being defended in a territory-like manner.

The burrow of each coterie consists of mounded entrance holes, with the mounds providing lookout points. There are actually two types of mounds, a domelike type and a cratered type, with connections that allow air to enter dome mounds and exit through crater mounds. These mounds

Fig 7. Black-tailed prairie dog, adults.

are organized in a radial manner around a central point. All the mound holes are interconnected with burrows so that air can flow through the system no matter the direction of the prevailing wind.

Sexual maturity and breeding requires two years, and the litter size is rather small, only three to four young, which may help to limit population densities. Unlike many plains rodents, prairie dogs do not hibernate, and they begin breeding early in the year. In spite of an effective warning system that spreads alarm calls and a clipping of vegetation near mounds to increase panoramic visibility, young and adults are often taken by raptors, coyotes, badgers, and the like. Additionally, females with young often kill the offspring of other females. Apparently to avoid inbreeding, young males leave the coterie before reaching sexual maturity, and females avoid mating with their fathers, brothers, or sons.

The black-tailed prairie dog has been considered for federal threatened status, but strong political pressures have caused the U.S. Fish and Wildlife Service to judge that such a classification is "warranted but precluded" in the eight-state administrative region that includes Nebraska. Such preclusion is based on pressures from ranchers and also because of limited funding for supporting other programs for more critically endangered species. Most of the species' remaining range is now affected by bubonic plague, though the largest remaining colonies (located mainly in South Dakota) are in regions still unaffected by this disease. The species' total current range consists of about 1 million acres, or approximately 1 percent of its presumed original range. Its Nebraska range was believed as of 1999 to consist of 90 to 125 square miles (about 70,000 acres), or about 0.1 percent of the state's land area.

Pronghorn

Probably no North American ungulate is more typical of the American plains than is the pronghorn. They are the sole surviving species of a large antelope-like group of grazing ungulates that has long occupied the ancient grasslands of interior North America, being most similar to but still different from the true antelopes of the African and Asian plains. They perhaps evolved in a semiarid landscape, where shrubs were as common as grasses, and thus evolved a dentition and digestive ability that allow them to shift from one type of forage to the other with ease.

Not only can pronghorns eat and easily digest sage and its associated arid-land shrubs, but they also consume other plants that are poisonous to livestock. Compared with bison, pronghorns forage more heavily on forbs (62–92 percent in one study) and on shrubs and eat a wider variety of plants that average higher in overall nutrient quality than those chosen by bison. They show more seasonal diversity in foods consumed, are likely to selectively eat leaves and flowers rather than stems, and generally prefer dicot plants such as forbs over monocots such as grasses and sedges.

Pronghorns have unusually small stomachs but have relatively large lungs and hearts, signs of a superb runner. Like cheetahs, pronghorns have a highly flexible spine, increasing their racing efficiency, and their scapulas (shoulder blades) are very loose so their legs can reach farther forward and increase their stride length.

Male pronghorns have larger and more impressive horns than females; they are unique among horned animals in that the black and horny outer portion is shed each year, but a tapering bony core is retained. Large males develop horns that branch and curve backward. The adult male also has a conspicuous black muzzle and large scent glands in front of the eye that are used for marking its territory. There are several blazes of white on the pronghorn's otherwise brown body, and the white area around the rump can be conspicuously raised when the animal is alarmed, producing a highly distinctive danger signal to others.

Pronghorns are highly mobile and may cover hundreds of miles when migrating seasonally. Like bison, they were hunted unmercifully, and the

Fig 8. Pronghorn, adult male.

20 to 40 million animals that once roamed the plains were reduced to about 30,000 by the early 1900s. In Nebraska, the species was almost completely extirpated, but reintroduction efforts in the 1950s and 1960s have brought the population up to about 7,000 animals.

Like other North American ungulates, pronghorns have a fall rutting season, during which males attempt to corral and control several females so that they can be fertilized during their short estrus periods. The males rub their cheeks and faces on brushes and twigs, scent-marking their territories, and if needed fight to defend them. In some areas definite territories as such may not actually be established and defended, but harems are nonetheless gathered and protected. Females become sexually mature before they are two years old, and their gestation period is about 250 days.

They typically produce twin offspring. These youngsters must be able to run within hours after birth, but most of their first few weeks of life are spent in a motionless hiding posture, rising occasionally only to nurse. Mothers usually stand about 70 yards away from their hidden young, far enough away to not reveal the location of the young but close enough to intervene should that become necessary. Shortly after birth the young seem to produce no bodily odors and are often overlooked by coyotes. Within a few days they can easily outrun a human and can put a coyote to a very good test. Nevertheless, coyotes are probably the pronghorn's most serious natural enemy, at least for calves.

Burrowing Owl

At one time, when there were hundreds of millions of prairie dogs on the Great Plains, there must have been tens of millions of burrowing owls. Even 40 years ago burrowing owls sometimes nested in the vicinity of Lincoln, where there then were no prairie dogs but where the burrows of woodchucks and other large rodents were available. Now burrowing owls are very hard to find anywhere in Nebraska; only in the Panhandle is one likely to locate any.

All owls exert a kind of strange attraction to humans; their large eyes, nocturnal behavior, and ghostlike presence, usually marked only by haunting calls in the night, appeal to our sense of the mysterious. Burrowing owls break this typical owl mold by being out in broad daylight, sitting on fence posts and beside rodent burrows like little feathered gnomes, swiveling their heads around to watch passing cars or animals, and sometimes flying from perch to perch like giant brown moths. Their eyes are a bright yellow, and unlike those of most owls their retinas lack the density of rod elements needed for efficient nocturnal vision. Thus they forage mainly during the day, mostly on larger insects such as scarab (dung) beetles, which are common around prairie dog colonies. They refurbish old, unused rodent burrows, nesting belowground, with an adult often standing guard at one of the entrances. Little modification is made to the burrow itself, but the birds often scatter dried fragments of cattle or bison dung around the entrance to their burrow. The function of this behavior is uncertain, but the dung perhaps helps mask the scents that might make the site attractive to inquisitive coyotes or badgers, both major enemies of these owls.

Unlike most Nebraska owls, burrowing owls are migratory and soon after their spring arrival set up housekeeping at any available burrow. Females lay rather large clutches of six or more eggs, each egg laid several days after the last, and incubation begins immediately. The result is that the clutch may hatch over a period of two to three weeks, so the youngest owl in a brood may be nearly three weeks younger than the eldest. As they grow, they increasingly venture out of the burrow; when several young of

Fig 9. Burrowing owls, adult and two young.

uniformly graded sizes are standing together near a nest, it may resemble a portrait of a family that never learned how to practice birth control. By the time of the fall migration perhaps half of the brood will have died because of starvation or other causes, and many of the rest will not make it through their first winter. Their first year is always a difficult time for young raptors, who must quickly learn to capture prey on their own after parental care ceases.

Vignettes of Endangered and Declining Species

Black-footed Ferret

The black-footed ferret is one of those near-mythic plains mammals; it was one of the last larger mammals of North America to be described, being painted by John J. Audubon in 1851 from a skin obtained in eastern Wyoming. At one time the species also occurred in the western plains of North Dakota, where I had been brought up to believe it was already extirpated in the 1950s. It had probably once ranged widely over the plains, feeding almost exclusively on prairie dogs. However, the decline of the prairie dog spelled disaster for the ferret, which needs a substantial prairie dog population to support a tiny ferret population.

There are few records of the black-footed ferret in Nebraska, the last being of an animal that was run over by a car near Overton in 1949. By the 1960s the only known surviving colony was found in southwestern

South Dakota. That colony finally disappeared, its last survivor dying in captivity in 1979.

Fig 10. Black-footed ferret, adult.

By 1967 the black-footed ferret had been recognized as an endangered species, and the Endangered Species Act in 1975 provided federal funds to try rescue it. A new wild population was discovered in Wyoming in 1981, which upon study was determined to contain some 130 animals. Then, its associated prairie dog population was decimated by an outbreak of sylvatic plague, and the ferrets were infected with canine distemper, perhaps contracted from coyotes, badgers, or skunks. By 1987 the last wild ferrets had been brought into captivity in order to try save the species through captive breeding. Seven different breeding facilities were used, including the Henry Dorley Zoo in Omaha. The resultant offspring were the progeny of 18 captive animals. The captive population now numbers in the hundreds, and attempts at reintroduction into the wild have been made in northwestern Montana, western South Dakota, and southeastern Wyoming. No reintroduction efforts have been made in Nebraska, which has little prime prairie dog habitat under governmental control and where until recently a state law required the annual control of prairie dogs on both private and state-owned properties.

The black-footed ferret is a close relative of the Asian and European ferrets and polecats and of the native North American weasels. It has a distinctive black "mask," as well as black feet and forelegs. It is also much larger than the long-tailed weasel, weighing up to 2.5 pounds, and about the size of a female mink. All of these animals are notable for their speed,

Fig 11. Swift fox.

agility, and effective killing ability. Ferrets are easily able to crawl down into prairie dog burrows and subdue prairie dogs by biting the neck and spinal cord.

Swift Fox

The swift fox, or kit fox as it is sometimes known, is a small, tan-colored fox of the American shortgrass plains with a silvery gray back and a similarly dark tail that reminds one of the somewhat larger black-backed jackal of the East African plains. The Great Plains version of the American species, the swift fox, has smaller ears and a shorter tail than the kit fox, a distinctive subspecies (some would argue a full species) found in the more arid southern semidesert and scrub-steppes. Both types are smaller than any of the other North American foxes and weigh only about three to seven pounds.

The prey of these tiny foxes is also small. In fact, during the summer up to half of the swift fox's total diet may consist of grasshoppers. In this it is a counterpart of the Swainson's hawk, which also specializes on grasshoppers while they are abundantly available. Otherwise, swift foxes feed on a variety of mammals and some ground-nesting birds but especially on rodents and rabbits. In spite of their small size they can run remarkably fast (up to 37 miles per hour) and thus are a fair match for a jackrabbit.

Swift foxes live as mated pairs throughout the year but often hunt alone, typically at night. The home ranges of an adult male and female greatly overlap, and sometimes a male may have two females within his home range. Each pair may have several available burrows, which they change fairly often, perhaps to help control the population of fleas and ticks. Breeding occurs once a year, during the winter, so that the pups are born in the spring (March and April) when an abundance of food is likely to be available. Males help provision the young with food, and the family remains together until the following fall. Through the fall and winter small mammals comprise their primary food, with carrion eaten whenever it might be necessary.

Swift foxes are now rare almost everywhere and have been extirpated from their one-time prairie habitats of western Canada. Fairly easily trapped and far too curious for their own good, they can be readily controlled by trapping, shooting, and poisoning. Swift foxes were almost eliminated from the Great Plains during the 1900s but at least in some areas have begun a recovery. Coyotes are a significant threat to them, as are golden eagles and larger hawks such as the ferruginous hawk.

Mountain Plover

The inconspicuous, shortgrass-adapted mountain plover should probably have instead been called the steppe plover; they see few if any mountains during their entire lifetimes. The High Plains, with grass so low and scattered as to be almost absent, best suit this bird. Like the similar-sized but much more widespread killdeer, the mountain plover has the large eyes and short bill that clearly mark a visually hunting, substrate-foraging shorebird. It lacks the conspicuous black breast-banding of the killdeer, and when crouched the bird can effectively disappear into the landscape, with its tan-above, white-below countershading pattern perfectly attuned to its pale, somewhat featureless environment. The dry plains of eastern Colorado and eastern Wyoming, instead of western Nebraska, best fit the needs of the mountain plover. Only in Kimball County near the Colorado border is it likely to be encountered, and then only by careful searching of the region's rather barren grassy plains. This "prairie ghost," like its equally hard to find shortgrass associate the McCown's longspur, is well worth searching for but not one whose photographic portrait will likely impress nonbirding friends. Like many inconspicuous steppe-adapted species, such as the swift fox, they are perhaps tended only by the God of Small Things. The ranges of the swift fox and mountain plover not only closely overlap, the plover probably often provides dinner for the fox.

Mountain plovers are migratory, returning to the plains of eastern Colorado and southwestern Nebraska in March. Territorial males soon begin advertisement displays, including "butterfly flights" with deep wingbeats above their territories as well as "falling-leaf" flights, in which they drop from a height of about 20 to 30 feet with their wings held in a nearly vertical

Fig 12. Mountain plover, head of an adult.

position while rocking from side to side, exposing their pale underwings and bellies. Pair bonds seem to be rather transitory; females may leave their first clutch of four eggs to the care of their mate alone so that they can find a second mate and begin a new clutch. Like the bird, the nest and eggs are very hard to find, as the nest consists of only a few bits of grass and the eggs match their backgrounds closely. The young are also well concealed and nearly impossible to see when they stop moving.

The mountain plover is considered an endangered species in Canada and has been under consideration as a nationally threatened or endangered species in the United States since 1993. It was officially listed as a threatened species in Nebraska in 1976, and the only recent records have been for Kimball and Box Butte Counties. The farthest east that it was reported breeding historically is 120 miles west of Fort Kearney, on the North Platte River. Its mostly likely remaining habitats are along the two westernmost tiers of Panhandle counties.

Ferruginous Hawk

If a single bird is ever proposed as a symbolic Great Plains species, one would be hard-pressed to find a better candidate than the ferruginous hawk or perhaps the prairie falcon. The sight of a ferruginous hawk majestically circling high in the atmosphere, with sunlight passing through its translucent wings and tail and surrounded by cumulus clouds and blue sky,

Fig 13. Ferruginous hawk, adult with rattlesnake.

can only be described as breathtaking. Not long ago I wrote that one could live a lifetime on the plains and never expect to see a ferruginous hawk, a prairie falcon, and a golden eagle all in the same day, only to have that very event occur a few weeks later on Colorado's Pawnee Prairie. Some days on the prairie can be that good.

This is the largest of the prairie hawks; adult males average about 2.5 pounds and females about 4 pounds. Almost twice the mass of a Swainson's hawk and nearly half that of a golden eagle, they fill an intermediate forging niche. With their powerful toes and a sharp beak having a nearly 2-inch-wide gape, these birds can readily kill large rodents and rabbits and can swallow smaller rodents whole. They are jackrabbit and prairie dog specialists, and their densities tend generally to coincide with the availability of these primary prey species. Where this prey base declines, as it has in the Prairie Provinces of Canada, the birds will also soon decline or disappear. Besides seeking out the High Plains, the birds favor those areas with elevated nest sites, such as knolls, buttes, rock outcrops, and low cliffs, but where such natural landscape features are lacking they may resort to nesting on haystacks, on transmission towers, in isolated trees, or even in small groves.

Ferruginous hawks have large home ranges, with the nests of adjoining pairs averaging about 8 miles apart. This spacing would suggest that their home ranges might be as large as 30 or 40 square miles, which roughly agrees with an average density estimate of a pair per 40 square miles in the Pawnee Grasslands of eastern Colorado. The birds are mostly migratory, with pairs returning early and sometimes starting egg laying by mid-March. If not permanently pair-bonded, such bonds are renewed or formed near the nest site. These birds sometimes displace common ravens from possible nest sites but surprisingly may themselves be displaced from nests by the smaller Swainson's hawk.

In 1991 the ferruginous hawk was nominated as a candidate for listing under the provisions of the Endangered Species Act, but this action was defeated. However, it is listed as a Category 2 species by the U.S. Fish and Wildlife Service and as a sensitive species by the Bureau of Land Management. In Canada, it was listed as threatened in 1980 but was upgraded to vulnerable in 1995, since its status there has improved there in recent years.

CHAPTER 5

Borne on the Wind

The Loess Hills and Mixed-Grass Prairies

Much of what we now think of Nebraska's typical visible topography is actually foreign. It came from Colorado, Wyoming, and sometimes even farther away in the form of gravel, sand, silt, and clay. Most or all of the larger particles, the gravels and sands, were carried in from the west by ancient rivers draining runoff from mountain highlands. These heavier materials tended to remain where they settled, along the edges and bottoms of streams that gradually gave them up as the streams lost velocity and became too slow to move them any farther. But the tiny silt and clay particles were not so easily discarded, and the terrible winds that came and went, especially during the interglacial periods, often tossed these materials about. Sometimes they carried them hundreds of miles, much like the dust carried by windstorms of the Dust Bowl years, when the red clays of Oklahoma and Texas were blown into Nebraska and even into the Dakotas in great, choking clouds.

Thus the Nebraska Sandhills, and perhaps other sandy regions of the central Great Plains, gave up their silts and clays during the interglacial periods. These materials were blown south and east, falling on and covering over the earlier deposits of sediments and other even older materials. Water and roots can easily penetrate these loosely packed and mostly silty deposits, and their surface layers were gradually knit together by a mixture of evolving perennial grasses and forbs but very few trees. Annual precipitation in this region now ranges from about 17 to 25 inches annually, the mostly summer rains providing for a rich and varied native flora.

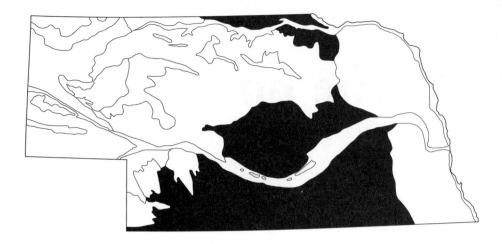

Map 7. Loess Mixed-grass Prairies subregion (lower two inked areas), Central Great Plains ecoregion. The Mixed-grass Prairies subregion of the Northwestern Great Plains ecoregion and the Sandhills—Mixed-grass Ecotone subregion of the Northwestern Glaciated Plains are also shown (two upper inked areas).

These often yellowish loessal soils are usually relatively low in fossil remains, but they sometimes contain fossil mammoths, such as the famous giant mammoth *Archidiskodon* of the late Pleistocene. They are sometimes also nearly riddled with the burrows of now-extinct rodents. Modern-day rodents such as the plains pocket gopher and ground squirrels (mainly thirteen-lined, more rarely spotted) find the soils equally desirable. Prairie dogs also use these same areas, now in sharply declining numbers, along with their commensal burrowing owl associates. Once immense herds of bison wandered seasonally over these soils, making wallows in the easily excavated surface layers and often spending their winters in the Republican Valley and other areas of mixed-grass prairies in what are now Kansas and Oklahoma. The bison could winter in areas receiving up to several feet of accumulated snow, using their massive heads and horns for pushing away the drifts to reach the sweet grasses hidden below.

Ecology of the Mixed-Grass Prairies

As one proceeds eastward across Nebraska's plains, the annual average rate of precipitation increases at a rate of about 1 inch per 25 miles, as the land increasingly comes under the influence of Gulf Coast moisture. Simultaneously, the rate of evaporation decreases as relative humidity increases, making it possible for plants and animals that are less drought-adapted to survive. It also means that taller plants requiring a greater degree of moisture during the summer growing season can survive. Much of Nebraska's mixed-grass prairies developed over loess soils, which are extremely uniform in particle size and typically are far deeper than any root system can ever penetrate. Water penetrates these soils easily, and they are highly fertile because of their calcium content. The upper level of calcium accumulation caused by downward leaching is generally from about 3 to

6 feet below the surface, or easily within the root depth of most native plants. Even low-stature grasses such as blue grama and buffalo grass may have roots that penetrate up to 6 feet in loess soil, and the roots of other grasses and some forbs may easily extend to a depth of more than 10 feet.

Because of original irregular deposition and subsequent erosion, the loess hills are fairly hilly, with intervening lower areas traversed by streams. Erosion by water and wind is rapid, causing steep gullies and small canyons, depending on the depth of the loess mantle. Loess characteristically erodes into nearly vertical slopes, which because of zones of differential slippage rates tend sometimes to slump slide sideways to form a series of small, steplike slopes and alternating nearly level areas of accumulation, or catstep, topography.

About 80 percent of the annual precipitation in the Loess Hills region of central Nebraska falls between April and September and indeed is often largely concentrated during a few heavy spring storms. June is usually the wettest month. The growing season lasts about 150 days. In the winter an average of 25 inches of snow falls, accumulating in ravines and valleys.

Because of their relatively rapid erosion rates, loess soils have little accumulated organic matter and thus do not retain moisture well, so water tends to drain away fairly rapidly after heavy rains. Like most of Nebraska, wind is a nearly constant presence in the Loess Hills, and thus leaf transpiration, surface evaporation, and wind erosion are all significant environmental influences. Trampling by cattle along regularly taken routes also often sets erosive forces into action. Because of the easily plowed and tilled soils, very little of the original Loess Hills prairie remains. Those sites that exist are nearly all badly overgrazed. Grazing, hay production, and dryland grain production all commonly occur on these soils.

Where erosion has cut down through the loess deposits to expose underlying substrate layers, such as locally along the Republican River and its tributaries, a chalky substrate known as the Smoky Hill member of the Niobrara geologic formation may be revealed. This layer of materials contains the toxic element selenium, which is of no known nutritive value to plants but which some plants accumulate nevertheless and store in their leaves and tissues. Examples of selenium-rich plants include various species of milk vetch legumes that are often called "locoweeds." Consumption of these selenium-containing plants may cause severe poisoning, perhaps thus serving as the plants' antigrazing biological defense.

Little bluestem, tall (prairie) dropseed, and needle-and-thread are the vegetational glue that hold mixed-grass prairie together; all are perennial bunchgrasses that tend to be spaced about 1 to 2 feet apart but send an umbrella-like network of roots down 4 or 5 feet into the soil that radiate outward to occupy the entire subterranean area between the grass clumps. These plants' roots, considering only the uppermost 4 inches of soil, may represent about a ton of air-dried organic matter per acre.

The prairie dropseed is a warm-season grass, with flowering and seed-set occurring in August or September. It drops its seeds shortly thereafter, but as with little bluestem, its long leaves remain attached to the stems all winter, providing accessible winter food for ungulates and cover for small mammals. Unlike prairie dropseed and little bluestem, needle-and-thread is a cool-season grass, its seeds maturing in mid-June and dropping by early July. The seed has a long, needlelike awn whose base is tightly coiled; periodic wetting and drying causes it to repeatedly uncoil and twist, gradually drilling the seed itself into the ground.

The mixed-grass prairies extend much farther south than Nebraska, into the Flint Hills of Kansas and beyond, and include some grassland species that extend northward only as far as southern Nebraska. Such is the Fremont's clematis, which barely extends into southern Nebraska. The Fremont's evening primrose and Fendler's aster also have ranges terminating along the Republican River valley.

Major Terrestrial Community Types

Loess Mixed-Grass Prairie

Widespread across the southern half of Nebraska and Kansas, the loess mixed-grass prairie also locally extends to northeastern Nebraska. It grows on undulating to sometimes nearly level soil that is usually deep and always well drained. Locally, limestone or sandstone outcrops may occur where the loess mantle has eroded away. The plant community of the this prairie type is significantly different from mixed-grass prairies growing farther west on drier, nonloessal soils (see the later discussion of the western mixed-grass prairie) and is dominated by tall- and mid-stature grasses, with an understory of low grasses. Side-oats grama and big bluestem are the most important of the taller perennial grass species on level uplands, and blue grama comprises the typical understory. On slopes, little bluestem replaces big bluestem, but on bottomlands big bluestem correspondingly gains in relative importance. In draws and ravines, shrubs such as chokecherry, buffalo currant, wolfberry, and poison ivy may occur, and on hilltops and uplands yucca may appear. There are many attractive perennial forbs, including prairie goldenrod, scarlet guara, prairie cone-flower, scarlet globe mallow, and plains yellow evening primrose.

Harold Hopkins identified 164 species of plants in various remnant Loess Hills prairies, which ranged in life-form from shortgrass to tall-grass types. Steven Rothenberger found 239 species on a single mixed-grass prairie site in the Loess Hills. On dry exposures, blue grama tends to dominate, but on typical mixed-grass sites side-oats grama and other midgrass species begin to shade out the blue grama. In the moistest sites, bluestems wrest dominance from both blue grama and other mid-stature grass species.

Loess Bluff Prairie

The loess bluff prairie is developed on the loess bluffs lining the Missouri River in eastern Nebraska and especially in Iowa. These bluffs show a great deal of slope-effect, with prairie developing on the warmer south-facing and west-facing slopes and hardwood forest on the opposite-facing, cooler, more shaded northeastern slopes. On the prairie areas, needle-and-thread is a common dominant, appearing fairly early in succession. It is later joined by yucca, blazing star, wild licorice, round-headed and purple prairie-clovers, and other prairie perennials. An ecotone to woodland and forest occurs at bluff crests and in shallow ravines in the form of sumacs, rough-leaved dogwood, wild plum, and hawthorn thickets.

Because of slope-effect, loess bluff prairies on the Nebraska side of the Missouri Valley are much more limited in extent but can be found in Indian Cave Park, in Ponca State Park, and in southern Thurston County. In these areas, they approach typical tallgrass prairie. Many of the Loess Hills plants of the Missouri Valley are Great Plains species that reach their eastern limits here or in western Iowa. These plants include Great Plains yucca, skeletonweed, scarlet guara, lotus and Missouri milk vetches, scarlet globe mallow, stiffstem flax, red threeawn, shell-leaf penstemon, and Flodman's thistle. Some Great Plains reptiles and amphibians also reach their eastern limits here, including Woodhouse's toad, Great Plains toad, Plains spadefoot toad, ornate box turtle, prairie rattlesnake, and Great Plains skink. Similarly, the white-tailed jackrabbit, Plains pocket mouse, and northern grasshopper mouse approach or reach the eastern ends of their ranges here.

Western Mixed-Grass Prairie

Over much of the Panhandle region and eastward along the Platte and Niobrara Rivers to Keith and Brown Counties, respectively, undisturbed uplands support a community that consists of a mixture of short grasses and sedges, along with some midgrasses, especially needle-and-thread. The typical sedge is threadleaf sedge, and blue grama is the common native grass, although with heavy grazing this may be replaced by annual brome grasses and western wheatgrass. Among the relatively uncommon shrubs are skunkbrush sumac and winterfat, a favorite winter food of cattle.

Juniper (Red Cedar) Woodland

On steep north-facing slopes of the central Platte Valley and on similar canyon, bluff, ravine, or other steep gradients farther west, an often fairly dense stand of junipers develops. These trees may be up to about 20 feet tall and when tightly clustered may have little vegetative undergrowth, the ground instead being covered by needle litter and a few annual species.

Under less dense canopy cover, some shrubs such as chokecherry, buffalo currant, and skunkbrush sumac may be able to survive, along with various herbaceous species typical of the mixed-grass prairie. The junipers of central Nebraska are largely introgressive hybrids between eastern and Rocky Mountain juniper, and these plants produce rich crops of berries that are eagerly consumed by birds such as cedar waxwings and, in winter, Townsend's solitaires and American robins. During summer, foraging orchard orioles and other insect-eating birds frequent the stands. Black-billed magpies often nest in the dense branches of junipers; their rounded accumulation of sticks and twigs provide a nearly impenetrable defense against possible mammalian predators such as raccoons.

In areas where fire protection is complete, cedars may reach a height of 60 feet and live a century or longer. However, cedars are intolerant of prolonged flooding and have traditionally been excluded from lowland floodplain forests because of recurrent floods. Unlike the other trees of eastern Nebraska, their evergreen leaves permit some photosynthesis to continue over winter, giving them an early growth advantage over deciduous species.

This community type has expanded phenomenally in Nebraska in recent decades, especially in the mixed-grass prairies of the Loess Hills, where a combination of cattle grazing, fire elimination, and expanded seed sources from shelterbelts planted during the Dust Bowl years have all greatly benefited cedars. Cedars are extremely fire-sensitive, but they are also highly drought-tolerant and rarely if ever are browsed by large animals. Limited grazing by cattle may favor the establishment of cedars by reducing grass competition. Cedar seedlings sometimes appear as a result of seeds carried in by birds that had consumed the fleshy fruits. The ingested seeds pass through their digestive tracts undamaged and are often soon deposited beneath fence-line perches, where they quickly take root. Within ten years the trees may themselves be producing seeds. The young trees also soon develop a reddish heartwood that is highly resistant to rot and to attack by wood-boring insects. For such reasons cedar chests discourage cloth moths and the like from taking up residence.

Profiles of Keystone and Typical Species

Little Bluestem

Little bluestem is the "shaggy" prairie grass of which Willa Cather wrote lovingly, whose English name refers to a bluish cast that is present on the lower leaves and stem nodes of growing plants. However, by midsummer much of the entire visible plant is starting to turn a rich reddish tint, and by fall one can easily recognize little bluestem by its combination of bunchlike or "shaggy" shape and its wonderful overall coppery red color, almost matching the colors of an autumnal prairie sunset. It and

Fig 14. Little bluestem. Illustration from Hitchcock (1935).

side-oats grama, whose equally distinctive florets hang down one side of the plant stem like the feathers of a Lakota brave's war lance (the plant was thus called "Banner-waving-in-the-wind grass") are two of the highly distinctive grasses of mixed-grass prairies.

Little bluestem is by far the most important plant of the mixed-grass prairie, and it extends eastward to dominate, along with big bluestem, the tallgrass prairie. It also penetrates the entire Sandhills region and locally may even find opportunities for survival in moist depressions of shortgrass prairie. Like big bluestem, it is a warm-season species, obtaining much of its growth in the warmest summer months and sending out graceful feathery flowering stalks in early fall, typically in late September and October. Its abundant seeds are soon dropped, but the upright stems and leaves persist over the winter. Cattle are not as fond of using little bluestem for winter forage as are bison. In good years little bluestem may produce 200 or more pounds of seeds per acre, or at least as much as big bluestem. This compares with about 100 pounds of seeds per acre produced by side-oats grama and 100 to 180 pounds for blue grama.

Bison

If ever there were an appropriate mammalian symbol of the Great Plains, it is certainly the bison. The entire economy, to say nothing of the mythology

Fig 15. Bison, adult male in flehmen display.

and religious symbolism, of the Plains Native Americans was inextricably bound up with bison. In a comparable way, early Christians liked to identify themselves as helpless sheep, being looked after by a benevolent godlike but also flesh-and-blood Good Shepherd. The bison Katonka was both flesh and spirit to the Lakota people, giving them a vital source of high-protein food and sturdy hides for blankets, clothing, parfleches, and canvases for recording personal histories. Dried buffalo chips provided fuel, the animals' paunches served as cooking pots, and utensils were made from their bones, horns, and hooves. It was only by decimating the bison that the Plains Native Americans were themselves finally conquered and their culture decimated by a concerted action of the American government.

Like cattle, bison are ruminate ungulates, but they are far better adapted to consuming native grasses without destroying them and more efficient in converting these foods to high-protein, low-fat carcasses. Compared to cattle, bison tend to eat a higher proportion of grasses and a lower proportion of forbs and browse. Bison also have a higher degree of digestibility of grasses and grasslike plants than do cattle, perhaps because their food-retention time is greater. They crop plants nearer to the ground than do cattle, are less selective in their foraging plant choices, and allocate less of their overall time for grazing. Bison can survive deep-snow winters and brutally hot summers. They are almost always on the move, so that no single area of grassland is likely to become too overgrazed.

Bison exerted strong influences on the grasslands they utilized, and some of these effects still persist. Their wallows often were filled with spring rains and, until they dried up in summer, became temporary ponds that could be used by shorebirds and other aquatic animals. These wallows are highly durable landscape depressions, often lasting decades or sometimes even for a century or more. Some wallows became extremely large when used by many animals, occasionally covering 2 or more acres and developing a unique plant community dominated by sedges, rushes, and tall dropseed. At least some of these plant differences relate to the higher soil phosphorus levels and more acidic soil chemistry associated with the wallowing behavior and to greater soil packing and altered moisture relationships. The abundant dung produced by bison also tended to spread nutrients and to fertilize the land over which they walked. The traditional Pawnee name for the Republican River translates as "Manure River," referring to the vast amounts of dung that were once deposited in it by wintering herds of bison.

Males are much larger and stronger than females, and during the late summer and autumn rutting season they become highly aggressive as they try to assemble harems through threats and actual fighting. At that time the males bellow often, and when near an estrus female they open their nostrils and retract their upper lips in a distinctive display. Especially during rutting bison must be accorded great respect; more people are killed in the national parks each year by bison than by any other wild animal. To illustrate their unpredictability and potential danger, I once made the mistake of getting too far from my car in trying to photograph a rutting male late one fall day in the Black Hills of South Dakota. I approached to within about 50 yards of a male standing on a hilltop and while focusing my camera noticed that his tail was raised, a certain sign of aggression. Perhaps the animal mistook my gunstock-mounted camera and its very long 500 mm telephoto lens for a rifle, and possibly the setting sun's image reflected off its front element, alerting it to possible danger. In any case, the animal suddenly lowered his head and headed straight for me. I dropped my camera and lens and quickly ran down the hill toward my car, where the door was open and the motor was still running. However, bison can run at nearly 40 miles per hour, and within seconds I could hear his hooves right behind me. In desperation I threw myself to the side, making a forward somersault and coming to a quick stop, just as the animal passed by me, its nearest horn within inches. I got up quickly and ran for the car as the bison began a circular turn back toward me. I had just enough time to close the car's

Fig 16. Coyote, howling adult.

door and speed away, with the bison still in half-hearted pursuit. Later that evening I went back with a flashlight to look for my camera and lens, expecting to find that they had been trampled into the ground. But they were still intact, and my film had captured a single, once-in-a-lifetime image of the male just before he began his charge.

Coyote

Coyotes are so much a part of the American scene it is hard to separate myth from fact, cartoon from creature. The Native Americans' elusive Trickster becomes Hollywood's Wile E. Coyote and then simply disappears into the background again, without ever revealing its true character. The actual coyote is part dog, part wolf, part fox, and part jackal, depending upon the time and place. It is smaller than a wolf, larger than a fox, and probably more abundant than either, in spite of a century or more of intense persecution. It can live in the literal shadows of a city or survive in the utter desolation of a desert. Probably part of the reason it is so universally disliked is because few humans are willing to accept the fact that they can be so easily outsmarted by an animal. No other American mammal is so often displayed as a corpse, often cruelly impaled on a barbed-wire fence, as if to prove that indeed a human actually managed to get the better of it. Yet the coyote has survived and has even expanded its range during the past century, occupying much of the range that wolves once controlled. In 1999 North Dakota spent over $800,000 of taxpayers' money in trying to control coyotes, which caused an estimated $126,000 in livestock damage during that same year. Apparently, coyotes are better at mathematics than government officials, and the state's coyote population remains healthy. Perhaps the God of Small Things has a soft spot for coyotes, too.

The secret of the coyote's success is its versatility. It can be a loner or a social hunter, be a predator or a scavenger, hunt diurnally or nocturnally, or make any number of other adjustments to fit its immediate needs. However, it is almost always a monogamous pair-bonding species, with the pair and their family the social unit. Small rodents are its primary prey; by eliminating other natural predators that might affect domestic livestock, such as wolves, bobcats, hawks, and owls, humans have stimulated the growth of a small-mammal prey base for coyotes, to which they have naturally responded. Like the often-persecuted buteo hawks and barn owls, coyotes are actually the farmer's closest friends and cheapest allies in fighting rodents and should be welcomed on one's land rather than used for target practice.

Sharp-Tailed Grouse

Of all the North American grouse species adapted to Nebraska's grasslands, shrublands, and woods, the sharp-tailed grouse has survived the best. Greater prairie-chickens have declined and retreated in their Nebraskan range, while the ruffed grouse, lesser prairie-chicken, and sage grouse have completely disappeared. But the sharp-tailed grouse's range in the state has hardly been affected by settlement; they have probably always been most abundant in the Sandhills region and in the cooler prairies to the north. That region closely corresponds to their basic habitat needs, namely large areas of grassland intermingled with scattered brushy areas. Other subspecies have somewhat differing habitat needs, especially in Canada and Alaska, but for the Great Plains race of sharp-tailed grouse this region comes close to Nirvana.

In the Sandhills, the sharp-tailed grouse and greater prairie-chicken overlap ecologically, but the prairie-chicken does best where small grains can be found and can be used to supplement their native diets of grasses, berries, and insects, whereas the sharp-tailed grouse has little use for such amenities. Sharptails in Nebraska have been found to consume more than

Fig 17. Sharp-tailed grouse,
adult male.

300 different types of foods, but a dozen or fewer plants comprise their basic diet. These include the fruits of ground cherries, wild rose, and poison ivy; the leaves and seeds of wild flax, knotweeds, and clover; and the buds of poplars, which together with rose hips are an important winter food. During the summer, protein is obtained by eating ground beetles, crickets, and short-horned grasshoppers. Although water may be drunk when it is available, the birds can probably obtain enough moisture from their foods to survive if needed. Like other grouse, sharptails can avoid extreme cold winter temperatures by burying themselves in snow drifts.

Sharptail males gather each spring on communal display grounds, or leks, during which time by postural displays and actual fighting they establish and contest territories. The older, more experienced males are able to control the most central territories, while younger and less experienced individuals must settle for positions on the periphery. Interior territories may not only be safest from roaming land predators but also often have the best panoramic visibility. Either by the relative lek position or by specific behavioral clues, females can rapidly identify these "master cocks" and selectively seek them out for mating. As a result, a very small number of males are likely to fertilize an entire local population. This behavior might lead to severe inbreeding were it not for the fact that few master cocks hold their exalted position for more than a year. Males are typically at least three years old before they can attain such competitive status, and very few male prairie grouse ever live longer than four or five years.

Vignettes of Endangered and Declining Species

Upland Sandpiper

In contrast to the mountain plover, the English name of the upland sandpiper is a good fit. To a greater degree than other American shorebirds this is a species of upland meadows and midgrass prairies, with little or no tendency to gravitate toward water. It is also America's most highly migratory grassland shorebird, traveling to the Pampas of Argentina every autumn, a distance of 8,000 miles each way. Although still hunted on its South American wintering grounds, it has been protected in North America since 1916, at which time it was in serious decline. Since then it has increased, but more recently it has again gone into a population decline as native prairies have disappeared throughout nearly all of its range. Such is the case in western Nebraska and also in the tallgrass prairies of eastern Nebraska, but in central Nebraska, and especially in and around the edges of the Sandhills, the species is still moderately common.

Few sights are more charming than watching an upland sandpiper alight, balletlike, on a fence post, always briefly lifting its wings to the vertical as if it were congratulating itself on its dexterity before tucking them gracefully into its flank feathers. During the spring and even well into the summer, males often call with a clear wolflike whistle while circling high above their territories in advertisement flights. Or they may flutter rather low above the ground, their wings cutting deep arcs in the air, as they whistle in a rolling or trilling manner.

Nest requirements of upland sandpipers are preferentially natural grasslands, with heavily grazed lands and lands converted to small-grain agriculture avoided. Hayfields and idle or abandoned fields are used, as are lightly grazed pastures and grassy rights-of-way. In general, the birds seek areas with grasses between 10 and 20 inches tall; both shorter and taller grasslands are avoided. Upland sandpipers also need rather large areas for their territories; few fields smaller than 250 acres are likely to hold any breeding pairs. Home ranges of a square mile or more for a pair are not unlikely.

There are many enemies of ground-nesting birds. For example, the nests might be trampled by large ungulates or destroyed by skunks, coyotes, or any number of other carnivores, and the eggs might be devoured

Fig 18. Upland sandpiper, adult landing.

by snakes or crows; no second nestings are attempted. Not long after the chicks gain their flying abilities at about a month of age, the birds leave their breeding grounds. They then begin their long autumnal migration, sometimes as early as late August.

American Burying Beetle

The inconspicuous and perhaps slightly repulsive American burying beetle

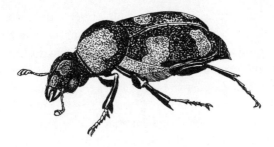

Fig 19. American burying beetle.

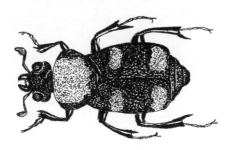

is one of Nebraska's few nationally endangered insect species. This beetle is part of a family of beetles also called carrion beetles or sexton beetles, names that give some indication of this beetle's ecological niche. Like vultures, they have undertaken the role of dealing with carrion effectively. In North America, there are numerous species in the burying beetle's genus, including at least 11 in Nebraska. However, the American burying beetle is the largest and by far the rarest of all. Although it once occurred over most of the eastern half of the United States, populations now are known only from Rhode Island, Oklahoma, and Nebraska. Recent Nebraska records are for west-central parts of the state, including Lincoln, Dawson, and Cherry Counties, and are from such diverse habitats as prairie grasslands, forest edges, and shrublands. All of these sites can be characterized by being fairly undisturbed by humans and of course having a supply of available carrion, typically small mammals such as mice or small songbirds.

Such potential foods, sometimes from as far away as 2 miles, are detected by the olfactory sensors located on the beetle's antennae. After flying to the site, the beetles, typically a pair, will try to lift the carcass and move it to a place where the ground is soft enough to allow for excavation. The insects do this by getting under the carcass and lifting it while crawling. This must require a great deal of strength as well as common knowledge by the pair as to the proper direction in which to pull it.

After completing the burial, the animal's fur or feathers are stripped away, and the remains are worked into a rounded ball. The beetles then cover the remains with secretions that help preserve the corpse and affect its decomposition. The female next constructs an egg chamber above the corpse and deposits up to 30 eggs. She then returns to the carrion and makes a slight depression at its top, in which both sexes regurgitate some partially digested food. When the eggs hatch a few days later there is a good supply of food for the small larvae. Parental care by both sexes continues through the entire larval development stages, a remarkable trait for any insect, especially any insect other than ants, bees, or termites. By the time the carcass has been fully consumed the young are ready to pupate, and the adults fly away. A month after pupation the young beetles emerge and the adults are likely to die, as they live only a single year.

CHAPTER 6

An Ocean of Sand

The Nebraska Sandhills and Sandsage Prairies

Yes, dear readers, there really is a Sandhills region in Nebraska. Many Nebraskans have been born, grown up, and perhaps even died of old age without even knowing this simple fact, much less ever having visited the region. This is in spite of the fact that the Sandhills cover an area of nearly 20,000 square miles, or about a quarter of the state's total area. They also represent the nearest thing to true wilderness area anywhere in the United States east of the Rockies, to say nothing of supporting the largest intact area of mixed-grass prairie that still survives south of Canada. The region's other superlatives include the remarkable fact that it rather inexplicably contains the largest hand-planted coniferous forest in North America (Bessey and McKelvie Divisions of the Nebraska National Forest), the most important waterfowl and shorebird breeding area south of the Dakota pothole country, and thousands of shallow, often small and unnamed wetlands that range from being some of the most alkaline waters in the western hemisphere to highly productive lakes and marshes that have an amazing diversity of plant and animal life.

The dunes themselves are an artist's dream, especially in early morning and late afternoon, when the low sun throws the entire countryside into a glacially slow theater of advancing or retreating curved shadows that are projected on a backdrop of rust-red grasses and grayish yellow sand. As evening slips into night, the stars gradually come awake in a pitch-black sky, providing one of the finest stellar shows imaginable. The dunes offer a multitude of lessons in survival, from the tiger beetles that quickly

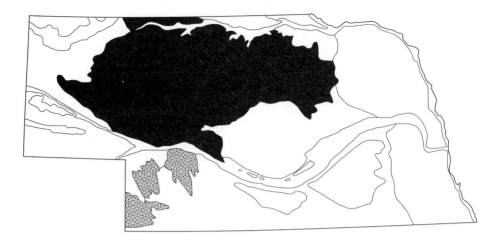

Map 8. Nebraska Sandhills and Border Prairies subregion (inked) of the Sandhills ecoregion. Isolated stippled areas to the south are Sandsage—Sandhills Ecotone subregions.

scamper over their surface in search of slower prey to the majestic long-billed curlew that would not hesitate to swallow a tiger beetle should one come into range of its long, decurved bill. Then there is the occasional prairie falcon patrolling the dunes that in turn could easily take down a curlew in full flight. Food chains tend to be short, and sometimes so are the lives of animals, in the Sandhills.

There are also the flowering plants, most obviously the yuccas that stand sentinel-like on dune slopes and burst into glorious white blossoms in June. All of this flowering effort is done to attract a few tiny white moths barely larger than an ant, to which the yucca is forever wedded in order to achieve fertilization. It is a two-way wedding, for the moth larvae must feed on some of the developing seeds in order to grow and pupate, but the larvae also must always leave enough seeds uneaten to insure the germination of another generation of yuccas.

Charles Darwin would have delighted in this story, just as he would have admired the Hayden's (blowout) penstemon, which will germinate and survive only in the transitory hardscrabble ecosystems induced by blowout disruptions of the dune community—an unstable, usually temporary environment of loose, drifting, or sometimes even blowing sand. As active dunes advance, they constantly cover the leafy parts of plants growing on their steep leeward faces and similarly expose the root systems of those growing on their more gently sloping windward slopes. Thus those plants that can survive repetitive burial by encroaching sand as a dune advances on them are unlikely to be the same species that are adapted to the constant loss of sand from their uppermost roots and the pummeling of their leaves by sand particles that gradually bounce up and over the dune crests through wind action. Sand-adapted grasses such as sand reedgrass and blowout grass are among the best plants for this kind of lifestyle; their rapid growth and narrow, flexible leaves are usually reinforced with silica, and their leaves often can roll up tubelike to avoid water loss and excessive abrasive damage to their surfaces by periodic assaults of sand. They also can use horizontal roots, or rhizomes, to gradually spread out and progressively capture the sand, often holding it in place long enough for other plants to take root.

To the south and west of the Sandhills proper are a few isolated islands of sand that consist of generally lower dunes reaching a maximum of about 150 feet high, or substantially lower than the dunes often found in the Sandhills proper. Geologic evidence suggests that these sands may be no more than about 10,000 years old and were deposited after the last glaciation. Like the Sandhills proper, the sand deposits here are hundreds of feet thick and are underlain by the Ogalalla aquifer, its edge not far below the ground surface. This easy access to groundwater and the generally more nearly level topography of the land have made this area especially vulnerable to center-pivot or other irrigation methods, which in contrast to the Sandhills have resulted in wholesale degradation if not complete destruction of much of this community type.

Ecology of the Sandhills and Sandsage Prairies

There are few if any natural communities in Nebraska more easily disturbed than the Sandhills; a horse that merely walks across them might leave hoofprints that can damage individual plants; a herd of cattle walking repeatedly across them will disrupt the thin mantle of vegetation and cause a blowout that can set a previously stabilized dune into motion, disrupting decades of previous plant growth. There are also few if any natural communities in Nebraska that are as compelling in their simple beauty or have such a sense of wilderness. The quietness and darkness of a moonless but star-filled Sandhills night remind one of sleeping in the Serengeti Plains of Tanzania, with the occasional yipping of coyotes replacing the distant roar of lions.

The Sandhills are notable for their relatively low species diversity of flora, a measure of the hardscrabble existence that plants encounter when trying to survive in this difficult environment. I have estimated that in six central Sandhills counties there are about 450 reported vascular plant species. By comparison, in Keith County, an adjacent county in the Platte Valley having about a fourth of the total area of these six Sandhills counties, there are at least 600 plant species. The entire state contains about 1,400 native and around 350 additional introduced vascular plant species, so the central Sandhills region has about 25 percent of the state's total vascular flora. The Sandhills do, however, support a scattered population of Nebraska's only endemic plant, the endangered Hayden's (blowout) penstemon. There are also an estimated 26 species of reptiles and amphibians (about half the state's total herp fauna), 50 species of mammals (about

60 percent of the state's total), and 50 species of native fish (also about 60 percent of the state's total). About 100 to 110 bird species breed in the Sandhills, which represents around half of the state's total breeding avifauna and from 10 to 20 fewer than breed in either of the nearby Platte or Niobrara Valleys. Nevertheless, the Sandhills are easily the most important breeding habitat for water-, marsh-, and shoreline-dependent birds in the state. It is also of great national and international importance to migrating waterfowl and shorebirds. There are about 2,000 square miles of Sandhills wetlands of all types, or nearly 10 percent of the region's total area.

Precipitation in the Sandhills region is no different in total (about 18–26 inches annually) from that supporting mixed-grass prairie to the north and south. The rain that falls is immediately drawn down through the sand to join the waters of the vast Ogallala aquifer. In some places the low interdune depressions are subirrigated, and in other places hundreds of permanent marshy wetlands are present as a result of the high water table. In still other locations artesian springs emerge from the bases of sand dunes to form small streams that eventually join others and flow eastwardly, southeastwardly, and northeastwardly out of the Sandhills in the form of rivers and permanent creeks.

Many plants and animals find long-term survival here almost impossible; the human population of the Sandhills has dropped by about 1 percent per year for the past half century. Prairie dogs and other large burrowing mammals such as pocket gophers that venture into the Sandhills soon discover that their burrows readily collapse in the soft sands, but kangaroo rats and pocket mice seem to do just fine. Perhaps their smaller-bore burrows are less likely to collapse under their own weight, or maybe they can dig new ones as rapidly as needed. Some small lizards such as the earless lizard also thrive in the sandy environment. The beautiful and harmless ornate box turtle somehow still survives there, even in the face of unscrupulous animal dealers who visit the Sandhills yearly to capture thousands to sell to the pet trade and even despite a few sadistic tourists who try to run over every turtle they see crossing country roads.

Major Terrestrial Community Types

Sandhills Dune Prairie

Over most of the 20,000-square-mile expanse of the Sandhills, the Sandhills dune prairie covers stabilized dunes that may attain heights of up to about 300 feet. Besides many of the usual species of the mixed-grass prairies, sand-adapted grasses such as sand bluestem (a close relative of big bluestem), prairie sandreed, sand lovegrass, sand muhly, and hairy grama are abundant. These are joined by arid-adapted sedges such as sun sedge and Schweinitz flatsedge, plus forbs such as stiff sunflower, lemon scurfpea, and bush morning-glory. The bush morning-glory is especially well adapted to this difficult environment as it has a greatly enlarged root that stores starches and other supplies.

Shrubs of the sand dune community include the distinctive yucca, which provides singing posts and lookout sites for songbirds; sand cherry; prairie rose; wild plum; and chokecherry. At times attractive flowers such as spiderwort, blazing star, and narrow-leaf penstemon break into blossom, along with bush morning-glory, stiff sunflower, and prickly-pear cactus.

With disturbance, the wind-formed and craterlike bare sand blowouts are initially invaded by blowout grass, which extends its long rhizomes into the open sand and begins to bind the sand together again. It is soon joined by sand muhly and prairie sandreed plus the leguminous lemon scurfpea. With enough time, the area "heals over" and reverts to typical Sandhills prairie.

Sandhills Needle-and-thread Prairie

Along the western, northern, and northeastern edges of the Sandhills, on level to slightly sloping substrates consisting of loamy sands, a community dominated by needle-and-thread, porcupine grass, and blue grama may replace the Sandhills dune prairie. The Sandhills needle-and-thread prairie has a fairly high level of species diversity, especially of legumes.

Sandhills Dry Valley Prairie

In dry interdune valleys, where the water table is at least about 5 feet below the ground surface, a prairie dominated by tall, mid-height, and short grasses develops. The tall grasses are mainly prairie sandreed and switchgrass; the mid-height grasses include needle-and-thread, little bluestem and western wheatgrass; and blue grama and sun sedge comprise the short grasses. There are many forbs such as prairie coneflower and globe mallow but few shrubs.

Sandhills Wet-Mesic Prairie

On interdune valleys where the water table is very close to the soil surface, providing subirrigation for many species, an assortment of tall grasses exist that otherwise are typical of tallgrass prairies. These grasses include big bluestem, Indiangrass, bluejoint, and prairie cordgrass. There are many forbs present, sometimes including prairie fringed orchid. This community is transitional to the Sandhills sedge wet meadow.

Sandhills Sedge Wet Meadow

In even wetter interdune valleys and around the edges of Sandhills marshes, fens, and lakes, a densely vegetated community of bluejoint, northern reedgrass, and several species of sedges and rushes predominate, as well as the taller prairie cordgrass. Shrubs, including some willows, are often present, too. There are some forbs, including many shoreline species such as water hemlock and swamp milkweed. Prairie fringed orchid has been reported from this community type, as has the equally rare small white lady's-slipper.

Sandsage Prairie

To the south and west of the Sandhills, beginning immediately south of the Platte River, the sandy substrate becomes gradually thinner and also somewhat loamier. In Nebraska, it is best developed in Chase, Dundy, and Perkins Counties, but it is also well developed in adjacent eastern Colorado and occurs disjunctively in southwestern Kansas. The sandy areas in southern Lincoln County and in Hayes County are transitional between typical Sandhills and sandsage prairie. Many of these small, is- landlike areas of sandsage are closely associated with rivers, which were probably the original source of the sand itself. The diagnostic species that identifies this community type is sandsage, which may vary from being rather sparse to fairly dense and may reach heights of about 3 feet. Its leaves are silvery green, and like big sagebrush at least some leaves persist through winter, making excellent winter food for deer, pronghorn, and jackrabbits, although cattle tend to avoid it. Nonetheless, its very presence traps winter snow, and its summer shade improves the microclimate for more palatable plants to sprout and grow.

Yuccas are often present on sandsage prairie, especially on steeper slopes. Between these and a few other shrubs, taller grasses such as prairie sandreed, sand bluestem, and needle-and-thread are typical, as is blue grama in an underlayer. Many of the forbs are those of Sandhills dune prairies, such as sand milkweed, prickly poppy, bush morning-glory, ten-petal stickleaf, wild begonia, and blazing star. The sandsage prairies also include a few plant species that occur in other sandsage prairies south and west of Nebraska but have not been found in the Sandhills themselves.

Nearly all of the birds of the Sandhills might be found also on the sandsage prairies, although the high level of degradation of the sandsage prairies has reduced their species diversity. Thus the upland sandpiper is now rare or absent. The lesser prairie-chicken was reported from the sandsage prairies from the late 1800s to the early 1920s but has since been entirely extirpated from the state. The greater prairie-chicken still occurs there but in a very precarious state. The Cassin's sparrow, probably was once also associated primarily with this habitat, but now breeds only very sparingly at the edges of the Sandhills proper. The pronghorn was once common but is now probably entirely gone, and the once-common jackrabbits (both species) are also little more than memories. Two snakes, the glossy snake and the western coachwhip snake, are known only from this corner of the state, and the plains black-headed snake has been recently reported only from a prairie in Chase County.

Profiles of Keystone and Typical Species

Great Plains Yucca

No Nebraska plant is more distinctive than the yucca; it can be recognized from as far away as one can see it. In June, when it is in full bloom, it lights up the Sandhills like Christmas trees illuminated by dozens of pendant white bulbs. The yucca has literally a rather prickly personality, and stumbling into one in the dark is almost as unpleasant as sitting down in a patch of cactus. But for these same reasons it survives the attentions of grazing animals, who usually give it a wide berth. It can also survive the rigors of life on a tall, dry sand dune, for it produces a long taproot that can find its way down to subsurface moisture without fail.

The Nebraska Department of Agriculture has unimaginatively classified yucca as an undesirable weed but nonetheless has noted that cattle, deer, and pronghorns will eat the flowers and fruit while avoiding the sharp leaves. The stems and leaves may be an emergency food that these animals can exploit in winter, and cottontail rabbits also may eat the leaves during winter. Native Americans used the strong leaf fibers to make rope, footwear, or various other items, and they used the tips of the leaves for needles. The baked or roasted flower stalks and young pods were eaten. A soapy extract could be obtained from the roots, thus the plant's alternative name "soapweed." Chances are good that Native Americans did not think of it as a useless weed.

The most famous aspect of the yucca is its extraordinary pollination mechanism, which depends on a tiny moth about a quarter inch in length. This white moth is attracted to the fragrant flowers and seeks out the pollen on the tip of the stamens. Using specialized mouthparts, she gathers together this into a large ball; the pollen from as many as four stamens may be used. After picking up the ball and carrying it to a separate flower, apparently looking for one that has not yet been pollinated, she then places the ball of pollen into the plant's stigma, bores a hole in the stigma with her ovipositor, and deposits her eggs. She typically lays a single egg in each of the three cells of the flower's ovary, usually repollinating the flower between each egg-laying bout. She then leaves, having insured reproduction both for herself and the plant. After the larvae hatch they begin to eat the developing crop of 300 to 400 seeds but never consume them all, assuring that the plant's chances of reproduction will succeed. Eventually, the larvae bore holes in the side of the seedpod and complete their own development outside the plant.

Long-Billed Curlew

There is something regal—if not surreal—about a shorebird that is almost as large as a grouse with a bill that may be nearly 9 inches long in females (shorter in the smaller males), and the bill of the long-billed curlew is indeed the longest of any New World shorebird. The generic name of curlews, *Numenius*, translates as "new moon," and the long-billed curlew's bill gracefully curves downward in the shape of a crescent moon. It would seem to be of little foraging benefit on the birds' prairie breeding grounds, where prey items such as grasshoppers are adroitly plucked from surface vegetation, but the bill instead is wonderfully adapted to probing in sand on the species' coastal wintering grounds. The English name "curlew" refers to the species' alarm call, a ringing "curlee" that carries far and wide across the Sandhills landscape, its sound alone enough to cause an instant pang of nostalgia for anyone lucky enough to have become attached to this wonderful wilderness.

The long-billed curlew needs little else to add to its aesthetic appeal, but when it raises is wings, as the male does in courtship display, or takes

Fig 21. Long-billed curlew, adult female in flight.

flight, it exposes cinnamon-tinted underwing coverts that contrast with its otherwise dead-grass patterning. This cryptic dorsal coloration provides perfect camouflage for nesting birds; I have approached to within 5 or 6 feet of an incubating bird, knowing almost exactly where the nest must be, and still have been unable to pick her out from the surrounding sparse grasses until she suddenly takes flight. Her wild alarm calls are then likely to attract not only her mate but any other curlews within hearing range, assuring the intruder of becoming the victim of diving attacks that come so close to one's head that it is impossible to avoid flinching.

By early May the curlews of the Sandhills are likely to be quietly sitting on their four-egg clutches, and the earlier territorial advertisement flights of males are a thing of the past. Males often take up watchful positions on a nearby hillside, alerting their females to any possible threat and trying to distract any intruders that blunder too close to the nest. Bullsnakes are a serious threat to curlew eggs, and coyotes are also predators for which no real defense is possible. Yet a considerable number of the nests make it safely through the four-week incubation period, and by the first of June there are many families of curlews to be found in the Sandhills. At this time the curlews are especially vigilant, and one can scarcely venture far from a car in the Sandhills without attracting the attention of a parent curlew. Fledging takes about six weeks, so by mid-July the birds begin to assemble in moist green meadows, taking on food in preparation for their migration. Shortly after that they simply disappear, leaving the Sandhills a more quiet and lonely place until they return the following spring.

Ornate Box Turtle

Somehow the ornate box turtle seems to be an ideal symbol of the Sandhills. It is independent, capable of surviving in areas that may seem desert-dry, hardy, long-lived, and even adventurous. Thus the box turtle frequently turns up in seemingly unlikely places, often hiking resolutely down the middle of a Sandhills road and reminding one of some out-of-state tourist who took a wrong turn and now is totally lost.

This, of course, is a misinterpretation of the facts. Chances are the turtle has a better idea of where it is than does the human encountering it. Chances are also good that the turtle is a male, which can be easily told by its red rather than pale brown eyes, most probably on the lookout for females. Both sexes have large home ranges, rarely up to 20 acres, and they have remarkable abilities for finding their way home after having been displaced. Of a sample of 100 turtles displaced from 1 to 1.7 miles in varied directions, 18 were recovered near their original point of capture. A male and a female returned from the maximum displacement distance, and in a later experiment one male returned to its home range by the following year, after having been displaced about 2 miles. These rather memorable examples suggest that the visual navigational abilities of these turtles must be quite remarkable, especially since they live in habitats notably lacking in conspicuous landmarks.

Nebraska is near or at the northernmost limit of this species' range (it also occurs in southern Wisconsin), and here the animals hibernate from about October to April. They are most active between late April and June, when males are likely to be in search of females and fertilized females are searching for nesting sites. During this period they expose themselves to being run over by vehicles, picked up by naive persons who mistakenly think they might make good pets, or being collected by greedy reptile dealers who know that Nebraska is the only midwestern state that does not protect its turtles and other herptiles adequately and who will try to sell them to out-of-state pet markets. All turtles may transmit salmonella, a serious bacterial disease to humans, when handled and thus are best left to their own devices.

The box turtle is named for its flexible lower shell, or plastron, which allows it to hide its head and tail when disturbed, providing an animal such as a coyote little opportunity to reach any vulnerable parts. Younger animals lack hinged plastrons, and their shells are weak enough that coyotes might be able to crush them. They also have somewhat higher back profiles than adults, especially adult females. Dexterous raccoons may be capable of pulling the plastron back long enough to kill young box turtles. However, by the time the turtles are mature at 10 to 12 years, they are fairly safe from most such dangers and may live for 20 years or more. It has been suggested that in a wild population the usual turnover time of a population may be about 25 years. Perhaps very rarely a turtle will live for 30 years.

The ornate box turtle population of the Sandhills is completely unknown, but thousands are now being removed yearly by commercial dealers. For a slow-growing and slowly reproducing species such as this, such exploitation is an unacceptable gamble with one of the most appealing symbols of the Sandhills. It should be terminated without delay, especially considering the fact that the animals make very poor pets.

Blanding's Turtle

The Blanding's turtle is a little-studied species that reaches the western edge of its overall range in the Sandhills of Nebraska. It is a much more aquatic species than the ornate box turtle, and it is likely that most Nebraskans have never seen one, nor would they recognize it if they did. It is about the same size as the ornate box turtle, but instead of having a radiating or starlike pattern of yellow lines on each large plate of the dorsal shell, there is a multitude of irregular tan to yellow dots. The underside is quite uniformly dark rather than crossed with many pale lines. The plastron is not hinged, and from the side the outline of the mouth turns upward near its base, producing a permanent, dolphin-like smiling expression that is quite distinctive.

The Blanding's turtle spends nearly all of its time in water, which is where courtship and mating also occur, but females, of course, must lay their eggs on land. This is usually done at night, between mid-June and mid-July in Nebraska, often in grassy habitats near marshes. Like many other turtles, sex determination is temperature dependent. If the eggs are incubated at 72.5 to 79.7 degrees F, nearly all of the young will be males, but an incubation temperature of 86 to 87.8 degrees F will result in nothing but females. Such a situation may limit the species' range to areas in which a mixture of sexes may be reasonably expected to hatch over a prolonged period of years. Blanding's turtles continue to grow until they are about 13 years old, and probably many live beyond 25 years. One wild specimen from Minnesota was known to attain at least 77 years of age and to have a reproductive period of at least 56 years, making it easily the greatest reported instance of longevity for this species. Raccoons, foxes, and skunks

Fig 22. *Above*, Western ornate box turtle, immature, and *below*, Blanding's turtle, adult

are known to be serious egg predators, and coyotes are also likely to be serious threats.

As with other reptiles and amphibians, essentially no information on the status of the Blanding's turtle in Nebraska is available. It is dependent upon the presence of permanent marshes in the Sandhills for its long-term survival

Ord's Kangaroo Rat

If Walt Disney had wanted to draw an especially appealing rat, he could have done no better than select the kangaroo rat for his basic stripped-down model. It has a gigantic head with a rounded face, luminous eyes that seem too large by half, tiny front feet, and oversized hind legs that make the animal look something like a frog in mouse's clothing. Its hairy tail is long and tipped with a decorative tassel, and its long whiskers stick downward and forward like a scraggly and worn-out artist's brush. Rather

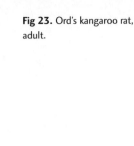

Fig 23. Ord's kangaroo rat, adult.

than walking or running like a typical rat, the animal bounces along, pogo-stick fashion, making its erratic way along the surface of dunes until it suddenly disappears into a small burrow opening. Chasing "roo-rats" with butterfly nets and a flashlight under pitch-dark conditions was an eclectic, elective sport at our biological field station for many years, until the fear of disease made it a more questionable activity. However, kangaroo rats proved to make wonderful pets, taming rapidly, requiring no water and thus producing little urine odor, and often living for years.

The Ord's kangaroo rat is one of the commonest small mammals of the Sandhills, but its nocturnal activities result in its presence being made known mostly by its tracks in the sand and by its burrow openings. Pocket mice (plains and hispid) are even more common, but they are smaller and their tracks are much less noticeable. Kangaroo rats and pocket mice are seed specialists, and they evidently are able to manufacture water metabolically, as they rarely if ever seek out surface water to drink. The usual gait of kangaroo rats consists of low and slow hops of from only a few inches to a foot or more, but when alarmed they can suddenly jump vertically nearly 20 inches or horizontally about 6 feet in a single bound. The may also kick sand in the face of an enemy. Such rapid and long jumps may be important when faced by a predator such as a coyote or a bullsnake; the air-filled middle-ear cavities of kangaroo rats are extraordinarily enlarged and allow for the detection of very low frequency and low amplitude noises such as those made by an approaching snake. The animals also communicate with foot drumming and low-frequency calls, which travel effectively over long distances.

Like nearly all other small rodents, kangaroo rats form no long-term pair bonds. Instead, the males are attracted to females in estrus by olfactory clues. Males may groom and follow such females for some time before finally having an opportunity to mate. Having achieved that, the male rather rapidly departs, as kangaroo rats tend to be nonsocial and would rather avoid one another than risk serious fighting. Males are able to mate at any time during the year, but female reproductive cycles are timed by the availability of green vegetation and associated seed production. Thus population density is likely to be set by rainfall patterns. Home ranges of adults extend up to about 2 to 3 acres, but these overlap, and densities of as many as 3 to 15 kangaroo rats per acre are not unusual in good habitats. Probably most individuals live no longer than a year under natural conditions, with owls, foxes, weasels, coyotes, and other nocturnal predators likely sources of mortality.

Vignettes of Endangered and Declining Species

Hayden's (Blowout) Penstemon

The Hayden's (blowout) penstemon, the sole plant species endemic to Nebraska, is confined to the Sandhills and now is only known from the western half of this region, from Cherry and Thomas Counties west. It was first listed as endangered in 1987 and indeed was believed to be extinct by the middle of the twentieth century. However, it was rediscovered in 1968 and is now known from more than dozen sites, mostly in Cherry County.

The limited range of this plant is a result of it being specifically adapted to the bare sand environment of dune blowouts and their immediate edges, where loose sand prevents all but a few species from taking root. It is a short-lived perennial, living up to about eight years, but can spread by rhizomatous roots. It can also spread by roots forming at the plant nodes if the stem itself becomes buried in sand. Although great numbers of seeds are produced, about 1,500 per plant per year, the seeds are often eaten, and those that are not are prevented from immediate germination by a thick seed coat that helps protect them from abrasion. In order for the seeds to germinate they must remain damp for at least two weeks, and the developing root must then reach a source of moisture before the seedling's stored food reserves are used up and it dies for lack of water. Because of these unlikely conditions, seedling establishment is rare, even where the

Fig 24. Hayden's penstemon.

plants are abundant. However, since the seeds may remain viable for up to 20 years, the chances of ultimate germination for a few are perhaps good if they escape being eaten, too deeply buried in sand, or blown into unsuitable habitats.

This species is one of Nebraska's most attractive penstemon species and the only one with fragrant flowers, which are usually blue to lavender. Similarly colored flowers are typical of shell-leaf penstemon, but this common species has leaves that are broadest beyond their midpoints. It has more lance-shaped leaves and larger blossoms than the similarly common narrow-leaved penstemon. All penstemons have asymmetrical flowers, with two petal lobes above and three below, the latter making a convenient landing place for bees, wasps, and other potential pollinators. There are four functional stamens and one long, sterile one (the staminode), which has a yellow hairy tip and is the basis for the common name "beardtongue" that is sometimes applied to plants of this genus. Dark magenta guidelines on the inner surface of the corolla help direct insects to the nectar, where the insects might also bring about pollination. It is one of the first plants to invade fresh blowouts but disappears by the time the blowout is fully "healed."

The Hayden's (blowout) penstemon is considered as endangered by both state and federal agencies, and a program of greenhouse rearing and reintroduction into Sandhills habitats is under way. It is hoped that eventually at least 10,000 plants can be established in the Sandhills. Currently, there are known to be about 3,000 to 5,000 plants in scattered locations in Cherry, Thomas, Garden, Morrill, and Box Butte Counties.

CHAPTER 7

Valley of Dreams

The Niobrara Valley and Its Transitional Forest

If there is a valley of dreams in Nebraska, the Niobrara Valley is certainly it. To drive north across the seemingly endless, grass-covered Sandhills and then suddenly come into sight of the deep, forest-covered canyon of the middle Niobrara Valley produces a feeling of awe and delight that must approach that felt by Brigham Young when he first saw the valley of the Great Salt Lake. The middle Niobrara Valley also reminds one of the Snake River canyon of Jackson Hole, Wyoming, except that instead of the Teton Range looming up on the horizon beyond there are likely to be giant cumulus clouds floating serenely in an ethereal sky. Unlike the indecisive Platte River, which never seems to know which channel it should use, the Niobrara River moves eastward with a certain business-like determination, only eventually to loose its identity in the much larger but also far more degraded and partially impounded Missouri River. In western Nebraska, for example, around Agate Fossil Beds National Monument, the river is so narrow one can nearly jump across it, and its associated valley scarcely makes a crease in the surrounding dry grasslands. Near its confluence with the Lewis and Clark impoundment of the Missouri River the Niobrara has a wide riverbed, many vegetatively overgrown sandbars, and a large floodplain with gentle valley slopes. Lewis and Clark described its mouth area in 1804 as being 152 yards wide and no more than four feet deep. They said its current was very rapid and referred to it as the *Qui courre* (from the French *L'Eau qui Court*), or the river that rushes. It no longer rushes there, but in the central part of Nebraska—where the river is

Map 9. Niobrara Valley Riparian Forest region, coniferous area (stippled) and deciduous area (inked). Based on maps by Kaul (1975) and Omernik (1987).

large enough and wild enough and its canyons deep enough to support the dreams and fantasies of vacationing canoeists, but where it is tame enough not to threaten their lives—the Niobrara is a perfect river.

Ecology of the Niobrara Valley Forest

There is perhaps no single river, or indeed any 100-mile stretch of river in all of North America, that is more interesting biologically than the Niobrara. The narrow Niobrara Valley is indeed where the East ends and the West begins, at least so far as its plant and animal life are concerned. In addition, it holds some unique relict species that reflect the glacial history of the northern Great Plains region, trapping a few such species that were left to survive within its cool valleys when the last glaciers retreated northward about 10,000 years ago.

In a detailed analysis, Gail Kantak and Steven Churchill determined that within the Nature Conservancy's Niobrara Valley Preserve, an area encompassing 1,126 square miles and about 40 river miles, there are at least 581 species of vascular plants, 80 species of mosses and liverworts, 278 vertebrate species, and 130 species of aquatic invertebrates. Among the vascular plants, most are also eastern and northeastern in their affinities, followed by northern and western affinities. A northern species of special interest is the paper birch, a Pleistocene relict whose closest populations now are in the Pine Ridge and Black Hills. Hybrid populations between the big-tooth aspen and quaking aspen are also indicative of mixed geographic influences; the bigtooth aspen now occurs no closer than about 250 miles away in the western Pine Ridge. There is also a population of hybrids of balsam poplar and cottonwood; the former species no longer occurs there in pure form and now is no closer than about 200 miles away, also in the Pine Ridge. The junipers in the Niobrara Valley range from the

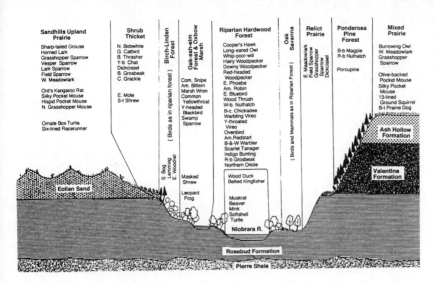

Fig 25. Profile view of typical middle Niobrara Valley habitats and some associated vertebrates. After Johnsgard (1995).

Rocky Mountain species in the Pine Ridge to a type that is closer to the eastern red cedar in the valley itself.

The mosses of the Niobrara Valley Preserve are mostly those with eastern affinities, but there are a few western forms and also some northern representatives that appear to be Pleistocene relict species. Of the eastern moss species, six reach their western limits here, and five are western species reaching their eastern limits. Two are northern species reaching their southern limits.

Considering the Niobrara Valley as a whole, the river extends for about 490 miles from eastern Wyoming to its confluence with the Missouri. Within Nebraska, its basin is about 300 miles long, with a gradient dropping some 3,400 feet, or an average of about 11 feet per mile. Since the Pleistocene period it has carved a deep canyon in northern Nebraska, which at maximum is about 450 feet deep. Since the time of Charles Bessey in the 1880s this area has been identified as a major transition zone for eastern and western biotas and significant shifts in species diversity. Robert Kaul, Gail Kantak, and Steven Churchill have determined that the total number of tree species decreases in the Niobrara Valley from 75 to 25 from east to west, the shrub species from 55 to 33, the woody and semiwoody vines from 14 to 7, the ferns from 33 to 6, and the mosses from 117 to 55. Much of the decrease in species diversity westwardly occurs close to the confluence of the Niobrara with the Missouri; west of that area there is mainly a substitution of western and northern species for eastern and southern ones. These authors report that almost 60 species (plus some additional subspecies or varieties) of Niobrara Valley vascular plants and 15 moss species have ranges mainly to the east of the central grasslands, whereas 15 vascular plants and 7 mosses have western affiliations. They also listed 2 butterflies, 3 fishes, 4 reptiles, 16 birds, and 1 mammal as eastern forms that occur in the valley, versus 17 butterflies, 1 fish, 4 reptiles, 5 birds, and 1 mammal as western species occurring there. There are also 5 northern fishes occurring as glacial relicts. This data suggests that the Niobrara Valley was colonized mostly from the east following the last glaciation, with different species and groups being able to colonize at varied rates and over differing distances.

There are at least 25 species of amphibians and reptiles in the Niobrara Valley Preserve. Most are associated geographically with the Great Plains, but there are also two eastern or northeastern forms (Blanding's turtle and eastern hognose snake). The western hognose snake is the eastern's western counterpart and overlaps slightly with it in the lower Niobrara Valley. The ringneck snake reaches its western Nebraskan limits here and also in the central Platte Valley, and the fox snake reaches its overall southeastern limits. There are also two lizards of southwestern affinities (many-lined skink and northern earless lizard) that are near or at the northeastern edges of their ranges.

Among the 24 documented species of fish, 2 species (brook stickleback and the threatened finescale dace) are probably glacial relicts. There is also a Pleistocene relict population of the eastern woodrat, which has since evolved into a unique endemic subspecies that is more than hundred miles removed from its nearest relatives, now located to south of the Platte. The olive-backed pocket mouse, an open-country western species, reaches its eastern Nebraska limits in the Niobrara Valley. The Franklin's ground squirrel reaches its western limits there, as does the northern short-tailed shrew.

I have estimated that 125 species of birds are probable or known breeders in the central Niobrara Valley. James Ducey also independently reported 125 known or likely breeders. He has further determined that 268 bird species are recorded in the valley, or about 60 percent of the total ever recorded for the state and about 75 percent of all the species likely to be present in any single year. The majority of the breeders are of eastern affinities, but there are a substantial number of western breeders (for example, long-billed curlew, common poorwill, lazuli bunting, black-headed grosbeak, Bullock's oriole, black-billed magpie, and red-breasted nuthatch). Two of these (curlew and poorwill) are grassland- or scrub-dependent birds; the rest are forest-edge or forest-dependent species. Species of northern affinities that breed in the valley but have rather limited ranges elsewhere in the state include the brown creeper, least flycatcher, common snipe, swamp sparrow, clay-colored sparrow, and tree swallow. Two of these are wetland-dependent forms (swamp sparrow and common snipe); the clay-colored sparrow is grassland-dependent; and the rest are more generally forest-adapted in their habitat traits.

In a 1997 survey of the Nature Conservancy's Niobrara Valley Preserve, 74 summering and presumed breeding species were found in a three-month period, as compared with 105 breeding or almost certainly breeding species found during a five-month survey in 1982. Considering both surveys, five warblers, four vireos, and four sparrows were reported, as well and six hawks and two falcons. A total of 231 species were reported from both surveys, or more than half of the state's known avifauna. During both years the mourning dove was the most abundant species.

The best evidence that this region is a major biological transition zone comes from the interactions and occasional hybridization of eastern and western species or races that occur in the central or upper Niobrara Valley. Examples of closely related east—west species pairs include the indigo and lazuli buntings, the Baltimore and Bullock's orioles (previously known as the northern oriole), the rose-breasted and black-headed grosbeaks, the eastern and spotted towhees (previously known as the rufous-sided towhee), and the eastern and western wood-pewees. It is likely that all of these represent gene pools that were split during Pleistocene times and have only rather recently come back into secondary contact. Hybrids of all these types have been reported from Nebraska. Eastern and western races (yellow-shafted and red-shafted forms) of the northern flicker also commonly intergrade here, as they do in the Platte Valley and widely elsewhere on the Great Plains. The western and scarlet tanagers meet and probably interact in the Niobrara Valley, since a possible hybrid has been reported from the central Niobrara Valley. The eastern and western meadowlarks overlap in the Niobrara and Platte Valleys but only rarely hybridize, as they tend to be ecologically isolated. However, their incidence of hybridizaton in the Platte Valley seems to be higher than anywhere else in their zone of overlap.

James Ducey estimated that four western bird species reach their eastern breeding range limit in the Niobrara Valley (prairie falcon, western wood-pewee, western tanager, and black-headed grosbeak). All but one of these species are associated with western forests; the prairie falcon is an open-country and cliff-dependent species. Because of recent taxonomic changes, the spotted towhee and Bullock's oriole should be added as western forms terminating in the Niobrara Valley and in the Platte Valley to the south. The violet-green swallow, Cassin's kingbird, and Say's phoebe should probably be added to this list, as might the ferruginous hawk and perhaps the golden eagle. The dusky flycatcher, gray jay, pinyon jay, and Townsend's solitaire are less likely but still potential candidates as occasional breeders in the western Niobrara Valley woodlands (they all breed commonly in the Black Hills of South Dakota), but specific regional breeding records seem to be lacking for all of them.

Likewise, at least 11 eastern species reach their western range limits or have western range extensions in the Niobrara Valley. They include the American woodcock, whip-poor-will, red-bellied woodpecker, eastern wood-pewee, blue-gray gnatcatcher, wood thrush, yellow-throated vireo, black-and-white warbler, American redstart, ovenbird, and scarlet tanager. All of these are essentially deciduous forest birds. To this list of eastern breeding birds approaching or reaching their western Great Plains limits one might add the wood duck, barred owl, ruby-throated hummingbird, willow flycatcher, eastern phoebe, Baltimore oriole, and eastern towhee. However, the blue-gray gnatcatcher has recently established a probable breeding population in northeastern Wyoming and thus may need to be deleted from the list of species having western range limits in the Niobrara Valley.

Major Terrestrial Community Types

Northeastern Upland Forest

Along the Niobrara Valley from Cherry County east to the river's confluence with the Missouri, a deciduous upland forest exists on valley bluffs and nearby uplands that is comparable to those of the upper Missouri, the lower Platte, and the Elkhorn Valleys. It is typically dominated by basswood and bur oaks, with some hackberry, green ash, and black walnut present. In the Niobrara Valley especially, the eastern red cedar is a frequent subcanopy species. Shrubs and vines are rather infrequent, but the herbaceous layer has fairly high species diversity, with some eastern and northern forbs such as bloodroot, ginseng, wild sarsaparilla, spikenard, and Canada lily. Breeding birds include the Cooper's hawk, American robin, Baltimore and Bullock's orioles, black-headed and rose-breasted grosbeaks, eastern wood-pewee, red-eyed vireo, red-bellied woodpecker, and wood thrush.

Northern Springbranch Canyon Forest

The springbranch forests of the middle Niobrara Valley are a highly localized and unique community type in the state. The canyons have formed around springs emerging out of the adjacent Sandhills, the waters flowing along bedrock shales until they break out into the soft, north-facing valley slopes and begin to cut lateral ravines and canyons into the steep bluffs of the Niobrara Valley. These shaded canyons provide a cool microclimate that has permitted the paper birch and a few other boreal plant species to survive for thousands of years, in spite of a generally warming macroclimate. Besides paper birch, deciduous trees of the eastern forest such as basswood, hop hornbeam, and green ash are prevalent, and eastern red cedar is common in the subcanopy layer. The herbs include wild sarsaparilla and grove sandwort, both of which may be quite common and are characteristic of this community type. The northern green orchid is also present, as are several other plant species of mostly boreal affinities.

Western Coniferous Forest and Mixed Coniferous Woodland

In Brown and Cherry Counties a well-developed coniferous forest is locally present, especially on steeper Niobrara Valley slopes just above the flood-plain forest but below the Sandhills prairie or mixed prairie of the level uplands. This forest is dominated by ponderosa pine but has a scattering of eastern floral elements. In cool, north-facing slopes or steep side canyons it may interdigitate with springbranch canyon forests. Eastward into Keya Paha County it often grades into mixed coniferous woodland, consisting of a dominant ponderosa pine canopy, a subcanopy of eastern red cedar (or possibly intergrades with Rocky Mountain juniper), and an underlayer of mostly eastern shrubs and herbaceous species. The red-breasted nuthatch and chipping sparrow are typical breeders. These general forest and wood-land types were discussed more completely in chapter 3.

Eastern Lowland and Floodplain Forests

The eastern lowland and floodplain forest communities are described in chapters 8 and 10. Typical birds of these communities include some east-ern warblers (American redstart, black-and-white warbler, and ovenbird), vireos (warbling and yellow-throated), scarlet tanager, white-breasted nut-hatch, and black-capped chickadee. Barred owls are found in mature floodplain woods, and house wrens are often abundant in floodplain forests having a supply of old woodpecker holes.

Eastern Floodplain Woodland and Willow Shrubland

The eastern floodplain woodland community is discussed in chapters 8 and 10, and the willow shrubland of the Niobrara Valley is little if at all differ-ent from the type described in chapter 8. Birds of the floodplain woodlands and shrublands include gray catbird, brown thrasher, loggerhead shrike, yellow-breasted chat, Bell's vireo, blue grosbeak, northern cardinal, and gray catbird. In wetter, willow-dominated communities the red-winged blackbird, common yellowthroat, and song sparrow are very common, and the swamp sparrow is a less abundant and more localized breeder.

Profiles of Keystone and Typical Species

Paper Birch

One normally thinks of paper birch as most typical of the north woods of the Great Lakes states, but it is a trans-Canadian species, occurring west to the Pacific and Bering Seas and east to the Maritime Provinces. It survives on a wide variety of soils but favors cool climates and well-drained, sandy loam soils. However, it also can grow around lakeshores in central Minnesota, its roots there probably reaching the water table.

Toward the southern edges of its range it is increasingly limited to generally cooler and often higher sites, such as north-facing slopes of ridges or mountains. Besides being present in the Black Hills of South Dakota, it extends south disjunctively into Missouri and Illinois as glacial relicts. It also grows as a probable glacial relict isolate near Boulder, Colorado. In Nebraska, the species occurs only in Cherry and Brown Counties, where it occurs on the south side of the Niobrara on the north-facing slopes of various shaded springbranch canyons, such as near Smith Falls.

Paper birch grows rapidly, producing seeds by the time it is about 15 years old. Like aspens, each year it produces catkins that generate large numbers of tiny seeds, which are dispersed by their winglike edges. Also like aspen and poplars, it is favored by periodic fires, as it can easily and quickly reproduce by suckers from its surviving roots. Paper birch usually does not grow very tall and seems to have a fairly short lifespan. Then, it eventually falls and subsequently decays rather rapidly, usually leaving remnants of its apparently more decay-resistant white bark as a memento to its previous presence in the forest.

Paper birch is probably best known for its papery white outermost trunk layer, called birchbark, which has long served Native Americans in making the waterproof, flexible outer skin of birch canoes or even lodges. It was also used for kindling, for making cutout patterns, as "memory scrolls" for learning songs and stories, for recording dreams, and for making containers. Typically the yellowish inner bark surface served as the primary surface of the object being made, as the white outer layers tend to curl back on themselves. The inner surface, however, makes a fine marking or writing substrate. The wood itself is fairly strong and hard and was used to construct items such as snowshoe frames and structural supports. The tree's sap can be boiled down to make a sugary syrup similar to maple syrup. In the winter, its buds are avidly eaten by grouse, as are its protein-rich catkins during spring. Beaver and deer relish its leaves and twigs, and these are important fall and winter foods for both.

Red-Tailed Hawk

The red-tailed hawk is one of those perfect hawks: a rodent-killing machine without peer, almost as much at home in the sky as a helium-filled balloon, and with eyes at least as keen as the best human eyes magnified by a resolving-power factor of four or more. It is powerful enough to kill a jackrabbit but flexible enough to live in nearly every North American habitat. What it especially needs is a good deal of open space, an adequate mammalian prey base, and sufficient security to nest and rear its family in peace. This it can find in such unlikely places as New York City's Central Park, on a Nebraskan farmstead, or in the deserts of the American Southwest.

Fig 26. Red-tailed hawk, adult.

Over this broad range the species' plumage patterns varies from a dark brownish black that may appear almost ravenlike at great distances to one so pale as to appear almost ghostlike. But among most of its plumage iterations a pinkish gray to rust red tint can be seen on the upper tail surface of adults, the basis for its common name. Most of Nebraska's birds are of intermediate plumage tones, with mainly white underparts and freckled or uniformly brown upperparts. However, during fall and spring a few very dark toned migrant birds from northwestern Canada and Alaska sometimes appear, and the very pale Krider's hawk variant is periodically seen among the breeding birds of the Sandhills. It was once thought that this latter type was a subspecies typical of the central Plains, but this is not the case. The plumage variations of such hawks as the red-tailed and Swainson's likely reflect the fact that the birds are powerful enough that any needs for concealing coloration that smaller and more vulnerable species might require have disappeared, so plumage variability is not a significant factor in influencing individual survival.

Red-tailed hawks are somewhat heavier than Swainson's hawks, with males averaging about 2.3 pounds and females about 2.7 pounds, as compared with about 1.8 and 2.4 pounds, respectively, for male and female Swainson's hawks. Together the two species occupy the middle-sized buteo-hawk niche in Nebraska, at least over the breeding season. To the west the even larger ferruginous hawk is added to this mix, and in the Missouri Valley the smaller red-shouldered hawk was a significant player until it became virtually extirpated during the twentieth century. Mammals make up nearly 100 percent of the prey biomass and the majority of individual prey items, with birds, reptiles (mainly snakes), and other prey progressively of less importance. These birds are strictly daylight hunters, with home ranges typically of a square mile or more during the breeding season. Their resultant population density may rarely be as high as a pair per square mile but more often is likely to be a pair per 5 to 10 square miles. During breeding they compete with the Swainson's hawk for similar-sized rodent prey and even more so with ferruginous hawks. However, the three species tend to have nonoverlapping home ranges where they occur together.

Swainson's Hawk

The Swainson's hawk was named for William Swainson, an English scientist and lithographer who certainly never saw one alive and thus missed out on becoming acquainted with one of the Great Plains's most wonderful and remarkable raptors. Like the upland sandpiper, it makes annual long-distance migrations to the temperate plains of southern South America. Starting on the Canadian plains, the birds gather early each fall into ever larger groups as they gradually move southward. This is probably not because they are strongly gregarious but rather simply because they are masters at detecting and riding thermal updrafts of warm air. This behavior allows them to glide for hundreds of miles a day, spending little energy more than that required for simply holding their wings outstretched and being able to remain within the primary thermal uplift zone. By the time the birds reach Mexico, and especially when they travel along the Isthmus of Panama, tens or even hundreds of thousands of these hawks may be riding the same winds. It seems likely that very little feeding is done during this migration, which may last for two months or more and may cover more than 6,000 miles each way.

During the time they on their wintering grounds, Swainson's hawks

Fig 27. Swainson's hawk, head of adult.

are distinctly insectivorous, often living largely on the abundant grasshop-pers. This is unusual behavior for a hawk that is almost as large as a red-tailed hawk, but it is a carry-over of similar behavior in North America, especially by nonbreeding individuals. Breeding birds eat a variety of small rodents, especially mice, voles, ground squirrels, and the like, but they occasionally eat mammals as large as rabbits plus some snakes and lizards.

Swainson's hawks probably initially breed at about three years of age, and adults begin to form pairs almost immediately after they have returned in spring from their wintering grounds. In Nebraska, this species begins to approach the population density of the resident red-tailed hawk in areas from about Grand Island westward. In open-country habitats with only scattered low trees, such as in the Sandhills and the shortgrass plains, red-tails are a relatively rare commodity as compared with Swainson's hawks. This may be in part because Swainson's hawks are willing to nest in low, solitary trees that may be little higher than a human can reach. Only rarely, human-made structures may be used. Males establish the center of their territory around potential nest sites and advertise them by soaring flights. When a female has been attracted, the two may soar together at a height of hundreds of yards. Then one bird, probably the male, will close its wings and dive, followed by a sudden pulling out of the dive, a regaining of elevation, and a repetition of this sky-dancing display. The flight ends with the male landing on a horizontal perch, where the female may soon also land, and perhaps mating will ensue.

Like the red-tailed hawk, the Swainson's hawk is a remarkably beneficial species to farmers and ranchers. Yet year after year several hundred hawks and owls are turned into raptor rehabilitation centers in Nebraska, a large proportion of the birds having been shot. It is heartbreaking to see these birds stoically enduring unimaginable pain from bullet holes or from dozens of pellets penetrating their heads and bodies and sometimes blinding them. Faith in the goodness of humanity is very hard to maintain under these conditions.

Baltimore and Bullock's Orioles

Two of the closely related species that come together and sometimes hybridize along the Platte and Niobrara Valleys are the Baltimore and Bullock's orioles. Studies on these birds in Nebraska during the 1950s led Charles Sibley and Lester Short Jr. to conclude that the birds had not been separated geographically long enough during the Pleistocene to develop a complete reproductive isolation between their respective gene pools. Then, when climatic and vegetational changes in the Great Plains allowed these two forms to again come into contact, hybridization began to occur and to threaten the genetic integrity of the two incompletely differentiated gene pools. More recent studies have indicated that the incidence of hybridization may have declined since then, causing taxonomists to again consider them as distinct, albeit occasionally hybridizing, species. It has even been suggested that they may not be one another's most closely related species.

Adult male Baltimore and Bullock's orioles are among the most attractive of Great Plains birds. It takes males two years to gain their full nuptial plumage, although first-year males are sexually fertile since they sing and display to generally unresponsive females whenever they get an opportunity. First-year males also occasionally manage to mate and raise young. However, their plumage is very femalelike, and their songs are also rather distinctive—so much so that a first-year singing male makes a wonderful field-quiz subject for elementary ornithology students. Most students can never imagine these drab males are exactly the same species as the much more brilliantly colored older ones that might be perched nearby and that they have learned to consider as a cinch for field identification.

Where both species are present, the male songs of the two are quite different, although males of either species will respond equally to the songs of

Fig 28. *Top and bottom*: breeding-plumage male Baltimore and Bullock's orioles. *Middle pair*: two hybrid males. The hybrid types are based on birds captured in Nebraska by W. C. Scharf.

their own or the other species. However, females can certainly distinguish them, and this might be an important means of making mate decisions, along with the rather strongly apparent male plumage differences. Females of both species are quite similar, and it is unlikely that males discriminate between them when courting. In an area of eastern Colorado where both species occur, pairs of Baltimore orioles were more successful in breeding than were pairs of Bullock's or hybrid pairs.

Although hybrids are certainly fertile, adult males have a variety of generally intermediate, odd-looking plumages, and they also may have maladaptive molt patterns. Thus, rather than molting their wing feathers once after breeding, they may molt twice, according to the different seasonal molt schedules of each. If generally true of hybrids, this might be a strong selective pressure against the production of second-generation hybrids. Observations suggest that Bullock's orioles may be more tolerant of heat and have lower water requirements than do Baltimore Orioles, a point of possible ecological significance in the West.

In western Nebraska, the Baltimore oriole is more likely to nest somewhat higher in trees than is the Bullock's oriole, a fact of no known significance and with no other obvious ecological differences. In Nebraska, both species are prone to nest in the same trees as eastern or western kingbirds, taking advantage of the kingbird's intense nest-protection behavior against jays, crows, or other possible aerial predators of eggs and nestlings.

Vignettes of Endangered and Declining Species

River Otter

Otters once were distributed along nearly all the major streams of North America, from northern Alaska to Texas and Florida. Like beavers, otters were often trapped for their lustrous pelts, and by the 1900s the animal had disappeared from many parts of the United States, including Nebraska. Improved protection in recent decades has improved the situation, and they now can be trapped in about half of the states and all of the Canadian provinces. Scarcely any records of otters in Nebraska were obtained until one was trapped in 1977 in the Republican River drainage. This animal probably represented a transient from Kansas or, more likely, Missouri. The river otter was declared an endangered species in Nebraska in 1986, and a restoration program was begun. Wild-trapped animals from a variety of geographic sites were released in several different Nebraskan rivers, including the Niobrara, Platte, North Platte, Calamus, Elkhorn, and South Loup. At least a few of these animals took a leisurely ride downstream, only to be recaptured 600 miles away in Missouri, but it is believed that most have adapted to their new surroundings and have become established near their release points.

Although a member of the weasel family, the otter seems to share few of the usual bloodthirsty weasel personality traits and instead is remarkably playful and at times seems almost carefree. Otters live almost entirely on fish, captured mainly using visual clues, but they also have a keen sense of touch. They are prone to prey on the slowest-swimming and most abundant fish rather than on faster-swimming and more agile game fish such as trout. They are rather short-legged and when underwater use their forelegs for power and let their hind legs trail behind, functioning as rudders. When running on land they are surprisingly fast, alternately arching and extending their long and flexible backbone to provide power and increase their effective stride, comparable to a running cat.

Fig 29. River otter, *above* running and *below* swimming.

Females care for their young for a prolonged period, and it is not a rare sight in Wyoming or Colorado national parks to see a family group playing together or sliding down wet mud banks or running and then sliding on the ice during winter. Sometimes their seeming playful activities can lead to trouble, as when a single inattentive animal gets separated from an escape hole in the ice and is trapped by hunting coyotes. Yet they are very agile and can defend themselves well against a single coyote. Probably it is not unusual for an animal to live for about 15 years under natural conditions or up to 25 years in captivity. In Nebraska, the otter was recently upgraded from endangered to threatened.

Ute Lady's Tresses

Nebraska has at least 18 species of native orchids but most are quite limited in distribution and some are known from only one or two counties. The Ute lady's tresses is one of those elusive species and is not even listed in T. M. Barkley's *Flora of the Great Plains*. It was placed on the federal list of endangered species in 1992, after having been first discovered in the early 1980s. Initially known from only a few locations in Colorado, Utah, and Nevada, it has more recently turned up in Montana, Wyoming, and western Nebraska. This is an unusually large range for such a rare species, but just because a species is known to occur in an area, it may not appear

Fig 30. Ute lady's-tresses and a likely pollinating bumblebee (*Bombus fervidus*). The inked-in part of the lateral view shows the stigma; the arrow indicates the pollinia. The dotted line shows a typical route up the flower stem made by visiting insects. In part after Nilsson (1992) and Cingel (1995).

every year. Like many orchids, this one may emerge only in very favorable years, then disappear again for many years. It also may germinate and remain underground for several years before appearing for the first time.

One reason this orchid was not discovered until recently is that it closely resembles another species, *S. romanzoffiana*, a common Rocky Mountain species. It requires an expert to distinguish the two. Like the Colorado butterfly plant, another endangered species in Nebraska, this orchid tolerates some grazing or mowing, and cattle seem to like its taste, which might prevent its blooming and setting seeds. It does best in moist or periodically flooded soil, thus its name *diluvialis*. Its generic name *Spiranthes* refers to the fact that the blossoms are arranged in a spiral up the stem of the plant. Each flower is built on a pattern of threes, including three petals and three sepals. The lowermost petal is elongated into a decorative lip, two sepals extend out sideways like wings, while the other two petals and the remaining sepal form a fused overhanging hood or tube that encloses the sexual parts of the flower. These parts are fused to form a central column, which in all orchids is highly specialized for insuring fertilization.

The fertilizing insects for this species are not yet well studied, but Sedonia Sipes and Vincent Tepedino have analyzed the plant's fertilization biology in Colorado and Utah. Small glands near the base of the column secrete a fluid that attracts insects, primarily bees and especially some long-tongued bumblebees. Of the two *Bombus* species (*B. morrisoni* and *B. fervidus*) observed at flowers there, only the latter is common in the Niobrara Valley, although other bumblebees (such as *B. pennsylvanicus*) are also present. Some long-tongued bees (*Anthophora*) were also observed carrying the plant's pollinaria, but their role as pollinators is uncertain. Only long-tongued insects are able to pollinate the plant; smaller bees cannot pick up the large pollinaria, or it may attach to an inappropriate location on their bodies. The insect lands on the lip and extends its proboscis to reach the nectar, thereby touching the sticky mass of pollen as well, which is carried away when the insect leaves the flower. The bee works spirally up the flower stalk, leaving the lower and oldest flowers that are functionally female and encountering the functionally male flowers toward the top of the stem. This differential rate of sexual development promotes cross-pollination as the insect visits one flower after another.

CHAPTER 8

Braided Streams and Sandy Shores

The Platte Valley and Its Riparian Forest

Perhaps few Nebraskans would claim the Platte River as their favorite river; it is slow, mostly knee-deep or even shallower, is sometimes somewhat unpredictable in its course, and often is no more scenic than your average Nebraska irrigation ditch. This is not the Platte's fault; for a century and more it has had its lifeblood water diverted, pumped, and polluted with fertilizers, pesticides, and animal wastes nearly all the way from its inconspicuous mountainous origins in eastern Wyoming and Colorado to its confluence in easternmost Nebraska with the Missouri. In May 1833, during his exploration of the upper Missouri Valley, Prince Maximilian of Wied reported that, even some 4 or 5 miles downstream from its confluence with the turbid Missouri River, the unsullied "clear and blue" effluent water from the Platte could still be easily distinguished from that of the much larger Missouri.

That the Platte River has survived at all in the face of almost two centuries of destructive influences is a small miracle. Had Michelangelo's statue of David been subjected to the same treatment as has the Platte, it would have long since been consigned to the scrap pile, so the Platte needs to be viewed with a gentle and forgiving eye. It must be appreciated for what it once was, like the once-pristine Parthenon, rather than what it has now become, a ghost of river and a frequent repository for trash. It is ironic that the 200,000 or so migrants who laboriously walked the length of the Platte on their way westward left only a few simple gravestones to mark their passing; modern humans prefer to leave abandoned cars to mark theirs.

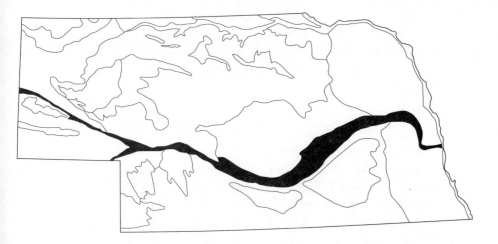

Map 10. Platte River Valley Riparian Forest region (inked). Based on maps by Kaul (1975) and Omernik (1987).

In spite of all this, the Platte Valley is still able to attract half a million sandhill cranes each March, and nearly 10 million waterfowl use it every spring, plus some 300 other bird species pass through the valley at various times of the year. Of these, at least 121 species breed in the central Platte Valley, including two nationally endangered species (least tern and piping plover). And the endangered whooping cranes reliably return each spring and fall, now in ever-increasing numbers, and sandhill cranes gather in almost uncountable flocks to rest and sleep beside the peaceful sandbars of the Platte. Through the night the cranes converse with the river, speaking in tongues that are both archaic and yet seemingly wise, and the river patiently listens. The voice of the river is even softer and possibly even older than that of the cranes; we would do well to try to hear and understand its plaintive message while it is still able to speak.

Ecology of the Platte River Valley

Major Terrestrial Community Types

Bur Oak Forest and Woodland

Along various stream valleys of eastern Nebraska, on well-developed and deeper loamy soils, and especially on steeper north- or east-facing slopes, a community largely or entirely dominated by bur oak develops. In the taller and denser forest version, bur oak shares dominance with hackberry and black walnut and to a lesser degree with somewhat smaller trees such as red elm, green ash, red mulberry, and hop hornbeam. In the drier woodland version, bur oak is likely to be the sole dominant, with only scattered individuals of green ash and elms. Both types have well-developed assortments of shrubs and both woody and herbaceous vines. Among the usual shrubs are gooseberry, coralberry, wolfberry, rough-leaved dogwood, and chokecherry. Smooth sumac, wild plum, and eastern red cedar may be present at more open sites. In the woodland version, big bluestem, little bluestem, porcupine grass, and sun-tolerant sedges may be found, but in the forest type there are fewer grasses, more shade-tolerant sedges, and a greater diversity of forbs.

EASTERN FLOODPLAIN WOODLAND

Closer to rivers, in the lower floodplain and on old river terraces that are only occasionally flooded, a distinctively midwestern gallery forest develops, which is dominated by cottonwoods. These trees may tower above most others, with older specimens up to 100 feet tall, and they often provide favored perches for wintering or migrating bald eagles. Lower-stature trees such as American elm, green ash, and red cedar usually surround these great trees. Box elders, silver maples, hackberries, mulberries, and other trees occur in various mixtures, and the shrub layer frequently consists of rough-leaved dogwood and coralberry. Woody vines such as poison ivy and Virginia creeper are common, but the forb layer is often affected by periodic or irregular flooding and may consist mostly of annual weeds that take root in the wake of receding spring waters. Although it has generally been accepted that the Platte was largely nonwooded during pre–Euroamerican settlement times, owing to periodic prairie fires, the most recent interpretation by W. C. Johnson and Susan Boettcher is that not only were many of the river's islands well wooded but a thin strip of gallery forest probably lined the riverbanks as well.

COTTONWOOD-DOGWOOD FLOODPLAIN FOREST

The cottonwood-dogwood floodplain forest in the lower Platte Valley, where it possibly but not certainly occurs, is probably little different from similar moist floodplain forests in the Missouri Valley, and is discussed in chapter 10. Cottonwood is the primary tree canopy species, and a dense growth of rough-leaved dogwood is the usual shrub layer, plus many climbing vines. It may differ from eastern floodplain lowland mainly in that it typically experiences more frequent flooding and correspondingly has a rather low level of species diversity.

SANDBAR/MUDFLAT AND PERENNIAL WILLOW SANDBAR

Along the streams and channels of many rivers throughout the state and especially along the braided channels of the Platte, sandbars and sandy islands are often formed. Sometimes the river's fluctuating currents and seasonally variable water levels quickly wash them away again. Historically, these changes in the Platte were extremely large. As meltwaters from the Wyoming and Colorado mountains made their way down the

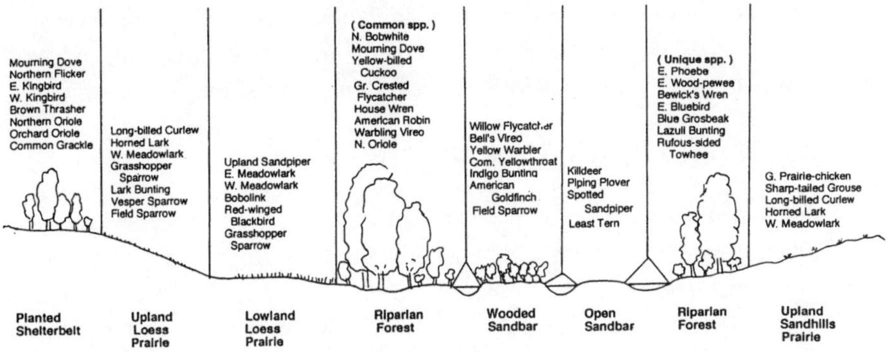
Mourning Dove
Northern Flicker
E. Kingbird
W. Kingbird
Brown Thrasher
Northern Oriole
Orchard Oriole
Common Grackle
Long-billed Curlew
Horned Lark
W. Meadowlark
Grasshopper Sparrow
Lark Bunting
Vesper Sparrow
Field Sparrow
Upland Sandpiper
E. Meadowlark
W. Meadowlark
Bobolink
Red-winged Blackbird
Grasshopper Sparrow
(Common spp.)
N. Bobwhite
Mourning Dove
Yellow-billed Cuckoo
Gr. Crested Flycatcher
House Wren
American Robin
Warbling Vireo
N. Oriole
Willow Flycatcher
Bell's Vireo
Yellow Warbler
Com. Yellowthroat
Indigo Bunting
American Goldfinch
Field Sparrow
Killdeer
Piping Plover
Spotted Sandpiper
Least Tern
(Unique spp.)
E. Phoebe
E. Wood-pewee
Bewick's Wren
E. Bluebird
Blue Grosbeak
Lazuli Bunting
Rufous-sided Towhee
G. Prairie-chicken
Sharp-tailed Grouse
Long-billed Curlew
Horned Lark
W. Meadowlark
Planted Shelterbelt
Upland Loess Prairie
Lowland Loess Prairie
Riparian Forest
Wooded Sandbar
Open Sandbar
Riparian Forest
Upland Sandhills Prairie

Fig 31. Profile view of typical middle Platte Valley habitats and some associated breeding birds. After Johnsgard (1995).

Platte each spring, rather brief but extensive flooding regularly occurred, sometimes sweeping away livestock and rarely even humans. An extended series of flood-control and irrigation river diversions has greatly reduced these changes in flow rates, and the once regular spring scouring by ice and high water is now a thing of the past. As a result, the sandbars and islands that are still being deposited have a better chance of becoming vegetated and stabilized before they are washed away in the following year's flood.

As spring and summer progress, the sandy bars and islands are increasingly colonized by annual weedy grasses and forbs, along with seedlings of smartweeds, cottonwoods, and willows, all of which are tolerant of the rather extreme conditions of sunlight, possible flooding, and an unstable substrate. Should the sandbar survive over the next winter and early spring, sandbar willow saplings and some taller perennial grasses such as prairie cordgrass may establish a footing. Many weedy annuals may still be present, but these soon are shaded out by the taller perennial grasses and seedling trees, which also include those of cottonwood and several additional willow species, among them diamond willow and peach-leaved willow. The root networks of these woody species increasingly help to bind the sandy substrate together, and eventually an almost impenetrable thicket may emerge.

Willow Shrubland

Given a few additional years of sandbar survival, the willow shrubland community gradually develops into a rather mature community that may be 7 to 12 feet tall and may consist of several recognizable zones, depending on their height above the water level. Nearest to the river and just barely above water level is a zone of sandbar willow, sometimes with a few examples of other shrub species such as false indigobush or diamond willow also present. At slightly higher levels sandbar willow shares its dominance with false indigo and with various water-tolerant forbs and grasses. On the driest sites the false indigobush assumes dominance over the sandbar willow. Saplings of cottonwood, peach-leaved willow, and diamond willow are also likely to be present in this zone.

WESTERN FLOODPLAIN WOODLAND

In central Nebraska and also westwardly, floodplains, river terraces, and filled-in sandbars mature into woodlands that are typically dominated by cottonwood and peach-leaved willows that may be up to about 30 feet in height. The substrate is still poorly developed beyond a stabilized sand, and drainage may be relatively poor, depending on the height of the land above the river level. Besides these two dominants, green ash may be fairly common, and the introduced Russian olive may invade, as might junipers. Sandbar willow manages to persist from earlier successional stages, especially near river edges, and the shrubs false indigo and rough-leaved dogwood are also likely to remain part of the community. Periodic flooding may stimulate the growth of annual weedy forbs such as thistles and ragweeds.

Profiles of Keystone and Typical Species

Eastern Cottonwood

I can never see an old, gaunt, and now often dying cottonwood along the Platte River without wondering if some immigrant may not have stopped to rest temporarily in its shade while traveling west along the Mormon or Oregon Trails. The trees carry a sense of history with them, and even when dying their dark skeleton-like forms reach higher into the sky than the other gallery forest trees, refusing for a time to give up their space on Earth to lesser trees. Whether dead or alive, their craggy branches provide lookout perches for wintering bald eagles, lofty sites for fox squirrel leaf nests, or rotted cavities for wood ducks to use for their own nesting purposes.

The cottonwood is really an overgrown poplar whose western limits correspond with the limits of the Great Plains, beyond which it is replaced by closely related species. Like other poplars, it grows remarkably rapidly, sometimes attaining 100 feet in 30 years or so. Its roots often penetrate to reach subsurface water levels, and thus it is especially typical of streamside woods. It likes sunshine and in wetter climates may extend beyond lowland floodplains onto moist upland slopes. Its fondness for water has made it undesirable in the eyes of some farmers, who think that it steals water from their nearby croplands and are thus inclined to cut them down or bulldoze and burn them.

Nevertheless, they are valuable trees for wildlife, with their buds, leaves, and twigs all being eaten by various bird and mammals, and

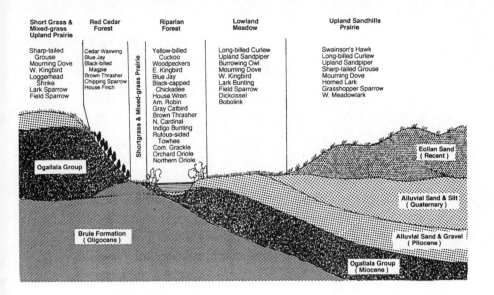

Short Grass & Mixed-grass Upland Prairie	Red Cedar Forest	Riparian Forest	Lowland Meadow	Upland Sandhills Prairie
Sharp-tailed Grouse Mourning Dove W. Kingbird Loggerhead Shrike Lark Sparrow Field Sparrow	Cedar Waxwing Blue Jay Black-billed Magpie Brown Thrasher Chipping Sparrow House Finch	Yellow-billed Cuckoo Woodpeckers E. Kingbird Blue Jay Black-capped Chickadee House Wren Am. Robin Gray Catbird Brown Thrasher N. Cardinal Indigo Bunting Rufous-sided Towhee Com. Grackle Orchard Oriole Northern Oriole	Long-billed Curlew Upland Sandpiper Burrowing Owl Mourning Dove W. Kingbird Lark Bunting Field Sparrow Dickcissel Bobolink	Swainson's Hawk Long-billed Curlew Upland Sandpiper Sharp-tailed Grouse Mourning Dove Horned Lark Grasshopper Sparrow W. Meadowlark

their cottony seeds may be used to line bird nests. The annual summer snowstorm of cotton, however, makes the trees unpopular as city trees, and their water-seeking roots are inclined to clog drains. Despite this, early pioneers often planted them in villages and towns because of their ease of planting and rapid growth. Probably because of this association with our pioneer history, the eastern cottonwood carries enough of a sentimental cachet to have been selected as the official state tree of Nebraska. As of the early 1990s there were about 100,000 acres of cottonwood forests in Nebraska.

Much earlier, the Pawnees of eastern Nebraska called it the "Dancing-leaf tree," a far nicer name than cottonwood. The Lakotas called it "Waga chun," the rustling tree, and considered it a sacred tree, in part because its triangular leaves are shaped like the outline of a tepee. They believed it to be the sacred tree that grew at the intersections of the two roads lying within the Sacred Hoop of the world. The red road of spiritual under-standing, passing from south to north and depicting the understanding that comes with increasing age, is represented by the cottonwood. The Lako-tas' most important tribal rituals for communing with the Great Spirit, such as the Sun Dance ritual, were performed around this tree. For this purpose a medicine man or the most virtuous young women of the tribe searched for the perfect cottonwood. The women would cut it down, strip its trunk of bark, and paint it four colors, each color representing one of the symbolic cardinal directions. The pole was then decorated and raised, and a variety of dances and exhibitions of personal bravery and tolerance of sacrificial pain were performed around it as a test of individual faith and as a thanksgiving.

Fig 32. Profile view of typical lower North Platte Valley habitats and some associated breeding birds. After Johnsgard (1995).

Fig 33. Sandhill crane, brooding adult.

Sandhill Crane

If the sandhill crane had been called the "sacred crane," something like the sacred ibis of Egypt, for example, it might have been accorded more respect by the general public. Instead, many Nebraskans still confuse sandhill cranes with great blue herons, a much more serious taxonomic blunder than, for another example, referring to some distant cousin as a "dumb ape."

Sandhill cranes were not, as some imagine, named for the Nebraska Sandhills, although the species once nested in the marshes there and they still frequent these same marshes in spring and fall. For nearly all of the twentieth century the cranes used Nebraska only as a way station to and from their breeding grounds, but in the late 1990s single pairs of cranes were known to have nested in the Rainwater Basin region.

Even if the cranes have not nested here continuously, they have used the Platte Valley for untold millennia as a prime staging area. Fossil remains of a sandhill crane dating back at least 8 million years have been found, suggesting the sandhill crane's love affair with Nebraska is a very long one indeed. From the time the Platte River becomes ice free in February until almost the middle of April, hundreds of thousands of sandhill cranes use the valley every year, by far the largest assemblage of any cranes in the world. Along dozens of the more remote stretches of river, roosts of up to 20,000 birds gather to spend the night standing in shallow waters. They select the widest stretches of river that offer nearly vegetation-free sandy islands and sandbars, which places them out of danger of land-based predators such as coyotes. Eagles may still pose a threat to sick or wounded birds, but a healthy crane is nearly an even match for an eagle, unless the crane is attacked in the sky when the eagle has the advantage of surprise.

We do not know what originally drew the cranes to the Platte, but the unique present-day combination of a wide, sandy river, adjacent wet meadows with a supply of invertebrate foods for a source of calcium, and an almost unlimited amount of waste corn in nearby fields for getting abundant carbohydrates that can be converted and stored as fat provide the magic attraction now. Millions of snow geese, Canada geese, and other geese join in on this feast, as do several million ducks, making March in Nebraska a bird-watcher's paradise. This alone is enough to warm the heart during the long days of winter, and the sounds of cranes filling the sky when they finally do arrive is at least as thrilling as hearing a massed choir singing the triumphant chorus to Beethoven's Ninth Symphony.

Barred Owl

The person who first described a woman's eyes as resembling "limpid pools" must have been unfamiliar with those of the barred owl, or such fine language would never have been wasted in describing a mere human. The barred owl's large and lustrous eyes seem very dark brown in one light and purplish black in another, but they always convey an intense sense of intelligence and mystery. As with other owls, its wide-set eyes are associated with wonderfully precise binocular vision under low-light conditions, and it has equally remarkable binaural hearing abilities. These highly nocturnal owls can home in on small rodent prey in total darkness, using only acoustic clues if needed. They can distinguish shapes and movements under light levels far too low for humans to use for any kind of visual orientation. Their ears also can detect the soft, rhythmic calls of another barred owl in the far distances; many a Hollywood movie scene set in a moonlit cemetery has been made more scary by the insertion of these evocative sounds. Except for the barn owl, the barred is the only Nebraska owl with brownish black rather than yellow eyes. Both are highly nocturnal birds; eyes make poor signaling devices at night, but the effectiveness of sound signals is undiminished

The barred owl is one of those woodland species that has benefited from the gradual forestation of the Platte Valley, allowing it to move west along its course. Nationally, it has also moved west, via the Canadian forests, all the way to Washington, Oregon, and, most recently, northern California. Throughout its entire range it is a small-mammal specialist, taking over the darkest hunting hours after the great horned owl and screech owls have essentially stopped hunting for the night. Barred owls, like other smaller owls, must also avoid contact with great horned owls, or they may end up as prey themselves.

Fig 34. Barred owl, head of an adult.

Barred owls begin breeding early in the calendar year, and their distinctive and repeated "Who cooks for you?" notes can be heard from late fall through winter in the woods of eastern Nebraska. Sometimes duets between paired birds can be heard, or a more general chorus may be generated by all the owls in the acoustic neighborhood. They are monogamous, the smaller male hunting for the incubating and brooding female and her dependent young over a several-month period.

Of all the owls that I see cooped up in rehabilitation cages, the barred owl affects me most. Their great dark eyes seem ineffably sad, and I find it hard to resist stroking their soft plumage in sympathy. One long-term captive owl, which had suffered the total loss of one eye as the result of a shotgun pellet and was essentially blind in the other, still could not resist trying to fly. It would take off and ascend to the ceiling of a large room and then fly along the ceiling surface, using its wingtips to feel its way, until it crashed into the wall at the far end of the room. Apparently it thought that, if it continued trying long enough, it would eventually find a way to freedom. It is not just humans who feel a need to be free.

Fig 35. Wood duck pair, *front* female and *back* male.

Wood Duck

Wood ducks are probably more beautiful than any duck has a right to be; they seem to belong more appropriately in the realm of fantasy art than reality. Explanations for their plumage splendor are hard to come by, but one possible clue is the fact that the birds are more active under the low-light conditions that occur near dusk than they are around midday, when most social interactions among other ducks occur. The wood duck's unusually large eyes are an indication of their abilities for seeing under these conditions. Thus the males may need their distinctive bold patterning and iridescent colors to attract females successfully. The courting males use a variety of head shaking, preening, and other body movements that effectively exhibit their bright colors to specific females.

Like barred owls, wood ducks have moved west along the Platte Valley during the twentieth century and now probably occur as breeders all the way to the Wyoming and Colorado borders. They have also greatly increased in numbers nationally as a result of improved federal protection, the setting out of nest boxes, and the growth of trees in once-lumbered areas. They use sites for nesting similar to those of hooded mergansers and goldeneyes and indeed also flying squirrels. For all these species, old woodpecker holes, especially those made by large species such as the pileated woodpecker, are probably preferred, but artificial nesting boxes will be accepted where natural cavities are rare or lacking. Sometimes two or even more females will try to nest in the same cavity, producing extremely large joint clutches that cannot be effectively incubated, as well as fights over ownership by the contending females.

The nearest relative of the wood duck is the mandarin duck (*Aix galericulata*) of China and Japan, the females of which are nearly identical to wood duck females, but mandarin duck males are even more elegantly plumaged. Together the two species represent some of the most beautiful birds in the world. The mandarin duck is also a forest-loving and tree-nesting bird and can often be seen in American zoos.

Northern Flicker

The flicker was seemingly named for the flickering visual effect of its wings in flight, since the word "flicker" originally meant a fluttering of birds. The word later came to refer to any general wavering, such as a wavering of candlelight. In either case, it seems to fit the flicker, whose yellow to red underwings flicker like a candle when the bird is in flight. In eastern Nebraska, nearly all the birds have bright yellow on their underwings and tail, but as one goes west along the Platte an increasing number show orange to red colors. These intermediate individuals represent a broad hybrid zone between the eastern yellow-shafted and western red-shafted races, with the zone of approximate genetic intermediacy being in the general vicinity of Bridgeport. The hybrids seem to survive just as well as the parentals, and mating among the various plumage types appears to be unrestricted. This hybridization zone is probably quite old, as known hybrids go back to the time of John J. Audubon. Because of the zone's breadth and length, extending from western Canada to Texas, it is unlikely that speciation will proceed any further in the flickers of the Great Plains.

Flickers differ from typical woodpeckers in that flickers frequently forage well away from trees, most often on open grasslands where they probe for ants and other insects. Their bills are just as sharp and chisel-like as other woodpeckers (their generic name *Colaptes* translates from the Greek as "a chisel"), and they can readily excavate nesting holes as do most other woodpeckers. These same holes may later be used for nesting by house wrens, eastern bluebirds, tree swallows, and other cavity nesters not so well equipped as woodpeckers for excavation.

Like other woodpeckers, flickers form monogamous pairs, their courtship involving loud calls, chases, and exposures of the male's yellow to red wing and tail feathers to the female's view. Males also have a distinctive "mustache" stripe behind the bill, which in yellow-shafted birds is black and in red-shafted individuals is red. Among hybrids this stripe can be either color but is often a mottled mixture of both. In some individuals the stripe may be black on one side and red on the other. Or an individual with yellow wings may rarely have a red stripe or one with pinkish wings a black stripe. The male yellow-shafted also has a red nape-patch, which is lacking in the red-shafted. The calls and postures of both types and their hybrids seem to be identical, at least superficially.

Vignettes of Endangered and Declining Species

Whooping Crane

The whooping crane is the very symbol of American conservation efforts and indicative of the value of the Endangered Species Act. Never common, the whooping crane was once almost hunted to extinction for trophies, its meat, its eggs, and its beautiful black-and-white plumage. By the late 1930s fewer than 20 birds were known to exist in the wild, which perhaps represented less than half a dozen breeding pairs. Aransas National Wildlife Refuge, near Corpus Christi, Texas, was then established to protect its limited wintering grounds. It was not until the 1950s that its remote breeding grounds were finally found, in the muskeg and forest wilderness of Wood Buffalo National Park in extreme northern Alberta, Canada. Since then the whooping cranes have made a sometimes painfully slow population recovery, the wild flock increasing by an average rate of only about two to three birds a year over the past six decades. After the start of the Endangered Species program, special efforts were made to rear some of the birds in captivity for release in other areas and to try to encourage sandhill cranes to cross-foster the chicks. The latter experiment proved unsuccessful, but there are ongoing efforts under way to start new flocks of whooping cranes in new habitats, such as on Kissimee Prairie, Florida, and at Necadah National Wildlife Refuge in central Wisconsin.

The Platte Valley has historically been the whooping crane's most important spring staging area between Texas and its nesting areas, and habitat modification has helped to bring some of the central Platte Valley back into an ecological state acceptable to the cranes. This, too, has been a slow and expensive process but has paid off in now-yearly use of the

Fig 36. Whooping cranes, adults landing.

region by a few whooping cranes. The birds are bigger and more space-demanding than sandhill cranes, and they also need more expansive areas of water for roosting and foraging.

One should not have to argue for the protection and conservation efforts that are being made on behalf of whooping cranes by federal, state, and private agencies; the birds are the most spectacular of Nebraska's avifauna. Cranes have an ancestral genealogy that is nearly as old as the Rocky Mountains, their ancestors probably having flown over what is now Nebraska while most of it was still a soggy marsh or an inland sea. In oriental cultures cranes are considered symbolic of longevity, fidelity, and happiness. We midwesterners should be grateful and should consider ourselves very lucky to be able to host them, even if only for brief periods and in still dangerously small numbers, each spring and fall.

Interior Least Tern

Of the several species of terns that migrate through or nest in Nebraska, the least tern is by far the rarest and the smallest. The state's nesting bird is part of a geographic race (the "interior" race) that once bred throughout the Mississippi, Ohio, and Missouri drainages, whereas two other races bred locally along the Atlantic and California coasts. Market hunters seriously

Fig 37. Least tern, brooding adult.

depleted all of these races for their attractive white feathers that made decorative additions to women's hats. Although the passage of the Migratory Bird Treaty stopped this frivolity, the coastal populations were then additionally impacted by development of shorelines for recreation and housing. The interior population was more influenced by river dredging, by channelization, and possibly also by water pollution. In any case, all populations were seriously affected, and by the 1970s it was determined that the species' population had declined by over 80 percent since the 1940s. In 1985 the species' interior population was federally listed as endangered. It is also considered endangered in Nebraska.

The nesting distribution of the least tern in Nebraska is now mostly confined to the lower Niobrara Valley, the central and lower Platte Valley, the upper Missouri Valley, the middle Elkhorn Valley, and the vicinity of Lake McConaughy. Nebraska's population, one of the largest remaining of the entire interior race, consists of about 1,200 to 1,400 birds, or 25 to 30 percent of the nation's total.

In Nebraska, the unvegetated sandbars and sandy islands that once were a typical part of the Platte River scene have largely disappeared, as river flows have declined and annual flow fluctuations have diminished. Thus nesting habitats for this sand-nesting tern have progressively disappeared. But, coincidentally, sandpit operations along the Platte Valley have produced hundreds of small wetlands, and the barren spoil piles around them have begun to attract least terns. If these small sandpits also support fish, terns may be able to hunt there; otherwise, they fly to the river for their daily fishing excursions. Like all terns, they have forked tails that are associated with slow-speed maneuvering and hovering, and sharply pointed wings for agile flight. They catch fish using vertical dives similar

Fig 38. Piping plover, chick and head of an adult.

to those of kingfishers and forage exclusively on minnow-sized fish such as shiners.

Nest sites are selected on bare sand, the birds often associating in a semicolonial manner, probably because of their limited site-selection opportunities. Both sexes help incubate the two- or three-egg clutch, and both sexes collect food for the developing chicks. Least terns nest in a high-risk environment, often losing their nests to flooding, human disturbance, or mammalian or avian predators. But they are persistent re-nesters, some-time making as many as three nesting attempts.

Increasing protection of nesting areas of the least tern have begun to show positive effects, and it would seem that the species is recovering slowly in Nebraska. With luck and persistence, the interior least tern can eventually be removed from the list of Nebraska's endangered species.

Piping Plover

In many ways the piping plover's story resembles that of the least tern; both species need barren sandy areas near water for nesting, and both have suffered similar population declines as a result of habitat alterations that are mostly the result of human activities. The interior breeding range of this species encompasses the drainage of the upper Missouri River and a few isolated outliers. There is also an Atlantic Coast population similar to that of the least tern but no Pacific Coast breeding.

The piping plover's breeding range once included all the larger rivers of Nebraska, including especially the Platte and Missouri plus parts of the Niobrara and Loup. Its current range includes the lower Platte, the Loup and Middle Loup, and the lower Niobrara. It also breeds very locally along the sandy edges of Lake McConaughy, where I have observed it nesting since the mid-1970s.

After first finding nesting birds near Kingsley Dam, I prepared and erected some signs indicating that this was a plover nesting site and that vehicles should avoid it. It was not long until the signs were run over. The birds continued to nest near a parking area of Arthur Bay, and it was frightening to watch tiny, newly hatched plovers running for their lives in trying to escape cars, motorbikes, and ATVs. Later, the local irrigation district erected more signs and wire fences, and effective monitoring of the nests is now under way. The birds often nest very close to the water's edge, and the color of the adult's back closely matches the color of dry sand. The eggs also perfectly match the color of sand and gravel, and the birds often place their nests where egg-sized gravel is nearby. The newly hatched chicks are just as cryptic and tend to crouch motionless when danger threatens. They then sprint away at the last possible moment, their tiny legs carrying them over the sand at a remarkable rate.

The species was first listed as endangered in the Great Lakes region and as threatened on the Atlantic Coast and the northern Great Plains. Nebraska's breeding population as of the mid-1990s was about 250 to 300 pairs.

Bald Eagle

It is a sad and ironic fact that our government's national symbol, the bald eagle, was one of the first birds to be placed on the list of endangered species in the 1970s. It had already received protection in 1940 with the passage of the Bald Eagle Act, which prohibited killing of the birds outside of Alaska. It continued to remain unprotected in Alaska until 1953 and until then had even been listed as a pest, for which a bounty was paid upon proof of killing.

Besides such slaughter, a major factor in the near demise of the bald eagle was the widespread use in the United States of "hard" pesticides such as DDT from the 1940s to the 1970s. These pesticides effectively sterilized much of the eagle population south of Canada, where pesticide use was greatest, and did similar damage to ospreys, peregrine falcons, and other raptors. It was not until the effective control of such pesticides was achieved that the eagle could have any hope for recovery. Lead poisoning, resulting from eagles eating birds that had been wounded or killed by lead shot, contributed to this problem and is still a significant source of mortality for eagles. Electrocution, illegal shooting, poisoning from pesticides

Fig 39. Bald eagle, adult.

or other poisons, and minor or accidental mortality factors contribute to the annual toll.

Bald eagles are potentially very long lived and do not attain their fully adult plumage and prime breeding condition until they are at least five years old. Until then they exhibit piebald plumages to varied degrees (the term "bald" refers to the all-white head of the adult eagle). Young birds, and especially first-year birds, are not as effective predators as are adults and are more prone to scavenge carrion than to try to kill live prey. Once paired, the birds maintain permanent monogamous pair bonds and return year after year to the same nest site. These large nests are added to every year, sometimes becoming so massive and heavy as to break their supporting branches.

Things have greatly improved for bald eagles in recent years. No known successful nesting of bald eagles had occurred during the 1900s until 1992, when a successful breeding occurred in Sherman County. Since then many other nesting pairs have been found, and there are now more than a dozen active sites around the state. Nebraska's wintering population has increased correspondingly, with about 500 to 1,100 wintering birds usually being found during the past decade, a significant proportion along the Platte River and especially around Kingsley Dam.

CHAPTER 9

Ghosts of Prairies Past

The Eastern Glaciated Plains and Tallgrass Prairies

Dreams do not die easily, nor did the tallgrass prairie. The roots of its typical plants drew deep into the glacial soil, and the plants repeatedly returned in the face of climatic adversity. But they could not fight the steel-bottom plow and the bulldozer. The spring after we moved into our several-years-old house in a fairly new subdivision in Lincoln, which had already been landscaped and carefully planted to Kentucky bluegrass, a few defiant stalks of big bluestem continued to spring up in our lawn. But bluestem does not fit the definition of "desirable plants" in the city's reference books, and after an official warning or two about having "undesirable weeds" on my property I gave in to convention and settled for a lawn with an exotic mixture of crabgrass-laced bluegrass, like those of all my neighbors. But I did refuse to use fertilizers, pesticides, or herbicides, and eventually my lawn devolved into one of the most disreputable assortments of greensward anywhere to be found in Lincoln.

Of all of the grassland types in North America, the tallgrass prairie has been the most ravaged. One estimate of its original extent, based on a map published by A. W. Küchler, was 221,375 square miles, as compared with 218,543 square miles for mixed-grass prairie and 237,476 square miles for shortgrass prairie. At least 95 percent of the tallgrass prairie is now gone; if the Sandhills prairies were classified as tallgrass prairie (they are usually considered as mixed-grass prairie) they would certainly be the largest remaining remnant in all of North America. However, the species diversity of Sandhills prairie plants is much lower than in true tallgrass

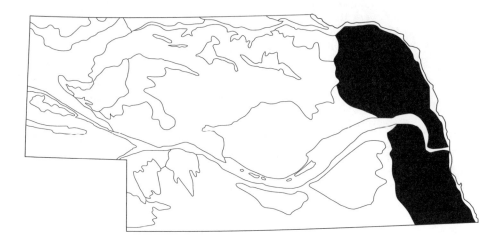

prairie. T. L. Stieger reports that at least 237 species of vascular plants occurred on various tallgrass prairie sites totaling about 10 square miles near Lincoln. One of the best studied of all tallgrass prairies is Nine-Mile Prairie near Lincoln, where over 290 species of plants have been detected over half a century of study by John Weaver and more recent botanists. It now comprises only 230 acres but was about twice that size when originally studied and before much of it was taken over for military purposes during World War II. Steven Rothenberger studied a mixed-grass prairie site in the Loess Hills that contained 239 species. These figures compare with a total of 194 plant species reported from the 2-square-mile Arapahoe Prairie in the central Sandhills that has been intensively surveyed by Kathleen Keeler, A. T. Harrison, and Linda Vescio.

Indiangrass, switchgrass, and, in wetter sites, prairie cordgrass are also important high-stature grasses of tallgrass prairie. All are well above 3 feet tall at maturity and have root systems that extend down 5 to 6 feet for Indiangrass, 7 to 8 feet for prairie cordgrass, and 8 to 12 feet for switchgrass. All are warm-season grasses that are strongly rhizomatous. Switchgrass and prairie cordgrass are continuous sod-formers, but Indiangrass is a bunchier species, mainly spreading from tillers that are produced from late summer rhizomes that overwinter and provide for early spring growth the following year. Seed production estimates for big bluestem, Indiangrass, and switchgrass perhaps average substantially less than that of little bluestem, but this statistic seems subject to considerable experimental variability. Additionally, all these species reproduce mainly by rhizomes rather than from seed dispersal.

Ecology of the Tallgrass Prairie

The tallgrass prairie is one of the most romantic concepts of the American West. The imagined view of endless bison herds plodding through grasses so tall that they half obscured the animals from sight is a powerful image and one that today must remain more in the realm of fancy than of fact. Quite probably most bison lived on prairies of shorter stature, and the taller grasses that were present were likely soon clipped by the hungry migrants, but it an attractive vision. One image that can still be enjoyed is the sight of tallgrass prairie in full bloom from June through September, when dozens of prairie forbs vie for the attention of bees, butterflies, and moths.

Such distinguished plant ecologists as Frederick Clements and John Weaver studied tallgrass prairies in eastern Nebraska for many decades. A study of several tallgrass prairies established that about 200 species of upland forbs were typically present and that 75 of these were found in 90 percent of the prairies examined. The most abundant and most consistently occurring upland forb was leadplant, which has a root system that can extend to more than 16 feet in length and possesses nitrogen-fixing nodules. The gayfeather or blazing star may have roots of a comparable length. Many species of goldenrods are also present, with roots up to 8 feet long. John Weaver once calculated that a square foot of big bluestem sod might contain about 55 linear feet and an acre about 400 miles of densely matted rhizomes from the surface to a depth of only a few inches. The strong roots of big bluestem have individual tensile strengths of 55 to 64 pounds, making prairie sod one of the strongest of natural organic substances. It is indeed strong enough to construct sod houses that have sometimes lasted a century or more in the face of Nebraska's relatively inhospitable climate.

Weaver also determined that the big bluestem has root systems up to about 3 feet in diameter that penetrate to a depth of nearly 7 feet. Some 43 percent of its underground biomass is concentrated in the top 2.5 inches of soil and 78 percent in the top 6 inches. The overall underground (root and rhizome) biomass of tallgrass prairies is usually two to four times greater than the aboveground biomass. The root component usually contributes about 30 percent of the annual primary production, or up to nearly 40 percent in the case of grazed prairie. Overall, annual primary production of organic matter in tallgrass prairie averages nearly 3,000 pounds per acre. Likewise, the total underground parts of tallgrass prairie may contribute more than a ton of new organic matter per acre annually. Annual turnover (decomposition) rates for the aboveground parts of tallgrass prairie average about 80 percent, resulting in an average turnover period for the aboveground component of about 1.25 years, whereas turnover periods for underground biomass averages about 3 or 4 years. As a result, prairie soils are constantly being refertilized by organic matter that has been produced during the past few growing seasons.

The soils of tallgrass prairie are among the deepest and most productive for grain crops of any on Earth. They represent the breakdown products of thousands of generations of annual productivity of grass and other herbaceous organic matter. Because of these organic materials and the clays usually present in prairie soils, such soils have excellent water-holding capabilities. In addition to the humus and related organic matter thus produced, many prairie legumes have nitrogen-fixing root bacteria that enrich and fertilize the soil to a depth of at least 15 feet. Earthworms and various vertebrate animals such as gophers make subterranean burrows that mix and aerate prairie soils, in the case of earthworms to a depth of 13 feet or more.

Major Terrestrial Community Types

Tallgrass Prairie

The five dominant grasses of upland tallgrass prairie are actually those of medium stature and consist of little bluestem, needlegrass, prairie dropseed, Junegrass, and side-oats grama. All are bunchgrasses, and of all these perennial native grasses, little bluestem is easily the most important. It alone may comprise 60 to 90 percent of the total vegetational cover, and in very favorable sites it may lose its bunching form and produce a continuous sod of interlocking roots. However, in most cases the major upland grasses occur in clumps spaced about a foot or more apart, with roots extending downward at least 4 to 5 feet. Weaver once calculated that a strip of prairie sod 4 inches wide, 8 inches deep, and 100 inches long held a tangled network of roots having a total length of more than 20 miles. In prairie the total weight of underground vegetation in the form of roots is likely to be as great as the aboveground parts, and much of this is recycled back into the soil on a yearly basis. In contrast, forests and woodlands store most of their productivity as woody aboveground parts, which recycle back into the soil only when the trees eventually die or perhaps are burned.

Most of the important grasses of the tallgrass prairie are 3 to 6 feet tall, with the higher slopes having a greater proportion of midstature species. One of the few large and bushy shrubs to be of significance on the uplands is wild plum, although the smaller leadplant is widely distributed.

Forbs of the tallgrass prairie are numerous, as mentioned earlier. In the uplands leadplant is usually the most important forb, although it has a woody base and might well be classified as a half-shrub. Other important prairie half-shrubs include prairie rose and New Jersey tea. The stiff sunflower is also one of the most widely distributed upland forbs and also extends to many lowlands. Other regular forbs are prairie goldenrod, prairie flax, wild alfalfa, heath aster, bastard toadflax, and daisy fleabane. Several sunflowers, such as the saw-toothed sunflower, Maximilian's sunflower, and Jerusalem artichoke, are important prairie forbs, especially in moister situations, and the Jerusalem artichoke has enlarged starchy tubers that can be eaten raw or cooked in various ways.

Wet Mesic Prairie

If the plants of upland tallgrass prairie are impressive, those of the lowland prairie are even more so. In this situation big bluestem may comprise 80 to 90 percent of the overall prairie vegetation. Big bluestem is substantially taller than little bluestem and where both occur together the shorter species may be shaded out. On slopes and drier hilltops the smaller species has an advantage over the larger one. The roots of big bluestem are about 6 to 8 feet deep, and those of little bluestem are about 5 feet deep, so big bluestem has an advantage in moister sites. However, its roots tend to grow directly downward, whereas those of little bluestem and other bunchgrasses tend to spread widely, intercepting a much broader area than the plant's aboveground parts. Like many prairie perennials, both bluestem species are believed to be long-lived. Both species are warm-season grasses and continue to grow through the summer. Big bluestem may rarely reach a height of 8 to 10 feet in some lowland sites by late summer, when it finally bursts into full flower. Other very tall grasses of lowland sites are prairie cordgrass, Canada wild rye, Indiangrass, and switchgrass. An additional 20 or more grass species are of importance in lowland prairie.

In typical lowland prairie, big bluestem is dominant, but Indiangrass, switchgrass, and Canada wildrye may also be abundant, and in wetter sites prairie cordgrass may take the place of big bluestem as the dominant species. Typical shrubs include wild plum, rough-leafed dogwood, and wolfberry. There are many summer- and fall-flowering composites, such as sunflowers, goldenrods, and asters, and prairie fringed orchids are likely in slightly moister ravines. Many taller forbs are part of the low prairie flora. Among these is compass plant, which grows to nearly 10 feet tall and has leaves that may be nearly 2 feet long. Younger plants especially have their leaves twisted vertically, and the leaf axis is oriented almost perfectly north-south (thus the plant's common name). This trait allows them to take advantage of early morning and late afternoon sunlight but not become too desiccated during midday hours. A related species, the cup plant, has opposite leaves united at their bases in such a way that a small cuplike structure is formed that holds water after rains.

Bur Oak Woodland

The bur oak woodland sometimes occurs near or is interspersed with tallgrass prairie, especially where recurrent burning or grazing helps to control the understory. See chapter 8 for a summary of this community type.

Sumac—Dogwood Shrubland

Over much of eastern Nebraska a shrubby community dominated by smooth sumac, rough-leaved dogwood, and sometimes wild plum, wolfberry, and coralberry is fairly common. It occurs along the edges of woodlands or in ravines of tallgrass prairies over well-drained soils. The shrubs may be fairly open or quite dense, at times even restricting human passage, but in the more open stands an understory of typical prairie plants may be present. Probably recurrent prairie fires once severely restricted this community, but fire suppression in recent times has encouraged its growth.

Profiles of Keystone and Typical Species

Big Bluestem

I remember seeing stands of big bluestem growing along the railroad track rights-of-ways near the little North Dakota village where I grew up. There it was simply known as "turkey-foot grass," and probably few people knew of its association with native prairie, which had largely disappeared from the Red River Valley for several generations before I had been born. Later I came to love its color and majestic stature. I still delight in letting my open fingers run through the grass as I walk through head-high stands of flowering big bluestem, until my shirt is tinted and spattered golden yellow by its abundant pollen. If there were a king of American grasses this must surely be it. Its companion tallgrass Indiangrass would be an appropriate queen; the bronzy fall color of Indiangrass is even more beautiful than that of big bluestem, and its bushy head of autumn florets seems to me to resemble a golden tasseled wand.

Big bluestem is a warm-season grass, often growing 6 feet or more during the hot summer months and finally bursting into blossom in September. By October it is starting to shed its seed crop, which in natural stands might reach 100 pounds per acre and much more in planted stands. By then its stemmy and rather rank foliage is not so attractive to large ungulates, but earlier in the season it is a highly preferred food for most grazing mammals.

Its generic name *Andropogon* translates as "man's beard," a fair description of its flowering head, which includes an equal mix of somewhat hairy and sessile but fertile spikelets and adjacent stalked but infertile ones. Although the undisputed dominant of moist tallgrass prairie, big bluestem has an overall range extending east to the Atlantic Coast, north in eastern Canada almost to James Bay, and south well into Mexico. Little bluestem, sometimes placed in the same genus, is also a warm-season grass with a range similar to that of big bluestem. The sand bluestem is an extremely close relative of big bluestem that is more sand- and arid-adapted but is otherwise nearly identical, and sometimes the two forms hybridize where their ranges overlap in Nebraska.

Western and Eastern Meadowlarks

Like those in many other states, Nebraskans chose the western meadowlark as the official state bird, not a bad choice considering that it is

Fig 40. Big bluestem.
Illustration from Hitchcock
(1935).

probably about the third-most common breeding species in the state, is firmly associated with the prairies, and has a spring song that would gladden the heart of even the most jaded listener. Growing up in North Dakota (whose residents also selected the western meadowlark as the state bird), when I heard the first spring song of a western meadowlark I knew that the winter was almost over and that better times lay ahead.

Nebraska is a bit unusual in that over much of the state both eastern and western meadowlarks can be seen and heard. Where they commonly occur together, as in the eastern fourth of the state, the eastern meadowlark is likely to be found in the lower, moister sites and the western on uplands and drier habitats. But often both can be heard singing almost simultaneously, and it is the difference in the males' advertising songs that makes field identification easiest. The western has a complex, melodious, and trumpetlike series of many short notes, uttered too rapidly to count them easily. The eastern has a more trombonelike series of a few more obviously sliding-scale notes. If one can see the singer, it may be apparent that the cheek ("malar") area of the western is more tinged with yellow, like the chin, whereas in the eastern this area has little if any yellowish color. Intermediate songs, as well as intermediate-looking birds, are sometimes present and may leave the observer in doubt as to their identity.

Individual male western meadowlarks sing a variety of unique song types, usually ranging from about 3 to 12. Some of these song types may be shared with other males in the local population, but no two males exhibit the exact same repertoire. A male may repeat one of his song types several times but will switch to a different type on hearing a rival, perhaps to reduce the likelihood of this other male becoming less responsive to a particular rival's song type. Song-switching may also be important both in territorial defense and in attracting a mate. Males having the largest song repertoires also tend to be among the first to obtain mates and have greater reproductive success than do less gifted males, suggesting that song is one of the effective ways to attract a mate, as humans have also more recently discovered.

Meadowlark nests are always extremely well hidden; those I have found have been more by accident than by design. When walking through prairies and looking for flowers or other things, I have at times been startled by the eruption of a meadowlark at my feet. A careful parting of the grasses will then reveal a roofed-over nest with four or five speckled eggs. Such nests are best left alone and carefully covered over again, for in spite of their concealment the eggs are often lost to predators.

Greater Prairie-Chicken

Greater prairie-chickens are well named. They are indeed closely linked to true prairie, and they are "great" not only in terms of relative size (at least as compared to the lesser prairie-chicken) but also in their aesthetic appeal. There are few other places in North America where one is, within about a hundred miles of Nebraska's largest cities, able to secrete oneself within a blind on a predawn spring morning and experience what is one of the most exciting avian shows imaginable. Some of my best memories have been formed in these locations. It is like being a front-row spectator at a play whose general plot one knows almost by heart, yet the outcome of every such experience is uncertain. Each viewing is like attending an opening-day performance, where the performers' roles may be unexpectedly altered. Add to this the sight of a golden sunrise on the eastern horizon and the sounds of meadowlarks and distant coyotes greeting the dawn, and the scene is complete. Or with the approach of a thunderstorm, sudden strong

Fig 41. Greater prairie-chicken, male.

winds, or the unexpected visit of a coyote or prairie falcon, the whole performance may suddenly disappear before one's eyes. They are truly "such things as dreams are made of."

Like the sharp-tailed grouse, prairie-chickens display sexually on traditional sites called leks, in which the social status of each male is the sole factor influencing his opportunity for mating successfully. This is Darwinian sexual selection in its clearest form; even a minor setback in status relative to that of neighboring males, such as repeated loss in fights over territorial boundaries, might be enough to exclude a male from hierarchical advancement toward the status of master cock. The master cock is the factor that molds the entire social structure into a working, coherent group. Should he be suddenly removed from a stable lek, the resulting fights over new territorial boundaries and associated disruptions over who might replace the dominant male will result in reduced fertilization rates among the females and a possible disintegration of the entire lek structure.

The sounds and postures of the greater prairie-chicken in display are quite different from those of the sharp-tailed grouse. A low-pitched, dove-like "booming" replaces the sharptail's "cooing," and a stately erect posture, with two vertical earlike pinnae and the lowered wingtips brushing the ground, replaces the frenzied dances and outstretched wings of the sharptail. Yet hybrids sometimes occur in areas where both species coexist. This is most likely to happen when females visit mixed leks and, for whatever reason, allow themselves to be mated by a master cock of the different species.

Once the females have been mated, the males play no further role in assuring the species' successful reproduction. The females make their nests, lay their eggs, and tend their broods alone, probably not encountering the adult males again until fall flocks begin to assemble.

Vignettes of Endangered and Declining Species

Massasauga Rattlesnake

Rattlesnakes of any type are not particularly appealing animals for most people, although the danger they pose to humans is considerably overrated. I know of only two students who were bitten by prairie rattlesnakes, a larger species than the massasauga, while I taught nearly 20 summers at the University of Nebraska–Lincoln biological field station. One student was grazed after picking up a snake that had been run over that the student thought was dead. A snake bit the other during a show of foolhardy machismo by another young man whose hormones had seemingly drowned any good sense that he might have otherwise possessed. Both victims recovered rapidly; fewer people are killed by snakebites each year in the United States (about a dozen, out of 6,000 to 7,000 bit each year) than are killed as a result of being stung by bees, wasps, or ants.

In part because of the universal human persecution of rattlesnakes and partly because of habitat loss in this species' original prairie range, the massasauga rattlesnake has nearly disappeared from Nebraska. One of its very few remaining haunts is around Burchard Lake State Recreation Area in Pawnee County of southeastern Nebraska, where greater prairie-chickens also survive in small numbers. When George Hudson surveyed Nebraska's reptiles and amphibians during the 1950s, he was aware of records from Lancaster, Fillmore, Gage, and Nemaha Counties. There were five records from Lancaster County, including one from the prairie remnant near Lincoln called Nine-Mile Prairie. John Lynch's more recent survey produced a few additional locations, but only in Pawnee County (Pawnee Prairie and Burchard Lake) is the species likely to be still present

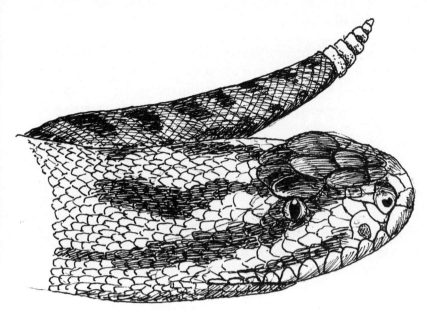

Fig 42. Massasauga.

and receiving some degree of protection. It was recently added to the list of Nebraska's threatened species.

No human fatalities have ever been reported from the bite of this small rattlesnake. It can be fairly readily distinguished from the larger prairie rattler by its rather thin tail and very small rattles, its several relatively large scales that are located on top of the head between the eyes (only three scales connect the eyes rather than two very large scales and many smaller ones), and the dark stripes on top of the head continuing back to the neck region without interruption. Snakes of this "pit viper" family can be easily identified in the field by their vertical, rather than round, pupils and by the round, heat-detecting openings, or "pits," near their nostrils.

Massasauga rattlesnakes tend especially to seek out grassy wetlands but also survive in open prairie and on rocky prairies, such as in the Flint Hills of Kansas. They range in length from 18 to 30 inches, and their rattles are so small that their buzzing can only be heard a few feet away. They are active only from about April to October and are nocturnal hunters. They prey on frogs, lizards, rodents, and some other smaller snakes but may also take birds' eggs, as well as birds, whenever they get the opportunity. Courtship and mating occur during both spring and fall, and the young are born during midsummer. From 3 to 13 offspring are usual, and these tiny snakes are venomous from birth.

Although small, this snake should be treated with due respect and left to its own for everyone's benefit. A little bit of danger on the prairie, whether real or only perceived, helps keep one alert.

Western Prairie Fringed Orchid

The western prairie fringed orchid is a lovely, all too ephemeral orchid that may remains hidden for years, suddenly appear in full bloom during late June or early July for a week or so, then disappear as quickly and quietly as it had materialized. Thus one must watch closely for it, usually in the wetter swales of tallgrass prairie. A farmer-photographer friend told me of once haying in a prairie meadow and seeing its blooms just as the plant was about to be mowed down. Before he could stop the machine the flower had gone into the mower. Returning in following summers, he was not able to find the plant again. The plants often remain unseen for several years in a dormant, subterranean state nourished by micorrhizae. They might then suddenly exhibit mass blooming, possibly stimulated by fire or by shifts in soil moisture that are associated with varied rainfall patterns.

There are many species of the genus *Plantathera*, most of which have whitish or greenish flowers and are pollinated by nocturnal or crepuscular moths. The white blossoms of the fringed orchid show up well under low-light conditions and no doubt help attract the moths. The enlarged and strongly fringed lower petal and sepals also might draw attention to the blossoms. Studies on the pollination biology by Charles J. Sheviak and Marlin L. Bowles have filled in the details for this species and a closely related but smaller one, the eastern prairie fringed orchid, which is fairly widespread in more eastern states. Both species creamy white to white blossoms, and in both the blossom fragrance is very sweet, intensifying after sunset. The blossoms of the western form are somewhat more creamy and their fragrance more spicy than in the eastern species. Their petal and sepal shapes also differ; in the western species the blossom heads are shorter and denser with fewer but larger individual blossoms.

Both species are specifically adapted to pollination by sphinx moths, being nocturnally fragrant, with extruded reproductive columns and extremely long nectar-bearing spurs. There is a very limited entrance access to the spur, and the pollinaria are situated in such a way that they will adhere either to the proboscis or eyes of the visiting moth. After the pollen-bearing structures have deposited their pollen on a moth, the columns rotate, so that they now fully expose their stigmas, ready to receive pollen from the next moth that visits.

Sheviak and Bowles estimated that any pollinating moths of the western species must have a proboscis length between 1.4 to 1.8 inches, and must also have an across-the-eyes distance that approximates the distance between the pollen-bearing viscidia. Five prairie-ranging sphinx moths seem to meet these requirements, all of which are native to Nebraska (achemon sphinx, white-lined sphinx, wild cherry sphinx, laurel sphinx, and vashti sphinx). Of these, the head measurements of vashti sphinx do not quite "fit" the proper requirements, and it may only be a nectar thief,

Fig 43. Western prairie fringed orchid and white-lined sphinx moth, a likely pollinator.

able to obtain nectar without carrying away pollen. The same is possibly true of the wild cherry sphinx.

Although it historically occurred all across eastern Nebraska, the current known distribution of the western prairie fringed orchid is limited to Lancaster County, eastern Seward County, Hall County, and east-central Cherry County. In 1989 the species was listed federally and by the state of Nebraska as a threatened species.

Small White Lady's-Slipper

The beautiful small white lady's-slipper once had a range similar to those of the eastern and western fringed prairie orchids combined. It extended west into eastern Nebraska and east into the southern New England states.

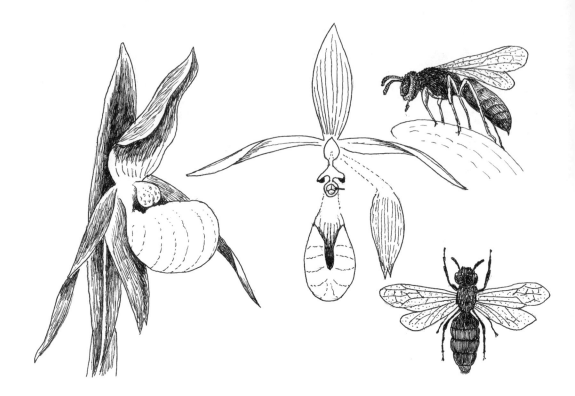

It favors damp soil but full sunlight, often occurring in wetter meadows where the prairie fringed orchid might also occur.

This little orchid blooms fairly early, in May and June, or about the same time as the yellow lady's-slipper, and before the fringed white prairie orchid. The blossoms may open before the leaves are fully unfurled, the flowers being mostly yellowish green except for the lower lip, which is glossy white with some flecks and narrow lines of purple. The conspicuous stamen-bearing structure is golden yellow with contrasting crimson spots, the colors probably serving as insect attractants. There is usually only a single blossom per stem, but there are sometimes two. However, the plants often grow in clumps, with stems up to 12 inches high and with the long, oval leaves wrapping around the stem at their bases. The white slipper-shaped pouch is up to an inch in length; the two lateral petal-like sepals are long, narrow, and rather twisted; and the dorsal hood is formed by a sepal that is also elongated and somewhat twisted.

The pollination ecology of this species is still little known but is perhaps much like that of a close European relative (*C. calceola*) that probably was separated from it during glacial periods. This species was one of the many orchids studied by Charles Darwin. He discovered that orchid flowers of this pouchlike type act as "conical traps, with the edges inwards, like the traps which are sold to catch beetles and cockroaches." Insects are perhaps attracted by scent or by the conspicuous white color of the

Fig 44. Small white lady's-slipper and an *Andrena* bee, a probable pollinator. The expanded front view shows the anthers (inked-in) and the triple stigma (arrow). The fused sepals at the rear are shown displaced to the side. In part after Luer (1975) and Dressler (1981).

pouch, with the crimson spots on the yellow staminode attracting further attention and the purple lines leading inward along the pouch perhaps acting as false nectar-guides. The plant produces a variety of fragrances, some of which are similar to sex-attractant pheromones used by bees for attracting females. Insects that crawl into the pouch become trapped and can only escape by exiting through one of the two rear openings. In doing so they must first brush the surface of the stigma and later one of the anthers. This sequence prevents self-pollination of the flower. Most of the visitors are bees, especially solitary bees of various genera such as *Andrena*, a large and widespread group of bees that dig nesting burrows in soil and are thus called "mining bees." Bumblebees may alight on the pouch but cannot enter, and some small bees and flies that do enter are too small to effect pollination.

Once very common in the wet meadows of eastern Nebraska, this orchid is now rare and is currently known only from Howard, Pierce, Platte, and Sherman Counties. It has been recently added to the list of Nebraska's threatened species.

CHAPTER 10

The Forests of Home

The Missouri Valley and the Eastern Deciduous Forest

When Lewis and Clark first explored the middle Missouri River along what is now the eastern border of Nebraska between mid-July and early September of 1804, they found a land that was still rich in plant, bird, and mammal life. They saw many broods of Canada geese along the river, hundreds of American white pelicans foraging around islands, and colonies of bank swallows on its steeper banks. Wild turkeys were numerous, as were American bald eagles. The explorers observed several grouse, almost certainly ruffed grouse, near present-day Council Bluffs. Deer were abundant in the extensive shoreline woods, and soon after the expedition reached what is now South Dakota they began to encounter elk and herds of bison on the open prairie.

The ruffed grouse is long gone from Nebraska, and in eastern Nebraska the elk is present only in place-names such as the Elkhorn River, Elk Creek (Johnson County), and Elk City (Douglas County). Until recent decades, Canada geese, bald eagles, and wild turkeys have been absent from the state's breeding avifauna, and most of the length of the middle and lower Missouri has been dredged, straightened, and otherwise manipulated to the point that it would probably have been nearly unrecognizable to these early explorers could they see it now. Lewis and Clark calculated that the Nebraska stretch of the Missouri River consisted of 556 miles of meandering river length, or about 115 miles more than at present. But a few tiny remnants of the river valley's original deciduous hardwood forest persist in preserves and parks such as Ponca State Park, Neale Woods,

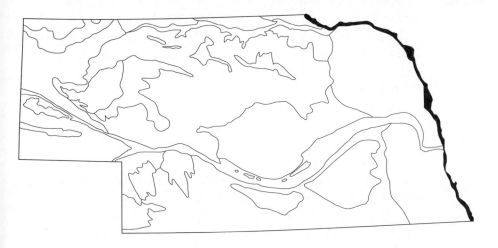

Fontenelle Forest, Indian Cave State Park, and Rulo Bluffs Preserve, so that at least some idea of how it must have once appeared can still be gained.

Ecology of the Missouri Valley Deciduous Forest

The deciduous forest of extreme eastern and southeastern Nebraska is a stripped-down version of the general forest type that mantles most of eastern North America. John Weaver has pointed out that, although in the heart of this great forest type there are about 200 woody species, only about 80 have made their way to southeastern Nebraska. Of these, about 31 are upland woody species. At the southeastern corner of the state the wooded area associated with the Missouri River may have been about 25 miles wide a century ago, narrowing to about 5 miles in the vicinity of Omaha and finally to less than a half a mile wide in northeastern Nebraska.

Other indications of the rich biodiversity of the lower Missouri Valley in Nebraska lie in the fact that 826 vascular plant species have been reported from Richardson County, based on an analysis by Robert Kaul and Steven Rolfsmeier, the largest total reported for any Nebraska county except for the intensively studied Douglas County. In two small areas of Missouri Valley forest totaling about 1,600 acres (Neale Woods and Fontenelle Forest in Sarpy County), nearly 75 percent of the state's 136 families of vascular plants are represented. Richardson County alone supports 25 species of reptiles and amphibians and 25 species of fish, larger totals than those given by Gail Kantak and Steven Churchill for any of the three Nebraska counties that they analyzed (Buffalo, Cuming, and Sioux).

There are an estimated 138 to 140 breeding species of birds in the southeastern corner of the state (the Missouri Valley from Douglas County south), judging from my own studies, which is also the largest total for any

Map 12. Missouri River Valley Riparian Forest region (inked). Based on maps by Kaul (1975) and Omernik (1987).

area in the state. Among the distinctly southern breeding species endemic to this part of the state are the red-shouldered hawk, broad-winged hawk, chuck-will's-widow, white-eyed vireo, Carolina wren, summer tanager, Louisiana waterthrush, and the northern parula, cerulean, prothonotary, Kentucky, and yellow-throated warblers.

Near the Kansas border the Missouri Valley is about 200 feet deep and about 1 mile wide, as compared with a much narrower canyon that is about 150 feet deep near the South Dakota line. As much as 36 inches of precipitation occurs annually at the southern end of Nebraska's Missouri Valley near the Kansas border, and this region's soils are fairly deep and fertile. The forest floor is well shaded, and a fairly high humidity is typical, allowing shade-tolerant forbs to thrive but often shading out those species that require more sun. As many as 40 species of shrubs occur in the floodplain forests of southeastern Nebraska, diminishing to about half this number in the northeast. There are also at least 50 species of forest-adapted herbs, none of which are part of the prairie flora. Many of these species bloom early in the spring, before the canopy foliage begins to shade out the overhead light.

Major Terrestrial Community Types

Southeastern Upland Forest

Along the edges of the Missouri Valley, from about Dakota County southward to the Kansas and Missouri borders, a riverine hardwood forest grows on the upper slopes of bluffs and sometimes also on the middle slopes of bluffs facing south and west. The forest is dominated by tall deciduous species such as red oak, bur oak, and bitternut hickory, with increasing amounts of chinquapin oak southwardly. Several other important trees can be found here, including shagbark hickory, basswood, green ash, and black walnut. Lower trees in the subcanopy include redbud, hop hornbeam, red mulberry, and red elm. Other oaks, such as black oaks, may also be present. The usual shrubs are rough-leaved dogwood and coralberry, and several woody vines are common. The species diversity of forbs is quite high and includes several rare and highly localized orchids, such as late coral-root, Wister's coral-root, and large yellow lady's-slipper. The rare forb ginseng is limited to this and the northeastern upland forest. Rulo Bluffs Preserve (Richardson County), Indian Cave State Park (Nemaha and Richardson Counties), and Fontenelle Forest (Sarpy County) are all good examples of the southeastern upland forest.

Northeastern Upland Forest

Along the upper Missouri Valley from the South Dakota border to Burt County and the lower portions of the Niobrara, Elkhorn, and Platte

Rivers, the northeastern upland forest is present, merging with the southeastern upland forest in Washington County. The northeastern upland forest is typically present on bluffs and lower slopes, especially on north- and east-facing slopes and above the floodplain forests. It is dominated by bur oak and basswood, with additional tree species often including green ash, red elm, and hackberry. The subcanopy is usually comprised of red cedar, hop hornbeam, and red elm. The same shrub layer as was mentioned for the southeastern upland forest is present, and the vines and forbs are also very similar. On drier upland slopes the northeastern upland forest is transitional to bur oak forest. Ponca State Park in Dixon County is an example of this community type, and on the Niobrara River it also occurs in the Niobrara Valley Preserve. As of the early 1990s there were more than 100,000 acres dominated by bur oak or oak-hickory forests in Nebraska.

Eastern Lowland Forest

On the upper floodplain terraces and bottoms of ravines along the Missouri River and other large to small rivers in the eastern parts of the state, a riverine forest dominated by elms (American and red), hackberry, and green ash is present. Most of these trees are 30 to 100 feet tall, and sometimes they are joined by silver maple or black walnut. Although cottonwood may be present, is it not a dominant species. Much the same shrub layer is present as that in the southeastern upland forest, and the vines are also similar. The herbaceous layer is typical of other floodplain forests and woodlands and includes a few of the upland forest species as well. As of the early 1990s there were about 200,000 acres of elm- and ash-dominated forests in Nebraska.

Cottonwood-Dogwood Floodplain Forest

At DeSoto Bend National Wildlife Refuge in Washington County, the cottonwood-dogwood floodplain forest, dominated by cottonwood, is present. Box elder and hackberry comprise some of the subcanopy trees. Shrubs include rough-leaved dogwood and some woody vines such as Virginia creeper and poison ivy. The community is little studied and perhaps is limited to areas exposed to frequent flooding.

Eastern Floodplain Woodland

The eastern floodplain woodland is mostly made up of lower-stature trees that must endure occasional flooding, although some fully grown cottonwoods up to 100 feet tall may be present as well. In the subcanopy, American elm is perhaps most common, but silver maple, box elder, hackberry, green ash, honey locust, and mulberry also occur. Shrubs and woody vines are common, and the herbaceous layer includes some forest perennials and a mixture of weedy annual forbs that are associated with periodic floods.

The loess bluff prairie is largely limited to the crests and slopes of river bluffs and is similar to the Loess Hills prairie. Perhaps best developed along the south side of the Missouri River between Knox and Dixon Counties, the loess bluff prairie generally is limited to slopes having a southern or western exposure. It also may be limited in northeastern Nebraska to areas having chalky limestone outcrops. In paintings executed by Karl Bodmer during the spring of 1843, the sharp ecological and slope-dependent breaks occurring between loess bluff prairies and the deciduous forests of the Missouri bluffs in the vicinity of Plattsmouth and Omaha can be clearly seen and appear to be little different from the same topographic patterns of Missouri Valley plant communities that are evident today. The loess bluff prairie probably merges with both typical loess mixed-grass prairie and with tallgrass prairie.

Profiles of Keystone and Typical Species

Bur Oak

Oaks are virtual synonyms for strength, and the bur oak is no exception. Its English name comes from the warty and burlike cap on its acorn, and its species name, *macrocarpa*, also refers to this large cap, which nearly covers the entire acorn. Another typical feature of the bur oak is its very thick and corky bark, which helps mature trees avoid being destroyed by the periodic prairie fires that once were a common aspect of prairie ecology. Even the small seedlings are fairly well protected from fire in this way. Because of this feature and an unusual degree of drought tolerance, the bur oak has managed to extend farther west away from the Missouri and other river valleys than have other upland forest trees. It therefore serves as a kind of advance guard for the woody species that comprise the eastern deciduous forest in its constant interplay with the prairie in their competition for dominance along the prairie-forest ecotone.

These are slow-growing trees, often not starting to reproduce until they are 30 to 35 years old or more. Very old trees may easily exceed 100 feet in height and may live well over 200 years. They do not produce large crops of acorns every year; instead, two or three years are needed to produce bumper crops. Squirrels eagerly gather the large acorns, and deer, turkey, rabbits, wood ducks, and various small rodents soon consume those that reach the ground. In this species the acorns mature in a single growing season and are thus produced on the current's year's branchlets. Because of their nutritious mast crop, oaks are among the most valuable trees in the forest for wildlife. Their strong branches resist breakage, and few if any woodpeckers are able to excavate nest cavities in them.

The bur oak is the most abundant and widespread of the several oak species in the Missouri Valley forest, but other oaks have similar habitats, and all produce crops of acorns, albeit of varying sizes and quantities. North American oaks evolved from ancestors in the mountains of Central America, and as they moved northward most of them evolved deciduous rather than evergreen foliage traits, allowing them to cope with winter temperatures and extremely dry periods.

Fox Squirrel

The fox squirrel, together with the much rarer southern flying squirrel, comprises a good deal of the mammalian biomass that occupies the canopy layer of the Missouri Valley forest. Although both the red squirrel and the gray squirrel are common in Iowa, the red squirrel does not occur widely in Nebraska, and the gray squirrel barely penetrates the southeastern corner of the state.

Fox squirrels are somewhat larger than gray squirrels (weighing up to 3 pounds versus up to 1.5 pounds) and might easily dominate them socially, but no obvious competition between them is evident. Both use leaf nests, which they build and occupy during summer as well as winter. Winter-built nests are more substantial than summer nests, and either type may last a few months or even up to several years if kept in good repair. They usually serve as supplement nests for the cavity nests that are the fox squirrels' primary homes; the cavity nest typically is used for nursing young, and the leaf nest is used when the primary nest is badly infested with parasites. The cavity nest may also be built close to a reliable source of food. Both types of nests are well lined with leaves that offer good insulation in winter.

Fox squirrels have home ranges of about 10 acres that are regularly used, but over the course of a year an animal might occupy as much as 40 acres, an area much larger than the home range of a gray squirrel. Thus fox squirrels are quite mobile, and individuals are known to have moved as far as 40 miles in a lifetime. Besides acorns, hickory nuts, walnuts, and the fruits of Osage orange are all eaten. Corn is taken wherever it is available. When gathering acorns or walnuts for storage, the fox squirrel removes the acorn caps and the walnut husks. Many nuts may be buried and their locations forgotten, resulting in a spreading of seedlings, which is the evolutionary advantage for trees to invest energy in producing such fruits. A single squirrel may consume well over 100 pounds food in a year, so the presence of nut-bearing trees is an important aspect of the animal's well-being. In very good forest habitats, the land might support up to three squirrels per acre.

Fox squirrels do not form pair bonds, so they tend to be solitary, except for a short period during mating. A female's young will remain with her through their first winter. In captivity fox squirrels might live as long as 15 years, but in the wild their lifespans probably average only a few years.

Some populations of fox squirrels around Lincoln and Omaha have individuals with all-black pelages. Such squirrels seem to be common only in towns; perhaps they are too conspicuous to survive well in the wild. Also, fox squirrels sometimes accumulate porphyrin compounds in their bones and teeth, a situation that in humans produces a disease that eventually leads to madness. Yet the squirrels do not appear to be affected by this malady, although the teeth of affected individuals will glow bright red in the presence of ultraviolet light.

Opossum

Opossums in Nebraska have always seemed slightly out of place to me; like the armadillo, they seem better to belong in the American tropics. Indeed, this species ranges south to Costa Rica and no doubt colonized the northern United States and southern Canada only in post-Pleistocene times. The species' overall fossil record goes back to mid-Pleistocene (interglacial) times, although marsupials of earlier types reach back nearly 100 million years in North America.

Marsupials have many remarkable features, of which the pouch (marsupium) is perhaps most familiar. While still a tiny embryo, each neonate must blindly crawl from the female's reproductive tract opening to the pouch and there become attached to a nipple, when less than two weeks have past since fertilization. There are about 13 such nipples; any individuals that get lost or for which there are no more available nipples soon die. After about 100 days the young are weaned and begin riding about on their mother's back, using their claws and prehensile tails for gaining a purchase. Their tails are surprisingly strong, and a baby can briefly (at least for a few seconds in my experience) hang upside-down from a small branch while holding on only with its tail.

Other reproductive oddities include the fact that the male has a forked penis and the female a forked vagina. Even the sperm occur in pairs, which swim parallel to each other. If separated, they tend to swim in circles. When mating, the female almost always lies on her right side; if copulation occurs with the female standing upright or lying on her left, fertilization is not successful, at least according to one report.

Amazingly, opossums are essentially immune to the venom of the rattlesnake family and regularly eat them. They also eat shrews and moles, which few other mammals ever do. Opossums seem quite resistant to plague and rabies, a characteristic typical of scavengers. They have more teeth (50) than any other North American terrestrial mammal, but these are only slightly differentiated into distinct functional types. Like primates, they have opposable digits that greatly help in climbing and clinging. Opossums are not noted for their intelligence. When threatened, they are unlikely to fight. Instead, they are prone to "play dead" while releasing a vile odor from their anal glands and possibly also defecating, a fairly effective deterrent to most potential predators.

Opossums are very common in southeastern Nebraska, especially in cities, where they raid garbage cans and gardens on a nightly basis, eating virtually anything they can find. Many get run over by motor vehicles, and some are probably killed by dogs. Yet they do well in cities, perhaps much better than in natural wooded habitats, where coyotes and great horned owls are a threat. They generally do not have long lifespans in the wild, maturing their first year, and few individuals live long enough to breed a second time.

Luna Moth

The nocturnal luna moth, named for the moon, is part of a group of giant silkmoths that otherwise occur in the Old World and are mostly large and pale green, with long and ornate tails on their hindwings and large eyespots on both their forewings and hindwings. The group includes Nebraska's largest, most ornate, and often very rare moths, as well as the Asian moth (*Bombyx mori*) that is the source of commercial silk. Like other silkmoths, lunas produce great quantities of silk when pupating. This silk envelope provides for insulation, protection against parasites, and especially disguise, as the cocoons usually closely resemble dead plant foliage or bark. The large, slow, and highly vulnerable larvae often mimic green leaves, but some have hairy tufts or stinging spines or may even resemble bird droppings. On disturbance some silkmoths regurgitate partly digested food or make clicking noises with their mouthparts.

Male moths have large, fringed, and feathery antennae and like many other moths can locate virgin females by following scent trails over great distances. Newly emerged females release a powerful pheromone and remain inactive, waiting for males to find them. Adults of both sexes live on stored fat reserves and devote their entire time to achieving reproduction. Fertilized females then scatter their eggs by depositing small groups over a wide area. The eggs of this species may include two types of larvae, one of which is fast growing and metamorphoses to produce a second brood during the season they hatched, and a second, slower-growing type that pupates and enters a state of dormancy until the next season. These spring-brood individuals are a darker green or more bluish green than the paler ones, which are more yellowish, and the spring-brood adult moths also have reddish purple wing margins, which are more yellow in later types. Males are also generally paler green overall than females and have larger antennae. They are highly mobile and may cover many miles in tracking down the location of virgin females.

The conspicuous eyespots of some moths have been proven effective in thwarting attacks by predatory birds, especially in those cases where the

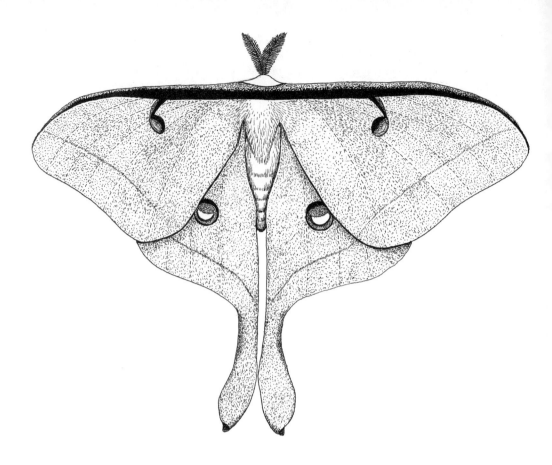

Fig 45. Luna moth, male.

eyes are typically hidden on the hindwing until the forewings are drawn forward. Luna moths and many other silkmoths have such eyespots, but this possible function for them has apparently remained unstudied.

The large larvae are bright green, and their irregular body outline suggests that of an indented leaf. They may be nearly 3 inches long just prior to pupating. The cocoon is usually placed near the base of the host plant in vegetational debris. It is papery brown, irregularly shaped, and easily overlooked. Upon being fertilized, the newly emerged female is stimulated to search for appropriate host plants on which to deposit her eggs. As with other silk moths, host plants tend to be trees or shrubs and often consist of plant species having compounds that help protect against their being consumed by herbivorous animals.

Nebraska is on the western edge of the luna moth's range, and the moth is almost entirely limited to the forested Missouri Valley, sometimes occurring west to Lincoln and beyond. The larvae usually are associated with hickory, walnut, sumac, and various other woody plants, depending on geographic region.

Southern Flying Squirrel

At times I have wished to have a flying squirrel for a pet, ever since one of my college mammalogy professors (who grew up in Georgia) told me that as a child he had kept one. It would sleep rolled up like a furry ball in his pocket while he spent the day at school but would become active at night, expertly gliding from one end of his room to the other and only occasionally doing some damage to curtains and window shades upon landing. Even today, I can't resist the idea of having such an amazing animal around to watch and marvel over. I have only seen a single flying squirrel in the wild, and it is likely that most Nebraskans go through life without ever seeing any.

Of course, flying squirrels do not really fly, but the loose skin between their forelegs and hindlegs provides an airfoil that, when stretched out, fully allows them to go from tree to tree without losing much altitude. Since they are only active in nearly total darkness, their large eyes must be wonderfully adapted to low-light vision. They are small squirrels, weighing only 2 to 5 ounces, with large ears, long whiskers, and a broad, flattened tail that serves as an aerial rudder. By varying the skin tension on one side or the other, the animal can control the angle of the glide and its speed, with the tail providing steering controls. The glide angle usually ranges from about 30 to 50 degrees below the horizontal, so typically a squirrel can glide laterally for about the same distance as it was above ground when starting its flight. Just before landing the tail is sharply raised, bringing the legs forward and allowing the animal to land facing upward. The hind feet absorb most of the shock, and all four feet have well-cushioned pads.

These little squirrels have fairly large home ranges of up to about 5 acres. In very good habitat their density may approach five squirrels per acre. Like other squirrels, they do not form pair bonds and are largely solitary, but several may huddle together in a nest to keep warm during cold weather. They only rarely make leaf nests and instead usually use old woodpecker holes. In fact, it was by once peering into an old woodpecker hole that I saw my first and only flying squirrel staring back at me, no doubt being just as surprised as I was. The animals store nuts and seeds in the nest and also rear their young there. They readily adapt to living in artificial nest sites such as nest boxes built for wood ducks, especially in places where natural tree cavities are lacking or rare. They eat a fair amount of animal material, including insects, carrion, bird eggs or nestlings, and insect larvae. They are especially attracted to the sap that flows from trees bored by sapsuckers, and they relish berries, fruit, and flower blossoms.

Flying squirrels probably never were common in Nebraska, since the state is at the very western edge of their range. They are difficult to survey accurately. The animals are known to occur north to about Nebraska City,

Fig 46. Southern flying
squirrel.

but only three populations have been found since 1985. Loss of habitat through lumbering in this part of the state has been substantial, and other previously wooded areas have been converted to cropland or have been affected by clearing and development for homesites. Probably places like Indian Cave State Park and Rulo Bluffs Preserve provide the best hope for maintaining a viable flying squirrel population in the state. It is classified as a threatened species in Nebraska.

Ginseng

Ginseng is a rather unusual plant whose use in herbal medicine and folklore has a long history. It is a perennial herb, growing less than a foot tall, with one or more stems supporting five broadly oval leaves that have serrated edges and radiate out like fingers. The leaf arrangement is similar to that of the Virginia creeper, a woody vine. Ginseng is limited in Nebraska

Fig 47. Ginseng.

to shady, rather moist forests along the Missouri Valley. It is part of a plant family that includes sarsaparilla (the basis for root beer) and other aromatic herbs. All have small, inconspicuous flowers with five petals and five sepals, which in the case of ginseng are greenish white. The flowers appear in a terminal, solitary cluster. Small, bright red berries are produced during June and July.

Ginseng's root is thick, tapered, and often curiously branched in such a way as vaguely to resemble a human form. The English name, "ginseng," derives from the plant's Chinese name, *jen-shen*, meaning "man-root." One Native American name for it was "garantoquin," which has the same meaning. Not surprisingly, this distinctive feature has resulted in a rich mythology and folklore being built up around the plant. There is an associated high level of its exploitation by herbalists, who have endowed it with supposed characteristics of improving male virility and, more recently, reinforcing the body's immune system. Neither of these attributes, nor others it is reputed to possess, has been scientifically proven, although the roots are known to be high in saponins, which are believed by many to have medicinal properties. The plant has long been very rare, and timber cutting as well as herb gathering has made it even rarer. Fresh roots from wild plants sell for as much as $150 to $500 per pound, the price varying with how large and how gnarled the roots are. Nearly all the roots from wild ginseng are sent to Asia, where the market for especially large and highly developed roots is incredibly lucrative, with single roots sometimes fetching thousands of dollars.

Cultivated ginseng roots bring only about $15 per pound. Such cultivated roots tend to be carrot shaped and pale colored rather than brownish and irregularly shaped, but they have about the same concentrations of active ingredients. Cultivating ginseng is now a big international business, especially in some states. More than 2.4 million pounds were produced in 1997, with Wisconsin alone accounting for 2 million pounds. The roots of wild plants are mostly obtained in the Southeast, with Kentucky, West Virginia, and North Carolina being areas where the digging is especially popular and where the plants are most highly prized.

Ginseng is a rather long-lived plant, and it may not bear fruit until it is at least seven years old. However, some plants barely more than a year old are dug up for sale, which results in little or no possible reproduction. Ginseng was recently added to the list of Nebraska's officially threatened species, and it has been listed as threatened by at least 31 other states.

Broad-Winged Hawk

Two hawks of the eastern deciduous forest, the red-shouldered hawk and the broad-winged hawk, were at one time probably regular breeders in Nebraska's forested valley of the Missouri River. Both are substantially smaller than the red-tailed hawk, but both resemble it in that they are primarily adapted for preying on small vertebrates, at least during the breeding season. Broad-winged hawk adults average about 0.8 (males) to 1.0 (females) pounds, whereas red-shouldered hawk adults average 1.0 (males) to 1.4 (females) pounds. Neither now breeds regularly in the state, but at least the southern part of Nebraska's Missouri Valley was once within their regular nesting range. Nesting in Nebraska by the red-shouldered hawk has occurred as recently as 1995, but there are no recent records for the broad-winged hawk.

Together with the Swainson's hawk, the broad-winged hawk makes a phenomenal migration each fall to South America. It takes a somewhat different route than the Swainson's and ends its migration in northern and west-central South America, rather than in southern parts of that

Fig 48. Broad-winged hawk, adult in threat display.

continent. Like the Swainson's, it increasingly congregates during the fall migration, funneling into eastern Texas during September and then passing down through Mexico, Central America, and southward, with the numbers of birds flying along the Panamanian isthmus during October almost uncountably large.

Broad-winged hawks return to their breeding grounds in late April, often seeking out stands of woods that are both close to water and near an open area or wetlands where hunting is good. Their home ranges are probably fairly small, at least as compared to red-tailed hawks, since population densities of a pair per square mile or less are fairly typical. Their foods seem to include a substantial number of amphibians, mainly frogs, with some lizards, snakes, birds, and mammals. In one study the average prey weight was less than 1 ounce, the size of a typical sparrow or mouse, whereas red-tailed hawks from the same area had an average prey weight of about 4 ounces, about the size of a ground squirrel or flying squirrel.

Territories are advertised by the male performing soaring and swooping flights, sometimes in company with the female, and copulation often occurs on a tree branch near the nest. Like other Nebraska hawks, there is a strong pair bond, with the male hunting and provisioning the female and her dependent young. Fledging occurs at about six weeks after hatching. The eggs and nestling young may be taken by raccoons, crows, and perhaps other predators such as large owls.

CHAPTER 11

A Wetness on the Land

Rivers, Lakes, Marshes, and Other Wetlands

The Ecology and Chemistry of Water

Although it has been written that we are but dust and to dust we shall return, it would be far more accurate to say that we came from water (our human bodies are nearly two-thirds water) and to water we must eventually return. Water is one of the most abundant compounds on Earth and one of the most critical components for sustaining life on this planet. It is also one of the most remarkable substances. No other common inorganic material is fluid at temperatures that support life, making it possible for floating and swimming organisms to have evolved and become mobile. Although fluid, the viscosity of water allows for rowing or undulating body movements to provide for easy mobility. Additionally, few natural compounds are so resistant to temperature changes (its so-called specific heat), which helps buffer living cells from rapid, perhaps fatal heating and cooling and also limits the rate of evaporation. Water dissolves essentially all the critical elements needed for life, including all the basic nutritional salts as well as vital gasses such as oxygen, hydrogen, and carbon dioxide. Its biochemical presence is also needed for facilitating all the basic photosynthetic and respiratory processes of living cells that involve carbon, oxygen, and hydrogen.

Although water becomes denser as it cools and approaches freezing, it actually expands as it freezes. Thus ice forms at the tops of lakes and rivers first, rather than from the bottom up. This curious fact allows aquatic organisms to survive in a fluid environment while separated from subfreezing atmospheric conditions above by a protective ice ceiling, thus enabling them to endure long winters or even more prolonged arcticlike conditions. Although water strongly resists freezing, it easily evaporates and by later condensation is able to spread life-giving moisture across the drier parts of the globe as various forms of precipitation.

Because colder water is relatively dense and sinks and warmer water correspondingly rises, oceans develop vertical cycles. As a result, warmer, low-latitude waters rise and replace sinking polar waters, producing upwellings of deep, dissolved ocean nutrients, access to which would otherwise be forever lost to plants and animals. These oceanic cycles, aided by the energy of planetary spin effects, produce enormous clockwise or counterclockwise surface-level currents such as the Gulf Stream, which may either warm or cool adjoining land masses, causing offshore or onshore winds and resulting in increased or reduced coastal precipitation. Oceanic currents therefore influence large-scale terrestrial wind as well as precipitation variables and thereby control continentwide climatic patterns, as recent cycles of La Niña and El Niño have so effectively proven in recent years.

Even in smaller ecosystems and communities such as lakes, marshes, and temporary wetlands, water strongly influences and ultimately controls local species diversity and the overall abundance of plants and animals. This control is largely brought about by the influence of available water in allowing photosynthesis to proceed and thus regulate the initial plant production and subsequent storage of organic matter such as carbohydrates. These materials are then successively funneled through food chains in a predictable and diminishing sequence of exploitation by consumer organisms such as herbivores and carnivores. The consumers in turn are ultimately exploited and transformed back again into their inorganic components by lowly decomposer organisms such as bacteria. In short, the water that helped to convert carbon dioxide to organic carbon molecules by green plants is again finally released as water in the processes of respiration and decomposition by both plants and animals. As stated earlier, water we are and to water we shall return.

Of all the wonders of Earth, nothing is more valuable than water; without it Earth would be as lifeless as Mars. Nevertheless, nothing on our planet seems to be wasted so flagrantly or polluted so recklessly by humankind as is water. Ultimately, we earthlings will have to decide if wish to share Mars's fate; postulated interplanetary attempts by NASA to reach and colonize it and to use its possible subsurface ice supplies for human consumption and industrial purposes would be roughly comparable to visiting Earth as it may have appeared 3 billion years ago and hoping to set up a profitable car-wash operation there. It would seem that conserving water on Earth is far simpler, much more profitable, and immensely more critical to our own survival than hoping to find and extract water on Mars.

Nebraska's Aquatic Ecosystems

Nebraska's surface water, most of which is in the form of reservoirs and lakes, covers less than 1 percent of the state's total area. These deep-water ecosystems differ from smaller and shallower wetlands by typically having at least some wave-washed shorelines that are free of permanent vegetation, producing sandy or gravely beaches. Deep enough to develop a seasonal temperature stratification (cooler and denser water below, warmer and less dense water above) during summer months, they may or may not fully freeze over during winter, depending in part on such factors as their size and depth and the temperatures and volumes of waters that flow into the lake from feeder streams. The relative degree of interface between land and water is not nearly so great in large lakes or reservoirs as it is in small wetlands and especially is far less than occurs in streams.

Other than lakes, Nebraska's most easily accessible contacts with surface water come from rivers, brooks, and creeks, which collectively represent more than 10,000 miles of interface between land and water. This is a dynamic and complex interface that provides the advantages and some of the disadvantages of both land-dwelling and aquatic living. There are no hard ecological distinctions between rivers, brooks, and creeks other than their relative sizes; all are stream-based ecosystems in which the continuous movement of water influences the distribution, abundance, and kinds of plants and animals that can survive within them. Moving waters, for example, tend to accumulate and hold more available oxygen than do still waters, but they tend to dislodge and remove rooted and floating plants downstream. They also tend to displace unattached animals, unless the animals can exert the energy needed to remain in place against the current. Currents may bring foods or nutrients to an animal but may carry these away, just as its eggs, sperm, or offspring may also be swept away.

In addition to these more conspicuous ecosystems, there are the small, still-water environments, including permanent marshes and fens. Both of these wetland types are usually surrounded by emergent vegetation. They may be too shallow to avoid being fully frozen during winter and are generally too small in surface area to develop enough wave action to influence shoreline vegetation growth. Fens differ from typical Nebraska marshes in that their bottoms are comprised mostly of variably decayed organic materials resembling peat, while marshes have sand or mud bottoms. However, in contrast to the somewhat similar bogs of the northern coniferous forests, fens provide a nutrient balance that allows for fairly high levels of plant and animal productivity.

There are almost innumerable temporary wetlands such as precipitation-dependent ponds, very temporary vernal ponds, and other only

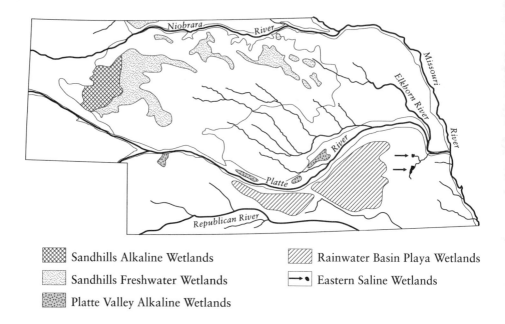

▨	Sandhills Alkaline Wetlands	▨	Rainwater Basin Playa Wetlands
▨	Sandhills Freshwater Wetlands	→•	Eastern Saline Wetlands
▨	Platte Valley Alkaline Wetlands		

Map 13. Major wetland areas of Nebraska. Modified from a map by Kaul and Rolfsmeier (1993)

slightly more permanent playa wetlands, the so-called rainwater lagoons of south-central Nebraska. There are also the wet Sandhills meadows fed primarily by subsurface irrigation. Such small wetlands may contain fresh water or may sometimes be highly saline. The water's chemical composition is often a more significant variable affecting plant and animal life than is the total amount or relative permanence of the water itself.

Nebraska has many rather small and often highly saline playa lakes, which are temporarily flooded lowlands that have gradually accumulated salts through evaporation. The accumulated salts exist because they are prevented from being removed through runoff, and they have not been leached down to substrate levels where they are no longer of biological significance. Most of the state's high saline wetlands of this type are found at the western edge of the Sandhills, where relatively low precipitation and high evaporation rates are typical. Clay plugging of the wetlands' bottoms may have also retarded or prevented significant subsurface drainage of these salts.

A small area of saline wetlands occurs in the immediate vicinity of Lincoln, especially along Salt Creek. Salt Creek flows in a general northeasterly direction from southwestern Lancaster County, passing through Lincoln and on to the northeast, where Little Salt Creek contributes its own smaller share of saline content. The water from both of these creeks flows through a layer of porous Dakota sandstone lying below much more recent loess deposits. The creeks apparently acquire their salt from passing over even deeper layers of shales that were deposited during Cretaceous times, when eastern Nebraska was still covered by a salty inland sea. In what was eventually to become western Lincoln, Salt Creek spread out to form Salt Lake, which was later developed for residential and recreational use and was more euphemistically renamed Capital Beach. These and similar salty wetlands in the Lincoln and Ceresco areas have hard clay, impermeable bottoms, preventing any leaching and allowing for the accumulation of so much crystallized salt that early efforts were made in the Lincoln area to gather and sell it commercially.

Marshy wetlands having permanent or temporary surface water grade into permanent or seasonal wet meadows, and these in turn are transitional to dryland habitats. Most of Nebraska's wet meadow habitats occur in the Sandhills, where an estimated 2,000 square miles of wetlands exist, from shallow lakes and permanent marshes to seasonally wet meadows.

Table 3 provides a list of wetland communities in Nebraska, based mostly on a recently proposed classification by Gerry Steinauer and Steven Rolfsmeier. A brief summary of each of the major wetland types occurring in Nebraska follows.

Rivers

In addition to the state's major rivers that have been described separately, there are several other rivers, especially the Republican, Elkhorn, and some stream-fed Sandhills rivers such as the Calamus and Loup, that are significant to the state as well. All of these rivers are rather similar to the larger ones already mentioned, except for the Sandhills rivers that tend to have very stable seasonal flow rates and sandy rather than muddy bottoms.

Spring-Fed Streams

Especially in the Sandhills, small brooks and seeps emerge from the base of tall dunes that at some time in the past may have blocked a natural waterway, causing the drainage to impound on the upstream side of the dune, forming a marsh or shallow lake; the streams then reappear on the downstream side. Examples include Birdwood Creek and Clear Creek along the southern edges of the Sandhills, both of which eventually flow into the Platte River. Other spring-fed streams occur in the northern Sandhills, where creeks emerge at the heads of "springbranch" canyons and flow into the Niobrara River. Such streams tend to be unusually clear and often support cold-adapted fish such as trout and glacial relict forms such as the rarer species of dace.

Rapids and Waterfalls

Practically the only rapids and waterfalls in Nebraska are found in the middle Niobrara Valley, where for part of its length the Niobrara River

Table 3.

Major Wetland Habitat
Types in Nebraska

Moving-water (Lentic) Habitats

    MUD-BOTTOM RIVERS

    SAND-BOTTOM RIVERS

    SPRING-FED (ARTESIAN) STREAMS

    RIVERINE SANDBARS AND GRAVEL FLATS

    RAPIDS AND WATERFALLS

Permanent Still-Water (Lotic) Habitats

    RESERVOIRS

    LAKES

    FRESHWATER MARSHES

        Sandhills Deepwater Marshes

        Shallow Freshwater Marshes

            Bulrush/Phragmites Marsh

            Cattail/Bulrush Marsh

            Cattail Marsh

    ALKALINE AND SALINE MARSHES AND MEADOWS

        Western Alkaline Marshes and Wet Meadows

        Eastern Saline Marshes and Wet Meadows

    FENS (ORGANIC SUBSTRATE WETLANDS)

Ephemeral and Semipermanent Aquatic Habitats

    PONDWEED (SEMIPERMANENT) WETLANDS

    PLAYA (RAINWATER) WETLANDS AND WET MEADOWS

    VERNAL POOLS AND MUDFLATS

---

flows over bedrock, producing local rapids. In the springbranch canyons of the valley's southern slopes, small waterfalls often can be found making their way down to the river, sometimes virtually hidden amid the rather dense wooded vegetation. The largest of these is Smith Falls, now part of a state park.

Riverine Sandbars and Gravel Flats

Along all of the state's rivers, sandbars and gravel flats form in areas where the water flow is slow enough to cause the solid materials to settle out.

These sandbars, sandy islands, or shoreline accretion areas provide important roosting sites for cranes and geese, nesting sites for plovers and terns, and opportunities for plants associated with early successional stages to take root and germinate, beginning the process of substrate stabilization.

Reservoirs and Lakes

Most of the surface water found in the state is present in the form of reservoirs. Scarcely any natural lakes are found outside of the Sandhills, and very few of the Sandhills wetlands are deep enough to qualify as lakes. Lakes and reservoirs typically are large enough for wave action to develop to the degree that barren, wave-washed shorelines occur along at least some parts of the wetland's perimeter. They are also deep enough so that a seasonal temperature stratification develops. Between the cool, deep water and the warmer surface water is a zone of fairly rapid temperature change, the thermocline. Such stratification strongly influences oxygen availability at deeper lake levels and the ease of vertical movement of dissolved nutrients.

Freshwater Marshes and Semipermanent Wetlands

Unlike lakes, marshes are typically surrounded by shoreline vegetation, which often consists of emergent cattails, bulrushes, reeds, or rushes. Marshes are too shallow for seasonal temperature stratification to develop, and their bottoms are sufficiently close to the surface that sunlight can penetrate and allow for rooted plants to survive. In deeper areas, these may include only submerged-leaf aquatics, such as various pondweeds, but in shallower waters floating-leaf pondweeds and water lilies are typical.

Fens

One is not inclined to imagine fens as common in Nebraska; their peaty or marly bottoms of undecayed organic matter would seem more characteristic of Minnesota or Canada. Indeed, fens might be regarded as glacial relics, for they sometimes contain plants that are much more typical of wetlands well to the north of Nebraska and that became marooned here with the retreat of the last glacier. The fens that do occur are largely limited to the Sandhills, mainly in Cherry County. A few small fens also are located in sandstone canyons associated with the Missouri River (Burt and Thurston Counties) and along the lower Little Blue River (Jefferson County).

In the Sandhills, fens generally are found along the edges of marshes or lakes and at the headwaters of Sandhills streams. The peat associated with them is sometimes quite thick, and unlike water in the bogs of Minnesota and Canada, the water is not especially acidic. Groundwater seepage to levels above the soil surface results in reduced air circulation, inhibits

Map 14. Rivers of Nebraska. Map courtesy University of Nebraska Department of Geography.

bacteria activity, and prevents the breakdown of organic matter, which builds up as peat. Mounds of peat up to 8 feet thick may form over time. Additionally, the cool groundwater allows plants characteristic of more northern climates to survive. Thus at least 12 rare Nebraska plants that otherwise are typical of Canada and the northern states have been found surviving on Nebraska fens. Most of the Sandhills fens have long since been drained and converted to hay meadows, but a few protected examples still survive, such as a fen at Dewey Lake in Valentine National Wildlife Refuge.

Alkaline and Saline Marshes and Meadows

Two other rare wetland types whose distribution in Nebraska is highly restricted are alkaline and saline wetlands. They occur in the western Sandhills, in the North Platte Valley, and to a very limited degree in eastern Nebraska, mainly the valley of Salt Creek and Little Salt Creek, Lancaster County (see map 13). In all these areas there are salt concentrations in the soils and wetland substrates that are so high as to restrict the kinds of

PAHA　BOYD

KNOX　CEDAR

Missouri

ROCK　HOLT　*Elkhorn River*　PIERCE　DIXON

DAKOTA

South Fork Elkhorn River　ANTELOPE　*North Fork Elkhorn River*　WAYNE　THURSTON

River

LOUP　GARFIELD　WHEELER

River

VALLEY　GREELEY

MADISON STANTON CUMING BURT

BOONE

PLATTE　COLFAX　DODGE　WASHINGTON

NANCE　*Loup River*

SHERMAN　HOWARD　MERRICK　POLK　BUTLER　SAUNDERS　DOUGLAS

River　*Big Blue*

SARPY

BUFFALO　HALL　HAMILTON　YORK　SEWARD　LANCASTER

CASS

River

Missouri

West Fork Big Blue River

PHELPS　KEARNEY　ADAMS　CLAY　FILLMORE　SALINE

OTOE

Little Nemaha R.

JOHNSON　*North Fork*　NEMAHA

HARLAN　FRANKLIN　WEBSTER　*Little Blue River*　THAYER　JEFFERSON　GAGE　*Big Blue River*　PAWNEE　RICHARDSON

NUCKOLLS　*Big Nemaha R.*

River

plants and animals that can survive there but allow for the relatively few saline-adapted species to thrive.

Playa Wetlands and Temporary Wet Meadows

Playa wetlands are transitory wetlands that develop in depressions as a result of snowmelt or excessive rainfall, forming wet meadows or shallow marshes that may persist for weeks, months, or sometimes an entire summer. The Rainwater Basin of southeastern Nebraska once supported hundreds of these wetlands, mainly because the region's soils are quite heavy and clay-based, which retards subsurface drainage. As a result, water accumulates, usually flooding the previous year's terrestrial plant life, making it easily available to swimming and diving birds. For such reasons, these wetlands are extremely important staging areas for migratory waterfowl in spring. By fall, when the birds return, they are usually dry but then offer thick, weedy habitat for many terrestrial birds and mammals.

Like the fens, most of these wetlands have been ditched and drained, although some are now under state and federal management and have been converted back to important wildlife habitat. Because of the relative scarcity of these wetlands, massive overcrowding of spring migrant birds such as geese is typical, producing opportunities for diseases such as botulism and fowl cholera to spread among the stressed birds.

Vernal Pools and Mudflats

Vernal pools are relatively small and temporary wetlands, sometimes only a few inches to a foot or so deep and often less than 100 feet across. They fill in spring from snowmelt or spring rains and for a time may attract frogs and toads. They may also host an abundance of invertebrates such as fairy shrimp or tadpole shrimp, whose eggs have lain dormant since the previous wetland cycle. Within weeks or at most a month or two, the water is gone, and an array of annuals soon carpet the bare ground.

Profiles of Keystone and Typical Species

Beaver

The beaver's history in North America is a long and significant one; trapping beavers for their valuable pelts provided one of the major early impetuses for exploring the West and for following the fledgling country's rivers to their mountain headwaters. In the course of such efforts, the most accessible mountain passes to the Pacific slope were explored, and an increasing comprehension of the continent's vast forest, wildlife, and mineral riches was attained.

The beaver is North America's largest rodent and the only nonhuman mammal that has a major influence on shaping its landscape. This has been achieved through the persistent, indomitable efforts of the animal to dam flowing streams, thus providing a nonfluctuating impoundment where lodges can be built or shoreline dens excavated without fear of recurrent flooding or the danger of being left dry and exposed to possible predation. Beavers are master engineers; I have seen beaver dams in the Teton Mountains that were at least 60 feet long with a shape gracefully bowing inward against the flow of water, thus probably reducing the chances of a washout. These impoundments provide wonderful habitats for fish and other aquatic life and stimulate the growth of pondweeds and other shoreline and aquatic vegetation. Prime environments are thereby created for open-water vertebrates such as muskrats, ducks, and grebes; shoreline-adapted species such as frogs, garter snakes, red-winged blackbirds, and song sparrows; and many other shoreline or aquatic amphibians, reptiles, birds, and mammals, not to mention the associated invertebrates.

Eventually these impoundments fill in with silt, converting the area to a wet meadow and thus generating other new and rich habitats. Then, a new dam must be begun, first by bringing in rocks and mud to provide a foundation and then by building up the dam itself with large and small branches that are firmly wedged into place. A well-made dam requires substantial amounts of dynamite to destroy, and unless the beavers themselves are removed, it will not take long before a new structure is begun on the remains of the old.

Adult beavers can be quite large; I have seen males that approached 70 pounds in the Teton region, but the average range of adult weights is from about 35 to 60 pounds. In spite of the animal's large size, few people actually see wild beavers, which tend to be crepuscular. The most common daytime indications of their presence are their dams and lodges and the presence of fallen trees. Or one might hear the slap of a beaver's tail should the animal suddenly be frightened and dive. They can remain hidden under water for very long periods, up to 15 minutes in extreme cases. Their large, flattened tail makes an effective underwater rudder (but is not used as a trowel, as folklore might suggest), and their hind feet are well webbed. While underwater, they can close both their nostrils and ears, and their eyes are protected by special membranes.

Beavers can even close their lips behind the enormous incisor teeth, so they can make dives with branches in their jaws without getting water in their mouths. These great teeth are extremely sharp, allowing the animals to gnaw through trees of substantial size and hardness, but they are especially fond of willows, poplars, birches, and similar fairly soft hardwoods. Much time is spent in collecting branches and depositing them near the lodge, by poking one end into the muddy substrate and leaving them there until they are taken into the lodge and used for food during fall and winter.

The beavers' social life is unusually complex for a rodent. The adults form monogamous pair bonds of indefinite length but probably usually lasting two to three years, with new mates taken as earlier ones die. The male's new mate is typically a young female that has been raised in another colony and is dispersing. Similarly, young males also disperse when they become sexually mature at two to three years of age. Until then, the lodge houses a nuclear family, with the pair, their current-year's brood, and any older immatures that have not yet dispersed. On average, a lodge contains about six individuals. Once the young are weaned, they become part of the working group, helping to keep the dam in repair and gathering stores of food for later use. In the wild, beavers often live for 10 to 15 years and perhaps longer where they are not subjected to trapping. In some areas, wolves and bears are significant enemies, but there are probably no significant predators for beavers in Nebraska.

Muskrat

There are few if any permanent marshes in Nebraska that lack a few haystacklike piles of vegetation scattered about, a sure sign that muskrats are present. Muskrats lack the beaver's trait of dam building, and their burrowing into human-made impoundments is more likely to cause leaks and damage than to improve their wetland habitat. But muskrats do regularly cut narrow paths through dense reedbeds, and these passageways are often used by ducks, grebes, and other swimming animals. Their rounded "houses" of reeds, rushes, and cattails also make convenient lookout points, nest sites, or loafing places for some of these same species.

Muskrats are less than 10 percent as heavy as beavers, the adults weighing about 1.7 to 3 pounds. On average, females are heavier than males. Because muskrats are much smaller than beavers, their population numbers are far greater, and under very good conditions their breeding density may at times exceed 30 animals per acre of marsh. However, there are often substantial annual variations in abundance, and at least in northern parts of their range the species may undergo regular cycles of abundance and scarcity.

Muskrats do not have all the adaptations for diving and social living that beavers exhibit, and unlike the beaver, they often supplement their herbivorous diet with various animal materials. But the two do share some similarities. Muskrats' hind feet are partly webbed, and their long tail is somewhat flattened vertically and is used as a rudder. They, too, can remain underwater for long periods, up to about 17 minutes. Like beavers, the animals produce a musky scent from glands in the anal region, which is especially typical of males during the breeding season and is used to mark territories. Also like beavers, they construct shoreline dens with underwater entrances that are difficult to detect visually. Abovewater "houses" are likely to be built only where banks are unavailable. They are usually built of dried cattails, bulrushes, and similar vegetation during the autumn and primarily serve as winter retreats.

Muskrats form monogamous pair bonds during the breeding season and are distinctly territorial. Females will kill intruding females, and males will fight with other males. Because of their fairly small size, muskrats are the favored prey of many predators, especially mink, which often enter their dens to capture and kill them. Otters are another serious aquatic predator. On land, coyotes and raccoons pose survival problems, as do mink and weasels. Larger raptors such as the great horned owl, red-tailed hawk, and northern harrier are potential aerial threats. Given all these enemies, the muskrat's average lifespan is rather short, probably about three to four years under natural conditions. Maturity occurs the first year or even as soon as six months after birth. The usual litter is of six to eight young, and there may be several litters per year, depending on the length of the breeding season.

Pallid Sturgeon

Two species of sturgeons occur in Nebraska; the state classifies one (the lake sturgeon) as threatened and one (the pallid sturgeon) as endangered. The pallid sturgeon is federally listed as endangered and thus is automatically listed similarly for the state. Both sturgeons are members of a very ancient group of fishes related to the sharks, in which most of the skeleton is cartilaginous, but unlike the sharks, there are numerous large, bony plates in the skin that provide for some protection.

There are 23 surviving sturgeon species, most of which are Eurasian. All sturgeons have five rows of bony plates extending down the back, sides, and underparts of the body. The head is also covered by plates and has a long snout. The tail is shaped in an unusual way, with the upper part larger and longer than the lower half, as in sharks. Sturgeons also have shovel-like snouts, with extensible lips used for sucking food from the lake bottom. Four tendril-like barbels hang down from the underside of the snout and are used as tactile sensory devices. Adults have no teeth, but, like sharks, there are spiracles, a pair of nostril-like openings, located in front of the eyes.

Commercial caviar comes from the eggs of the huge Russian beluga sturgeon of the Caspian and Black Seas, and historically the eggs of all the American species were also used for this purpose. The paddlefish of the Missouri River (and a single closely related species found in China) is a relative of the sturgeons, but it has no bony plates and only a few small scales on the skin. Both groups are suction-feeders, adapted to living at the bottom of large, turbid rivers.

The pallid sturgeon is paler throughout than the lake sturgeon, and in the pallid species the barbels are of unequal lengths, the inner ones much shorter than the outer pair. All sturgeon species can be quite large, but the pallid is smallest. It can grow to a length of 6 feet long, and record-size individuals reportedly have weighed more than 80 pounds. Today, individuals weighing as much as 20 to 30 pounds are very rare. By comparison, the lake sturgeon can reach 7 feet long and may rarely weigh up to 300 pounds. Both are slow growing and very long lived, sometimes living for more than a century.

Both sturgeons are bottom-dwellers, the pallid spawning over gravel beds at the mouths of tributary streams or in the main channels of the Missouri River. Spawning in this species occurs only every three to ten years, and they require fairly fast-moving water. Their highly adhesive eggs become attached to gravel or rocks, where they remain until the young hatch some five to eight days later. The spawning time has evolved to take advantage of the rapid water flows associated with spring runoff,

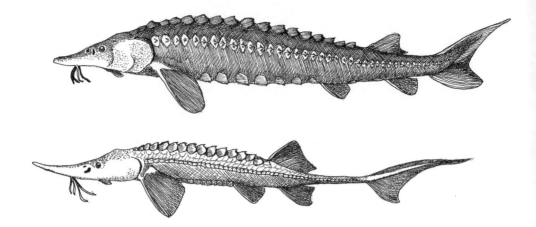

Fig 49. *Above,* lake sturgeon and *below,* pallid sturgeon.

but changes in river flow have made this runoff unpredictable, and little or no natural spawning now occurs in Nebraska.

Lake Sturgeon

The lake sturgeon is almost as rare as the pallid sturgeon. To the north and east of Nebraska, the lake sturgeon has had a historic range extending north to Hudson Bay and east to New England. It occupies both lakes and rivers but in Nebraska is limited to the Missouri River and mainly occurs in the unchannelized ports of the Missouri River around Gavin's Point Dam and some stretches below Sioux City, Iowa.

This species differs from the pallid sturgeon in being darker, with four barbels of about equal length that are not fringed along their edges. The plates along the animal's body become smoother and smaller with age and may almost disappear in old adults. The largest lake sturgeons may exceed 8 feet in length and weigh over 200 pounds. Like other sturgeons, this is a bottom-feeder, and it sucks up a mixture of invertebrate life as it feeds, as well as sand and some plant materials.

During the spring spawning period females lay enormous numbers of eggs that are adhesive and stick to the substrate. Some estimates suggest that a large female may lay up to 3 million eggs, which must require a great deal of energy and perhaps helps explain why spawning occurs only every two or three years. The eggs are deposited at random and receive no parental care. Young hatch in a week or longer, depending upon water temperature. The young are slow growing, and it may take about 20 years to reach sexual maturity and 70 or more years for the animals to reach a length of 5 feet.

Northern Redbelly Dace

The northern redbelly dace is very similar in its Nebraska distribution to the finescale dace, and both species are now classified as threatened in the state. Nebraska lies near the southern edges of both species' ranges, which extend north well into Canada and east to the Atlantic Ocean. Both species are glacial relicts, trapped in the cooler Nebraska streams when the last glaciers retreated. This species is smaller than the finescale dace but like it has very small scales. It also is olive green dorsally, with some dark flecking, and has a lateral black stripe, above which is a broad golden stripe and below which is a bright red abdominal band, at least in breeding males. Nonbreeding fish may be yellow or silvery white here. A second narrow black line occurs between the yellow lateral stripe and the dark olive back. Breeding males also develop tubercles on their pectoral fins and elsewhere.

This little minnow prefers quiet, small beaver ponds that have slow flow-rates, silty bottoms, and well-vegetated shorelines. It is a summer spawner, and males are like finescale dace males in that they chase fertile females. Sometimes several males gather around a single female, but it seems likely that only one male is successful in fertilizing her. In streams where both species occur, hybridization between the two sometimes has been reported, the hybrids being mainly female and closely resembling females of the redbelly species. Such hybrids may be fairly common in some areas.

Finescale Dace

The finescale dace has a distribution in Nebraska that is very much like that of the northern redbelly dace, now being largely limited to streams in Keya Paha, Rock, Brown, Sheridan, Wheeler, and Cherry Counties and in the Niobrara River in Sioux County. Small populations also occur in Custer, Logan, Lincoln, and Keith Counties. At one time both species had much wider distributions in central and western Nebraska.

Adults may be up to about 5 inches long. This little fish is dark olive green above and silvery white below. A dark stripe runs along the side of its body. Above the dark stripe is a narrow golden line, and below it is a bright red band in breeding males. Breeding males develop tubercles on their scales, especially near their pectoral fins. The scales are extremely small, thus the name "finescale" dace.

Like the redbelly dace, this species is found in clear, pristine streams that are usually spring-fed and have a sandy or gravelly bottom and a fairly slow current. The finescale dace may also occupy beaver ponds or similar standing-water habitats. These carnivorous fish eat a variety of small invertebrates.

During the spring and early summer breeding season, males follow females closely, the females stimulating courtship by swimming in a zigzag manner. While courting, the male holds his large, scooplike pectoral fins behind one of the female's, which helps control the pair's swimming. The

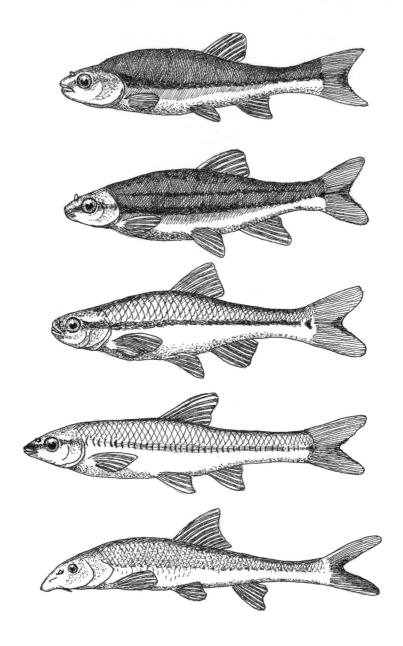

male then tries to direct the female against some solid object, where he can
curl his tail over the female's, placing his anal fin and its tubercles against
the vent of the female. This evidently stimulates spawning. The males hold
no territories, but they do fight among themselves.

Blacknose Shiner

The blacknose shiner is a typical example of the several species of endangered fish in Nebraska. It, along with some similar-sized dace minnows,

is a fish that is mainly found to the east and north of the state in cool, well-oxygenated waters. Its southern limits roughly correspond with the southern edges of the glaciers that once extended down to Kansas, Missouri, and the upper Mississippi-Ohio drainages. As the glaciers retreated, populations of cold-adapted fish were left in the cooler and clearer streams. Because of such factors as increased soil erosion and associated water turbidity and declining water flows, many of these populations disappeared during the early 1900s. Since 1939 this species has been collected in only a few Nebraska streams, including Gordon Creek and Brush Creek in Cherry County, the Niobrara River in Dawes County, and Holt Creek in Keya Paha County. All but one of these streams are in the Niobrara drainage; the exception is in the North Loup drainage. The species has recently been shifted from threatened to endangered status in Nebraska.

The black-nosed shiner is a tiny fish, with adults usually no longer than 2 to 3 inches. A single black or dusky stripe extends from the snout through the eye and back along each lateral line to the base of the tail. On a very similar species, the blackchin shiner, the black area of the snout extends down to include its lips and chin.

Like many endangered or threatened species, this one can only be saved by protecting its habitat, namely small, clear-water streams with abundant aquatic vegetation and with substrates ranging from sand to mud. Enough overhanging shoreline vegetation is typically present to help shade the steam and keep it cool. Pesticide runoff from adjacent land must be prevented in order to avoid killing the invertebrate food base for the fish.

Topeka Shiner

The Topeka shiner is on the federal list of endangered species. It is a small minnow, up to about 3 inches long. Its back is olive yellow, its sides have a dusky stripe against a silvery background that extends from the upper lips through the eyes and back to the base of the tail, and its belly is silver. There is a small triangular or arrow-shaped black spot at the base of the tail fin. Breeding males develop orange-red tint on their fins; their head (especially cheeks) and body (especially ventrally) are likewise tinged with orange. Males then develop tubercles, especially on the snout and the top of the head.

The species was named for Topeka, Kansas, where the first specimens to be described were collected. The minnows live in quiet pools of permanent streams that have a substrate of sand, gravel, or larger stones. Deposition of silt will eliminate the minnow and perhaps has been the cause of its widespread disappearance from a much wider range in the upper Midwest. It once occurred in all the major drainages in Kansas, but agriculture gradually polluted these streams with silt, and perhaps the same is true of Nebraska. Not much is known of the species' life history, but Topeka shiners probably normally live for two to three years. Reproduction begins in the first year of life and lasts from late June to August, at least in Kansas.

Sturgeon Chub

The sturgeon chub is a small and slender minnow (about 3 inches long) that is largely confined to the Missouri River and a few tributaries between Montana and Missouri and to the lower Mississippi River below the mouth of the Missouri. It has a long and somewhat flattened snout (thus the name sturgeon chub), small eyes, and abundant chemoreceptors (taste buds) scattered over its head, its body, and even its fins. These features help adapt it to a life in turbid water, where eyesight is of little value. The fish inhabit fairly fast flowing riffles in shallow and turbid streams, especially those with gravelly substrates.

The body has a somewhat wedge-shaped and downward-curving profile, which may help to hold the animal at or near the bottom in strong current. Many of the scales on the sides and back are ridged, an adaptation of somewhat uncertain function, but it has been suggested that they might serve as current detectors. More likely they help stabilize the body in currents, reducing turbulence as water flows backward over the fish's body. The sexes are similar in appearance, but males develop small tubercles along the rays of their anterior pair of fins during the breeding season.

In Nebraska, this species is little studied but is thought to be limited to the Missouri River downstream from Fort Calhoun and to the lower Platte downstream from Fremont. It has been recommended for addition to the list of Nebraska's endangered species. In Kansas, the species is also highly limited in range but is believed to breed in late spring and early summer, based on the reproductive condition of males. Its reproductive biology is still unknown.

Salt Creek Tiger Beetle

Tiger beetles are small, very active diurnal and predatory beetles that occur in many habitats. Many are notable for their colorful dorsal patterning. They can run very rapidly and can easily take flight. Tiger beetles hunt in broad daylight, and on dark days or at night they hide effectively in holes or under various types of cover. Among the many habitats used in Nebraska are sand dunes, roadsides, unvegetated or slightly vegetated riverbanks, and saline flats. Of the 85 species known of the family's largest and most typical tribe (Cicindelini) from the United States, 30 forms (28 species and 2 additional subspecies) have been found in Nebraska. Of these, the rarest is the Salt Creek tiger beetle, a highly restricted race of a much more widely distributed species, of which a second race also occurs in the Nebraska Sandhills and High Plains and widely elsewhere in the American

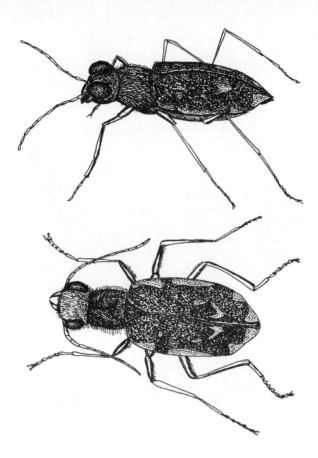

Fig 51. Salt Creek tiger beetle.

Southwest. The Salt Creek tiger beetle is dark brown or dark olive green on the thickened coverings (elytra) of the forewings as well as on the head and thorax, with a slight amount of lighter yellowish to whitish spotting. The underside is shiny and uniformly dark green.

This tiger beetle is active for only a few months in late spring and early summer, from May to July. A single reproductive cycle per breeding season is typical. Although once much more common and widespread in Lancaster County, it is now known only from a few wet mudflats that drain saline flats. These sites include a wetland along I-80 near the Lincoln airport and another near North 27th Street, just beyond the city limits and near I-80. This is one of the most restricted distributions known for any insect, and the species' population perhaps numbers only in the hundreds. It has been recommended for inclusion on the list of Nebraska's endangered species.

Three other tiger beetle species (*C. circumpicta*, *C. togata*, and *C. fulgida*) also occur in these same areas but are much more common. They generally are found on drier substrates and remain active much later in the summer.

Fig 52. Colorado butterfly plant and clouded crimson moth, a possible pollinator.

Colorado Butterfly Plant

The little-known Colorado butterfly plant is a close relative of several other plants of the same genus (*Gaura*) that are well-known wildflowers and cultivated garden flowers. Their name does not refer to any butterfly-attracting ability but rather to the fact the four somewhat teardrop-shaped petals of the flower are oriented toward the upper half of the blossom and are arranged in such a way as to suggest a small butterfly, perched with its wings outspread. In addition to its conspicuous white petals, there are eight long stamens with yellowish to reddish anthers hanging outward and downward and a similarly long white pistil with a four-lobed yellow stigma at its tip. The four narrow and tapering sepals are directed backward.

This biennial herb grows to about 10 inches tall. Known in Nebraska only from the floodplain of Lodgepole Creek in western Kimball County near the Wyoming line, this small population is an outlier of 22 known populations known from Laramie County, Wyoming. Small populations of the same subspecies are also known from Colorado. It has been recommended (Category I) for listing as endangered in Nebraska. Its ecology and reproductive biology are largely still unstudied. Although typically a biennial—maturing, flowering and dying in two years—it sometimes delays flowering until at least four years. Each flower may produce only one or two viable seeds, and there is no special seed-dispersal mechanism.

The Colorado butterfly plant's pollinating organism is still unknown, but it is part of a large group of evening primroses whose flowers often open toward evening. Considering its conspicuous pinkish white petal color, one might assume that its likely pollinators are seminocturnal insects having long tongues, since a rather long style is present. Hawk moths or other seminocturnal moths are possible candidates. In the case of the more common species of other evening primroses, various hawk moths, nocturnal moths, and "millers" are all attracted to the plants' odor and color and are likely pollinators. Most evening primroses have white to yellow flowers; many open late in the day, the blossoms usually closing again soon after sunrise on the following morning. In this species, the petals are pale pink and have gland-tipped hairs that probably help attract pollinators. The clouded crimson (*Schinia gaurae*) (fig. 52) is known to be associated with other closely related species of Gaura but is not yet known to be associated with the Colorado butterfly flower. However, it is mostly pale pink and matches the color of several gaura species (including the Colorado butterfly plant) almost perfectly and may even spend the day hiding among their flowers.

Saltwort

Like the Salt Creek tiger beetle, saltwort is a saline-adapted species that is now largely or entirely limited to a few small saline wetlands in the northern third of Lancaster County, where it is often the only plant growing on the salt-encrusted flats. One can immediately recognize this plant by chewing one of the succulent and rather scalelike leaves, as they have a distinctly sharp and salty taste. The plant's generic name, *Salicornia*, refers to salt, and has the same Latin origin as "salary" (working for one's allotment of salt, thus also the basis for the expression "not worth one's salt"). The plants grow only 3 to 8 inches tall, and their almost invisible flowers,

Fig 53. Saltwort and
long-billed curlew chick.

clustered in groups of threes, are likely to be overlooked. At maturity, the stems turn reddish, the basis for its specific name *rubra*.

Saltwort is the most salt tolerant of the Nebraskan alkali flats plants. Lancaster County almost represents its southernmost range limits; the only more southerly known location is one in central Kansas. It is more common in the Dakotas, especially North Dakota. A member of the goosefoot family Chenopodiaceae (named for the distinctive shape of the leaves in many species), saltwort is an annual herb. Its tiny flowers are nearly hidden, the stamens barely visible. It is likely to be found in the most alkaline soils, tolerating levels of alkalinity that no other species can withstand. In some places the plants grow close together, forming a colony, while in others they are more widely spaced and shrublike.

CHAPTER 12

Sad Stories
and Short Visions

Vanishing Habitats and Changing Times

Lives are short and human memories are even shorter. Virtually none of the people who have experienced the passing of the twentieth century will have remembered the turning of the nineteenth. For us, stories of vast herds of bison on the central Nebraska plains are as remote a concept as is the idea of hunting woolly mammoths or even of living along the Olduvai Gorge in Africa in constant fear of attack by large predatory mammals. Likewise, for people of the late twenty-first century it will be almost impossible to imagine the great flocks of sandhill cranes that now still visit Nebraska each year or visualize a time when the sky was not criss-crossed with contrails of jet airplanes, or of other kinds of aircraft yet to be developed. I have often thought that we of the early twenty-first century have been lucky to have lived when efficient means of transportation were already highly developed but not so highly developed that there were no more areas of wilderness left or no undiscovered species yet to find. During my lifetime the ivory-billed woodpecker has disappeared from the mainland of North America, and the Eskimo curlew is perhaps forever gone as well. In my parents' lifetimes the passenger pigeon and Carolina parakeet also became extinct. During my grandparents' lifetimes the last Labrador ducks vanished off the coast of Long Island without a trace. And long before that the last surviving great auks were bludgeoned to death by hungry sailors. For such reasons it is worth recounting some of the events of the twentieth century before they are too badly blurred by failing memories.

Gray (Prairie) Wolf

The prairie wolf is one of those half-mythic creatures of my early childhood memories. My mother talked of hearing wolves calling on her father's homesteaded land in North Dakota's Red River valley when she was a small child. She also told of homeless American Indians camping periodically along the nearby Sheyenne River and of finding arrowheads and other Native American artifacts on newly ploughed prairies. The wolves disappeared from the northern plains as quietly, and almost as quickly, as snow melting in spring. Most of the wolves were gone from the central and northern plains by the 1890s, contemporaneously with the disappearance of the bison herds. In Nebraska, they were nearly gone by 1900, and they had completely vanished by 1920. At that time there were still some wolves in North Dakota, but these were largely restricted to the Badlands region. The last one killed in South Dakota was taken in the Black Hills in 1934.

With the disappearance of the wolves, coyotes took over the role of large mammalian predator of the plains. Previously, coyotes had been subordinate to the wolves and probably had been limited to denser brushland habitats and badlands topography, to which the wolves were seemingly less well adapted. Stock ranchers, who initially had to contend with wolves preying on their large domestic livestock, now had to deal with coyotes taking their chickens, calves, and lambs. Poison baits and leg traps, which had worked well in controlling wolves, were used on the coyotes, although with somewhat less success. Perhaps in part these differences in susceptibility resulted from the coyote's greater flexibility in diet; its capacity for adapting to diverse habitats, including living in close proximity to humans; and its reduced dependence on large prey obtained by hunting in packs. Coyotes have adapted so well in Nebraska to living on farmlands that they sometimes even interbreed with domestic dogs, producing "coydogs" that are less wary of people and thus are more prone to enter farmsteads and steal domestic livestock or poultry than are wild coyotes.

Both coyotes and wolves are no more (or less) than completely wild dogs; they all have similar social lives. Indeed, domesticated dogs are nothing other than domesticated wolves, reshaped through selective breeding to suit various human predilections. Adult wild dogs of all species (foxes, jackals, and the like) are monogamous, with the male remaining with his mate to help protect and provide food for her while she is nursing their young. All species of dogs are playful, especially as puppies, with strong social bonding. They also have a strong sense of social structure. Each wolf pack has a dominant male and a reproductive female, with the other pack members serving subsidiary roles.

As befitting large, mobile predators, wolves and coyotes have large territories and home ranges and need effective means of long-distance communication as well as in absentia territorial marking devices. Urine markings serve for such semi-permanent signposts, and loud, far-carrying calls provide real-time locational information to those within hearing range. All of the members of the canid family are highly vocal, with the yipping serenades of coyotes still a common nocturnal treat for residents and campers in outstate Nebraska. Those lucky enough to hear the ghostly wail of wolves on a cold night in northern forests are likely to remember this hair-raising sound for a lifetime. A city dog senselessly barking at the moon at night may seem to be nuisance to neighbors, but the dog may be simply responding to innate urges it probably cannot control and does not understand itself. It is perhaps similar to our own strange fascination with fire; staring into a campfire is a hypnotic experience that in effect takes us back a million years or so, when fires kept wild dogs and large predatory cats at bay and provided enough warmth and security for enduring the long, dangerous nights of the East African savanna.

Snow Goose

Snow geese, like pasque flowers, have always been my personal symbol of spring. North Dakota winters are so long and dreary that as a teenager I literally began counting the days in March until the first snow geese arrived in the just-melting marshes near my hometown at the south end of the Red River valley. Hearing their wild, excited voices overhead was enough to make me forget everything else and plead with my father to let me borrow the family car long enough to witness the arrival of the incoming flocks. Then, standing knee-deep in the near-freezing water of glacial-age marshes and hidden by head-high phragmites, I would exult in the sheer joy of the moment, imagining that I was alone with uncountable wild geese and headed with them for unknowable lands still much farther to the north. It was another 20 years before I finally traveled to the Canadian tundra to see the birds on their nesting grounds and was able to feel that I had at last begun to understand some of their secrets.

When I was a boy in the 1950s, biologists had little knowledge of the total North American population of snow geese, as federal efforts in management of our waterfowl resources were still in their relative infancy. One might well guess that the population then was in the range of a million birds, most of which migrated back and forth across the Great Plains, nesting broadly throughout the Canadian arctic and wintering mostly along the coasts of Texas and Louisiana. This estimate excludes the population of greater snow geese wintering on the Atlantic Coast and other generally still smaller populations of lesser snow geese wintering in the Pacific coastal and Rocky Mountain states. Midwinter estimates by the U.S. Fish and Wildlife Service during the early 1970s suggested a total population of at least a million lesser snow geese in the Central and Mississippi Flyways.

Fig 54. Snow goose flock.

Early fall counts in eastern Canada for that same time period approached 2 million birds, with more than half of the total associated with breeding colonies on Baffin Island. By eliminating the annual fall harvest of about 300,000 birds in Canada and the United States, a compromise figure of about 1.5 million birds returning north in spring might be estimated.

The 1960s and later decades were marked by an expanding federal refuge system, milder springs that improved breeding conditions on the tundra breeding grounds, and probably greater quantities of available fall and spring foods associated with increased and improved grain crop production. These factors all interacted to benefit snow geese, which naturally responded by increasing their annual productivity. The midcontinental population of about 1 million snow geese of the 1960s doubled in the next ten years, and by the mid-1990s the fall population hovered somewhere between 3 and 6 million birds. This may have spelled good news for goose hunters, but it represented very bad news for the tundra where the colonies bred, which was quickly degraded and eventually even virtually destroyed by the increasing avian biomass that it had to support.

Larger flock sizes also placed strains on the resources of traditional fall and spring stopover points. In Nebraska, the returning spring snow goose flocks gradually spread west out of their traditional and restricted Missouri Valley Flyway into the Platte Valley, where abundant quantities of waste corn were to be found and where other goose species such as white-fronted geese and Canada geese had already learned to exploit this free largess. At its peak, during the late 1990s about a million geese eventually funneled into the Platte Valley in February and early March, about half of which were snow geese. Additionally, nearly half a million sandhill cranes joined the flocks, arriving slightly later in the spring and remaining longer. Their dependence on corn is great, and the prior use of it by great flocks of geese raises questions of how that species might be affected.

The crowded spring conditions in the Platte Valley and the nearby Rainwater Basin set the stage for disaster, which in part has taken the form of repeated outbreaks of fowl cholera, a usually fatal avian disease associated with crowding, limited supplies of fresh water, and associated stress. Repeated outbreaks of this disease have occurred in recent years, sometimes killing tens of thousands of geese and probably weakening many more.

There is no easy answer to this problem. Spring hunting seasons have been tried in the Dakotas and Nebraska, but relatively few birds are killed, and one can only guess at the stresses associated with goose hunting on nontarget birds such as cranes and other waterfowl. Additionally, many nontarget species are shot, including some hawks and other protected species that probably happened to pass too close to the blind of a goose hunter.

As with the white-tailed deer, there can be too much of a good thing in nature, and there are no simple answers to complex ecological problems. However, if humans are unable to find an answer, we can be certain that one will come about anyway through the forces of natural selection.

Peregrine Falcon

Few birds carry so much emotional baggage attached to them as does the peregrine falcon. The following descriptions might apply: fastest-flying bird; rarest of the North American falcons; a regal bird appropriate only for royalty and sheiks to possess. Perhaps not all of these perceived traits

Fig 55. Peregrine falcon, stooping.

are entirely true. The peregrine's speed during its near-vertical stoop on its prey has often been wildly overestimated, and happily the species has now begun a long comeback from near extinction, or at least extirpation, in North America. But the bird still carries the unmistakable imprimatur of royalty, and the sight of even one coursing above the horizon in the far distance provides a heart-stopping moment for all nature lovers.

Like osprey and bald eagle numbers, the peregrine falcon population in North America crashed during the 1950s and 1960s with the wide-spread use of hard pesticides such as DDT. It was not until the publication of Rachel Carson's *Silent Spring* in 1962 that most Americans awoke to the damage that had been visited upon our country and a political groundswell arose for combating the well-entrenched lobbyists for the commercial pesticide interests. Even though it was well known that raptor populations were declining all around us, the insidious mechanism by which DDT affected the reproductive mechanisms of birds was not understood for some time. Eventually, it was discovered that, during the metabolic breakdown of DDT, intermediate compounds were produced that affected egg production. That is, the mobilization of calcium stores by the female, needed to produce normal eggshells, was blocked in such a way that only thin-shelled eggs were laid. These eggs either were crushed by the incubating parent or did not allow for enough available calcium for the developing embryo to manufacture its skeleton properly. In either case, reproduction failures ensued, and before long there were no peregrines breeding successfully anywhere in the eastern states, where DDT use was most prevalent. The peregrine was first federally listed as an endangered species in 1970 and was listed again in 1984.

Only the forced elimination of DDT sales (but not its manufacture) in the United States and Canada and the passage of the Endangered Species Act in the early 1970s enabled the peregrine to have a chance of remaining part of the North American scene. This process was greatly aided by the use of unaffected and captive birds as a source of new blood by conservation groups such as the Peregrine Fund. By placing fertile eggs or newly hatched chicks in the nests of pairs that, in spite of their infertility, were still attempting to nest each year, the tide of population loss gradually was halted and reversed. Many of these efforts were made in cities, where high-rise buildings provided an artificial nesting cliff for hand-raised adults and where domestic pigeons and other city-adapted birds offered a source of fairly easy meals. As a result, an entire population of city-adapted peregrines began to emerge. One such effort was made at the Woodmen Tower in Omaha beginning in the late 1980s. Over the next several years, many young peregrines were fledged successfully. The population of city-breeding peregrines in the United States is still growing and has greatly helped to restore the birds in the eastern half of the country.

It was believed that, prior to the peregrine's national decline, as many as 800 pairs bred in the western states, but by 1965 fewer than 20 pairs were known to exist west of the Great Plains. There is only one known historic nesting record in Nebraska, dating from about 1904 in Sioux County. Efforts at captive breeding and "hacking" of young birds into the wild have gradually increased these numbers, and by the late 1980s some 200 pairs were known to be present in the American West. However, peregrines are highly mobile (*peregrinus* means wandering), and pesticide use on their Latin American wintering grounds is still uncontrolled, so the species is by no means out of danger.

White-Tailed Deer

Like the other large mammals of the plains, the white-tailed deer was once almost wholly eliminated from Nebraska by uncontrolled hunting. So few animals were left by 1900 that in 1907 laws were passed that made deer hunting illegal in the state. At that time perhaps fewer than 100 deer still survived, and it was not until 1961 that the population had recovered to the point that hunting of white-tailed and mule deer was again allowed.

By 2000 the state's deer population approached 300,000 animals, and deer are now so common in many areas that they pose a threat to suburban gardens and especially to motorists who are inclined to drive too fast at night near wooded areas. The watchword for evening recreational driving along the Platte River is no longer "Look for deer" but instead is "Look out for deer!" Deer may also damage crops and haystacks and have been implicated in helping to spread Lyme disease–infected deer ticks in the state, with a consequent increase in this seriously debilitating disease among humans.

In spite of such usually minor inconveniences that deer present for Nebraskans, they offer great visual benefits and provide hunting opportunities for almost 100,000 hunters. In 1997, 78,000 hunters took nearly 59,000 deer, which represents about 2.8 million pounds of venison. The monetary effect of deer hunting on the state's economy was estimated at $40 million in 1996. This figure is similar to an estimate of the annual monetary effect of spring bird-watching along the Platte River on the state's tourism economy. The number of hunters is also very close to the estimated number of birders (80,000) who visit the Platte Valley each spring. Thus birds and deer are renewable resources that individually feed more money into Nebraska's outstate economy than college football generates each year for the city of Lincoln.

Although the white-tailed deer population has continued to increase in recent years, the mule deer population of western Nebraska is essentially stable, so white-tailed deer now compose about 75 percent of the state's total deer population. Reasons for these differences are not at all clear, but it is known that the majority of white-tail does in Nebraska begin reproducing when they are only six months old, compared to only about 7 percent of mule deer does who do so. This differential reproductive rate gradually compounds the white-tail population more rapidly than that of the mule deer by an annual factor of 1.4 to 1. Additionally, white-tails seem more at home close to humans and thus can cope better with increasing human encroachment on once undisturbed lands.

Eskimo Curlew

It may well be that we shall never know exactly when the last Eskimo curlew disappeared from the planet Earth. The last convincing photographs of a wild bird (taken during spring on the south Texas coast) date to the 1960s, but there have been repeated accounts of sightings since then. Like other extinct species, such as the passenger pigeon and Carolina parakeet, this little curlew was once so common that its decline and probable extinction would have at one time seemed inconceivable. Market hunters

Fig 56. Eskimo curlew, adult in flight.

slaughtered them each spring, along with larger and even more highly prized shorebirds such as the whimbrel and long-billed curlew, sending them to high-priced culinary markets in midwestern cities.

The Eskimo curlew differed from its larger relatives that are still surviving in one important respect; it had a transequatorial migration. Beyond the sheer distances involved and the attendant stresses of such long flights, which involved flying to the Argentine pampas each fall, the fall flight was especially risky. Following breeding, the birds moved eastward and southward from their northern Canadian nesting ground to the vicinity of Newfoundland and Nova Scotia. During August and early September, the birds gorged on the ripening crop of crowberries and the abundant coastal invertebrates, fattening themselves for the long trip southward. In early September, the flocks would head south over the Atlantic Coast, making a nonstop journey to the Lesser Antilles about 2,000 miles away. After a stay of just a few days, they again took off, first crossing eastern Brazil and finally arriving in Argentina by about mid-September. This route had the advantage of being rapid and allowed the curlews to exploit the early fall food resources of the Canadian maritime coastline. But it did not allow for making major navigational mistakes or for the possibility of hurricanes.

Thus, although this journey must have evolved over millennia, it occasionally placed the birds directly in harm's way. Tropical storms in the Caribbean are common in the fall, and a small shorebird is no match for wind velocities of a hundred miles per hour or more. Such storms sometimes pushed the birds eastward, forcing them to make emergency landfalls in Bermuda or even in the British Isles. But since these situations were probably rare and the birds once so abundant, the overall population was not put in danger.

The Eskimo curlew was one of the many previously hunted species that first gained complete United States and Canadian protection by the Migratory Bird Treaty (approved in the United States in 1918), but by then the population had already been decimated. Additionally, hunting has continued in South America, where conservation remains unpopular.

The spring migration northward was much safer. The birds journeyed up along the interior of Central America and Mexico through the Great Plains, where they gradually traveled along a fairly broad front that encompassed Nebraska. A flock of 75 to 100 birds, one of the last flocks reported anywhere, was seen in Merrick County in April 1900, while smaller groups were seen in York County in 1904 or 1905 and in Madison County in 1909 or 1910. In 1911 seven birds from a flock of eight were shot during April in Merrick County. Finally, on April 17, 1915, a single bird was shot near Norfolk, perhaps being the last curlew to be shot in the United States before the signing of the Migratory Bird Treaty. A group of eight was seen near Hastings on April 8, 1926, which represents the last known sighting of Eskimo curlews in Nebraska.

In September 1932 a single Eskimo curlew was shot near the Strait of Belle Isle on the coast of Labrador, the last one known to be killed on the North American continent. Only a few weeks later, the noted American ornithologist Robert Cushman Murphy saw a flock of four at the tip of Montauk Point, Long Island. These were the last birds to be seen along the Atlantic Coast for nearly 30 years.

Since the several Texas sightings of 1959 to 1962, there has been relatively little news. Between Murphy's sighting in 1932 and 1976 there were at least six sightings of Eskimo curlews on the Texas coast, seven from the Atlantic Coast, and one from James Bay. There have been a few unconfirmed reports during the 1990s, including one from Nebraska that subsequently was retracted, but silence has been typical.

We will never know for certain what caused the final death knell of the curlews. Perhaps it was a fall storm that fatally swept all the remaining flocks out to sea; possibly it was a succession of poor breeding years on the Canadian tundra for the few remaining pairs; or maybe the last birds ended up as stew for some hungry Latin American family. We can only imagine that, as I once wrote, "The sight of these great shifting flocks of birds, combined with the distant sound of avian sleigh bells or wind whistling through a ship's riggings, must have been a magical vision that, like the great herds of bison on the Great Plains, has forever vanished from the American scene."

An Ecological Afterword

One should perhaps never finish a book in a minor key, and at the start of a new millennium it may be worth casting a hopeful eye forward as well as a more jaundiced one backward. It is certainly true that we have lost much in Nebraska during the past few generations, but during my tenure here two great sanctuaries by been developed along the central Platte River, one funded and developed by the National Audubon Society and the other resulting from federal litigation over habitat losses associated with dam building upstream. Additionally, the heart of the jewel-like central Niobrara Valley has been saved by the actions of the Nature Conservancy in establishing the Niobrara Valley Preserve. A new if small national wildlife refuge (Boyer Chute) has been established along the third of the state's abused and variously endangered rivers, the Missouri, and the dire ecological condition of that river is finally gaining a degree of attention.

The Nature Conservancy has also played a key role in saving wetlands in the Rainwater Basin and rare saline wetlands in the Sandhills and the Salt Creek Valley. The natural wetland marshes of eastern Nebraska have greatly declined in the past 40 years, whereas artificial impoundments have proliferated there, mainly for flood-control reasons. Thus it is now common to see snow geese and Canada geese flying over Lincoln each spring and fall, a sight that was very unusual when I arrived here in the 1960s. I could not have imagined that Canada geese would eventually come to be seen as a "problem" species in town or that bald eagles would become relatively common migrants along the area's many impoundments. Remarkably, sandhill and whooping cranes have not only become more common in the Platte Valley since the 1960s, they are now finally recognized as representing a major tourist attraction, even being scheduled to appear on the state's license plates. Happily, Nebraska has refrained from making sandhill cranes legal hunting targets, thereby helping to protect not only whooping cranes but also great blue herons and almost all other large flying creatures.

Relict prairies in Nebraska are becoming increasingly identified and designated for some degree of protection, usually by private agencies and individuals. The Nebraska Game and Parks Commission is slowly, if sometimes grudgingly, recognizing that even nongame birds, mammals, and fish might deserve some respect and conservation attention. Perhaps even someday the ornate box turtle, doughty symbol of the Nebraska Sandhills, will be given some protection from unlimited commercial exploitation before it becomes completely eliminated.

The state of the forests in this "Tree-planting State" is slowly changing. The total forest acreage in Nebraska increased substantially during the 1980s and 1990s, but some native prairies are being increasingly overgrown with red cedars and other invasive vegetation. Moreover, the increased logging of the state's relatively rare mature hardwood and coniferous forests needs to be carefully monitored.

When I came to Nebraska in 1961, no national conservation organization had a state or field office. Now the Nebraska section of the Nature Conservancy has over 5,000 members and six field offices. The National Audubon Society operates two important sanctuaries in the state and has over 3,500 state members, with active chapters in four cities. Both organizations have been powerful beneficial influences for conservation in Nebraska, and the state would be far poorer without them.

Other national conservation-oriented groups with a significant presence in the state are the National Wildlife Federation, Ducks Unlimited, and Pheasants Forever. The Sierra Club, Wilderness Society, American Rivers, and Friends of Wildlife provide a smaller influence. The U.S. Forest Service, the Bureau of Reclamation, Army Corps of Engineers, and various private in-state agencies such as the Central Nebraska Public Power and Irrigation District are all to varying degrees legally charged with overseeing the ecological effects of their activities and mitigating any adverse effects that might result. Nebraska's Natural Heritage Program, partly funded by the Nature Conservancy; the Nebraska Environmental Trust, funded by the state's gambling proceeds; and the Wildlife Check-off Program, funded by voluntary state tax contributions, have been initiated in recent years and have provided valuable support for inventorying and protecting Nebraska's natural treasures. These programs are especially valuable for the study and conservation of nongame species that tend to "fall between the cracks" of state agencies' primary responsibilities.

The Plains Prairie Resource Institute in Aurora, Nebraska, manages remnant native prairies across the entire state and has an active educational program. There is also the Outdoor Education Center at Schramm State Park and a relatively under-developed one at the Wildcat Hills State Recreation Area. At the local level, Lincoln and Omaha have important conservation education programs and interpretive centers (at the Fontenelle Forest Preserve, Neale Woods, and Pioneers Park Nature Center). It is a rare, almost sensual, pleasure to be able to stroll among the almost cathedral-like trees of Fontenelle Forest during the warbler "fallouts" of early May or to gaze across nearly unbroken vistas of tallgrass prairies such as those of Nine-Mile Prairie or Spring Creek Ranch in October when the bluestems turn coppery red and the fall sky turns azure blue. And it is an almost surreal experience to share a prairie sunrise solely in the company of prairie grouse, whose archaic language is that of the Oligocene or even earlier and whose endless story has been repeated every spring, year after year, century after century, never tiring in its urgency or intensity. Experiencing this event for the first time must be almost akin to initially landing on the moon or discovering some new phenomenon—there are no words within one's personal experience to describe it adequately.

There are thus still places in Nebraska where one can lie back on a fragrant bed of last-year's bluestem in early April, with the half-intoxicating odor of freshly germinating grass invading one's nose, with the shrill but majestic music of cranes almost constantly overhead, and with occasional harmonies added by arctic-bound if nearly invisible geese. There is then a true sense of belonging to and being a part of the land, and one can only offer an unspoken prayer that such treasures will still be there for the next generation to savor and love. At such times one realizes that, although there may be places with higher mountains, with magnificent rock-bound coastlines, or with misty cloud forests, it really doesn't matter. Nebraska is our spiritual home, our own self-chosen Nirvana, our prairie-born paradise, and the natural surviving legacy of long-forgotten winds, immense amounts of water, now-vanished glacial ice, and unfathomable eons of time. It has been freely bestowed upon us, either to keep or to destroy. May we choose to keep it.

PART THREE

Appendixes

Checklists of Nebraska's Flora and Fauna

Vertebrates

Mammals

This list is based largely on Jones and Choate (1980) but excludes some species of uncertain current occurrence and includes recent taxonomic changes. Nebraska's species were described by Jones et al. (1985) and also by Jones et al. (1983). Current accounts, range maps, and photos of all species are provided by Wilson and Ruff (1999), whose taxonomy supplants these regional works. Recent range changes have been documented by Benedict et al. (2000). An early but still useful field guide (Burt and Grossenheider 1976) illustrates all species, although several bat, shrew, and rodent species have since undergone confusing nomenclature changes. Familial and generic sequences are arranged in traditional taxonomic sequence; species are listed alphabetically within genera.

FAMILY DIDELPHIDAE—NEW WORLD OPOSSUMS
Virginia Opossum. *Didelphis virginiana*. Common, cities and woods (mainly eastern); rare in Panhandle but increasing.

FAMILY SORICIDAE—SHREWS
Masked Shrew. *Sorex cinereus*. Locally common (especially north), rare in southern counties; mainly in moist grasslands or somewhat brushy to wooded mesic habitats; increasing southwardly.
Merriam's Shrew. *Sorex merriami*. Very rare (northwest), arid grasslands and sage.

Northern Short-tailed Shrew. *Blarina brevicauda.* Common (east and central), diverse habitats, from grasslands to woods.

Elliot's Short-tailed Shrew. *Blarina hylophaga.* Common (south), diverse habitats.

Least Shrew. *Cryptotis parva.* Uncommon (east and central), rare in Panhandle; open grassy or weedy habitats, especially tallgrass prairies.

FAMILY TALPIDAE — MOLES

Eastern Mole. *Scalopus aquaticus.* Widespread, mostly subterranean, loamy soils and taller grasses.

FAMILY VESPERTILIONIDAE — VESPERTILIONID BATS

Western Small-footed Myotis. *Myotis ciliolabrum* (previously *M. subulatus* and *M. leibii*). Panhandle (Niobrara and White Rivers), rocky habitats; hibernator.

Little Brown Myotis. *Myotis lucifugus.* Pine Ridge and eastern quarter only, coniferous and deciduous forests; hibernator.

Northern Myotis. *Myotis septentrionalis.* Southeastern, west along Niobrara and Republican Rivers, wooded habitats; hibernator.

Fringe-tailed Myotis. *Myotis thysanodes.* Pine Ridge and Wildcat Hills, coniferous habitats; hibernator.

Long-legged Myotis. *Myotis volans* (previously part of *M. keeni*). Pine Ridge only, open forests, rock ledges; hibernator.

Silver-haired Bat. *Lasionycteris noctivagans.* Widespread, forests and meadows; migratory.

Eastern Pipistrelle. *Pipistrellus subflavus.* Southeast only (Cass and Sarpy Counties), woods and woodland edges; hibernator.

Big Brown Bat. *Eptesicus fuscus.* Widespread, diverse habitats, including woods; hibernator.

Eastern Red Bat. *Lasiurus borealis.* Widespread, diverse habitats; migratory.

Hoary Bat. *Lasiurus cinereus.* Widespread, wooded areas; migratory.

Evening Bat. *Nycticeius humeralis.* Eastern third, deciduous woods; probably migratory; expanding north and west.

Townsend's Big-eared Bat. *Plecotus townsendii.* Panhandle only (rare, Sheridan County); hibernator.

FAMILY MOLOSSIDAE — MOLOSSID BATS

Brazilian Free-tailed Bat. *Tadarida brasiliensis.* All but northwestern corner; rare migrant.

FAMILY DASYPODIDAE — ARMADILLOS

Nine-banded Armadillo. *Dasypus novemcinctus.* Southern and eastern counties only (where rare), dry scrub in most of range but reported along river valleys.

FAMILY LEPORIDAE — HARES AND RABBITS

Desert Cottontail. *Sylvilagus auduboni.* West and central, drier grasslands, especially mixed-grass and shortgrass prairies.

Eastern Cottontail. *Sylvilagus floridanus.* Widespread, woods and taller grasslands.

Black-tailed Jack Rabbit. *Lepus californicus.* More common in southern half in grasslands, especially mixed-grass prairies and disturbed habitats.

White-tailed Jack Rabbit. *Lepus townsendii.* More common northwardly in grasslands, especially mixed-grass or Sandhills prairies; becoming rare.

FAMILY SCIURIDAE — SQUIRRELS

Eastern Chipmunk. *Tamias striatus.* Southeast (very rare), deciduous woods; possibly extirpated (western edge of species' range).

Least Chipmunk. *Eutamias minimus.* Panhandle only, coniferous woods where affected by increased logging.

Woodchuck. *Marmota monax.* East and central, deciduous woods; expanding in range.

Wyoming Ground Squirrel. *Spermophilus elegans* (previously part of *S. richardsonii*). Extreme west only; possibly extirpated in Nebraska but present in adjoining southeastern Wyoming in sandsage grasslands.

Franklin's Ground Squirrel. *Spermophilus franklini.* East and central, tall-grass prairies.

Spotted Ground Squirrel. *Spermophilus spilosoma.* West and central, grasslands, especially mixed-grass prairies; widespread but uncommon.

Thirteen-lined Ground Squirrel. *Spermophilus tridecemlineatus.* Widespread, especially mixed-grass prairies.

Black-tailed Prairie Dog. *Cynomys ludovicianus.* West and central, drier grasslands; declining over entire species' range.

Gray Squirrel. *Sciurus carolinensis.* Southeast only, deciduous woods; probably declining.

Fox Squirrel. *Sciurus niger.* East and northwest, deciduous woods.

Southern Flying Squirrel. *Glaucomys volans.* Southeast only, deciduous woods; rare and declining, state threatened.

FAMILY GEOMYIDAE — POCKET GOPHERS

Northern Pocket Gopher. *Thomomys talpoides.* Southwest and northwest only, grasslands, especially mixed-grass or sandsage prairies; probably declining.

Plains Pocket Gopher. *Geomys bursarius.* Widespread, especially in taller grasslands.

FAMILY HETEROMYIDAE — HETEROMYID RODENTS

Olive-backed Pocket Mouse. *Perognathus fasciatus.* Northwest and north, shortgrass prairies.

Plains Pocket Mouse. *Perognathus flavescens.* Widespread, Sandhills and sandsage or mixed-grass prairies.

Silky Pocket Mouse. *Perognathus flavus.* West to central, drier grasslands, especially mixed-grass prairies; rare.

Hispid Pocket Mouse. *Perognathus hispidus.* Widespread, sandy grasslands.

Ord's Kangaroo Rat. *Dipodomys ordii.* West and central, sandy grasslands.

FAMILY CASTORIDAE — BEAVERS
Beaver. *Castor canadensis.* Widespread in aquatic habitats.

FAMILY MURIDAE — RATS AND MICE
Western Harvest Mouse. *Reithrodontomys megalotis.* Widespread, taller grasslands.

Plains Harvest Mouse. *Reithrodontomys montanus.* Widespread, shorter (mixed-grass) grasslands; rare.

White-footed Mouse. *Peromyscus leucopus.* Mainly east and central but increasing westwardly, all habitats.

Deer Mouse. *Peromyscus maniculatus.* Widespread, all open habitats.

Northern Grasshopper Mouse. *Onychomys leucogaster.* Widespread, especially sandy and mixed-grass grasslands, moving eastwardly.

Hispid Cotton Rat. *Sigmodon hispidus.* Southeast and south-central only, dense herbaceous vegetation and taller grasslands; possibly now declining.

Bushy-tailed Woodrat. *Neotoma cinerea.* Panhandle, rocky slopes and crevices.

Eastern Woodrat. *Neotoma floridana.* Southern counties and (*N. f. baileyii*) Niobrara Valley; wooded areas, buildings, outcrops.

Southern Bog Lemming. *Synaptomys cooperi.* East and central, also southwest; fenlike and dense grassy habitats.

Prairie Vole. *Microtus ochrogaster.* Widespread, taller grasslands.

Meadow Vole. *Microtus pernnsylvanicus.* Widespread, increasing southwardly, moist grasslands.

Woodland Vole. *Microtus pinetorum.* Southeast and Missouri Valley, wooded habitats.

Muskrat. *Ondatra zibethicus.* Widespread in aquatic habitats.

House Mouse. *Mus musculus.* Introduced; statewide near human habitations.

Norway Rat. *Rattus norvegicus.* Introduced; statewide near human habitations.

FAMILY ZAPODIDAE — JUMPING MICE
Meadow Jumping Mouse. *Zapus hudsonius.* East and central, grassy or herbaceous habitats.

FAMILY ERETHIZONTIDAE — NEW WORLD PORCUPINES
Porcupine. *Erethizon dorsatum.* West and central, rare in east; in or near trees.

FAMILY CANIDAE — COYOTES, WOLVES, AND FOXES

Coyote. *Canis latrans.* Widespread, especially in grasslands.

Gray (Prairie) Wolf. *Canis lupus.* Extirpated by 1920; nationally endangered.

Swift Fox. *Vulpes velox.* Rare, shortgrass prairies; reintroduced in west; state endangered.

Red Fox. *Vulpes vulpes.* Widespread, especially near woods.

Gray Fox. *Urocyon cineroargenteus.* Mainly east but increasing north and west, deciduous woods.

FAMILY FELIDAE — CATS

Mountain Lion. *Puma concolor.* Very rare but several Panhandle records recently.

Lynx. *Felis lynx.* Extremely rare; less than a dozen modern records; proposed for national threatened or endangered list.

Bobcat. *Felis rufus.* Increasingly common statewide, most common in west in rocky topography; recently increasing in southeast.

FAMILY PROCYONIDAE — RACCOONS AND ALLIES

Raccoon. *Procyon lotor.* Statewide, near trees.

FAMILY MUSTELIDAE — WEASELS, BADGERS, SKUNKS, AND OTTERS

(Skunks are sometimes separated into Mephitinae)

Long-tailed Weasel. *Mustela frenata.* Statewide, grasslands and woods; declining.

Black-footed Ferret. *Mustela nigriceps.* Probably extirpated; nationally endangered.

Least Weasel. *Mustela nivalis.* Statewide, diverse habitats.

Mink. *Mustela vison.* Statewide, near rivers and marshes.

Badger. *Taxidea taxus.* Statewide, grasslands, especially drier grasslands.

Spotted Skunk. *Spilogale putorius.* Widespread but rare, forest edges.

Striped Skunk. *Mephitis mephitis.* Statewide, diverse habitats.

Northern River Otter. *Lutra canadensis.* Very rare; reintroduced in some rivers; state threatened.

FAMILY CERVIDAE — DEER, ELK, AND MOOSE

Elk (Wapiti). *Cervus elephas.* Extirpated; reintroduced in Pine Ridge and Fort Niobrara.

Mule Deer. *Odocoileus hemionus.* West and central, mainly grasslands; declining.

White-tailed Deer. *Odocoileus virginianus.* Widespread, forests and grasslands; expanding westwardly.

FAMILY BOVIDAE — CATTLE, SHEEP, AND GOATS

Bison. *Bison bison.* Extirpated but reintroduced locally, usually in confined locations.

Mountain Sheep. *Ovis canadensis.* Extirpated but reintroduced in Pine Ridge.

FAMILY ANTILOCAPRIDAE — PRONGHORNS
Pronghorn. *Antilocapra americana.* West, open grasslands.

Birds

This list excludes nearly all accidental, extinct, extirpated, and hypothetical bird species. Species status mainly adapted and simplified from Johnsgard (2000) and Mollhoff (2001). Breeding status based on number of total numbers of probable and confirmed state records for each species (1984–88) listed in Mollhoff (abundant = 300+, common = 200–299, uncommon = 100–199, occasional or scattered 50–99, rare or local = 10–50, extremely rare or highly local = 1–9). Family sequence as well as common and Latin names follow the most recent American Ornithologists' Union listing. All species are illustrated in Dickinson (1999), the most useful and complete field guide available.

* indicates known breeding species.

FAMILY GAVIIDAE — LOONS
Red-throated Loon. *Gavia stellata.* Extremely rare migrant.
Pacific Loon. *Gavia pacifica.* Extremely rare migrant.
Common Loon. *Gavia immer.* Common migrant.

FAMILY PODICIPEDIDAE — GREBES
Pied-billed Grebe. *Podilymbus podiceps.* Common summer resident.*
Horned Grebe. *Podiceps auritus.* Regular migrant, rare breeder.*
Red-necked Grebe. *Podiceps grisegena.* Extremely rare migrant.
Eared Grebe. *Podiceps nigricollis.* Common summer resident (west).*
Western Grebe. *Aechmophorus occidentalis.* Common summer resident (central and west).*
Clark's Grebe. *Aechmophorus clarkii.* Regular summer resident (west).*

FAMILY PELECANIDAE — PELICANS
American White Pelican. *Pelecanus erythrorhynchos.* Common nonbreeding summer resident.

FAMILY PHALACROCORACIDAE — CORMORANTS
Double-crested Cormorant. *Phalacrocorax auritus.* Common summer resident.*
Neotropic Cormorant. *Phalacrocorax brasilianus.* Very rare migrant.

FAMILY ARDEIDAE — BITTERNS AND HERONS
American Bittern. *Botaurus lentiginosus.* Common summer resident.*
Least Bittern. *Ixobrychus exilis.* Uncommon summer resident.*
Great Blue Heron. *Ardea herodias.* Common summer resident.*

Great Egret. *Ardea alba.* Regular nonbreeding summer resident, rare breeder.*

Snowy Egret. *Egretta thula.* Irregular summer resident, rare breeder.*

Little Blue Heron. *Egretta caerulea.* Irregular nonbreeding summer resident.

Cattle Egret. *Bubulcus ibis.* Regular summer resident, rare breeder.*

Green Heron. *Butorides virescens.* Common summer resident.*

Black-crowned Night-heron. *Nycticorax nycticorax.* Common summer resident.*

Yellow-crowned Night-heron. *Nyctanassa violacea.* Rare summer resident.*

FAMILY THRESKIORNITHIDAE—IBISES AND SPOONBILLS
White-faced Ibis. *Plegadis chihi.* Local summer resident (west).*

FAMILY CATHARTIDAE—AMERICAN VULTURES
Turkey Vulture. *Cathartes aura.* Common summer resident.*

FAMILY ANATIDAE—SWANS, GEESE, AND DUCKS
Greater White-fronted Goose. *Anser albifrons.* Common migrant.

Snow Goose. *Chen caerulescens.* Common migrant.

Ross's Goose. *Chen rossii.* Regular migrant.

Canada Goose. *Branta canadensis.* Common summer resident.*

Trumpeter Swan. *Cygnus buccinator.* Regular summer resident (west).*

Tundra Swan. *Cygnus columbianus.* Casual migrant (mainly east).

Wood Duck. *Aix sponsa.* Common summer resident (mainly east).*

Gadwall. *Anas strepera.* Common summer resident.*

Eurasian Wigeon. *Anas penelope.* Casual to rare migrant.

American Wigeon. *Anas americana.* Common summer resident.*

American Black Duck. *Anas rubripes.* Regular migrant (east).

Mallard. *Anas platyrhynchos.* Common summer resident.*

Blue-winged Teal. *Anas discors.* Common summer resident.*

Cinnamon Teal. *Anas cyanoptera.* Regular summer resident (west).*

Northern Shoveler. *Anas clypeata.* Common summer resident.*

Northern Pintail. *Anas acuta.* Common summer resident.*

Green-winged Teal. *Anas crecca.* Common summer resident.*

Canvasback. *Aythya valisineria.* Uncommon summer resident.*

Redhead. *Aythya americana.* Common summer resident.*

Ring-necked Duck. *Aythya collaris.* Common migrant, rare summer resident (central).*

Greater Scaup. *Aythya marila.* Regular but uncommon to rare migrant.

Lesser Scaup. *Aythya affinis.* Common migrant, rare summer resident (central).*

Surf Scoter. *Melanitta perspicillata.* Regular but uncommon migrant.

White-winged Scoter. *Melanitta fusca.* Regular but uncommon migrant.

Black Scoter. *Melanitta nigra.* Casual migrant.

Oldsquaw. (Long-tailed Duck). *Clangula hyemalis*. Regular but uncommon migrant.

Bufflehead. *Bucephala albeola*. Common migrant.

Common Goldeneye. *Bucephala clangula*. Common migrant.

Barrow's Goldeneye. *Bucephala islandica*. Casual migrant.

Hooded Merganser. *Lophodytes cucullatus*. Regular migrant, casual summer resident.*

Red-breasted Merganser. *Mergus serrator*. Regular migrant.

Common Merganser. *Mergus merganser*. Common migrant, casual summer resident.*

Ruddy Duck. *Oxyura jamaicensis*. Common summer resident (central and west).*

FAMILY ACCIPITRIDAE — KITES, HAWKS, EAGLES, AND ALLIES

Osprey. *Pandion haliaetus*. Regular migrant.

Mississippi Kite. *Ictinia mississippiensis*. Casual migrant, rare summer resident (west).*

Bald Eagle. *Haliaeetus leucocephalus*. Regular overwintering migrant, rare resident; nationally threatened.*

Northern Harrier. *Circus cyaneus*. Regular summer resident.*

Sharp-shinned Hawk. *Accipiter striatus*. Regular but rare summer resident.*

Cooper's Hawk. *Accipiter cooperii*. Regular migrant, rare resident.*

Northern Goshawk. *Accipiter gentilis*. Regular but rare wintering migrant.

Red-shouldered Hawk. *Buteo lineatus*. Rare (extirpated?) resident (southeast).*

Broad-winged Hawk. *Buteo platypterus*. Regular migrant, rare (extirpated?) summer resident (east).*

Swainson's Hawk. *Buteo swainsoni*. Common summer resident (west).*

Red-tailed Hawk. *Buteo jamaicensis*. Common resident.*

Ferruginous Hawk. *Buteo regalis*. Regular but uncommon to rare resident (west).*

Rough-legged Hawk. *Buteo lagopus*. Regular wintering migrant.

Golden Eagle. *Aquila chrysaetos*. Regular but uncommon resident (west).*

FAMILY FALCONIDAE — FALCONS

American Kestrel. *Falco sparverius*. Common summer resident.*

Merlin. *Falco columbarius*. Regular but uncommon migrant.

Prairie Falcon. *Falco mexicanus*. Regular resident (west).*

Peregrine Falcon. *Falco peregrinus*. Regular migrant, rare summer resident (east), previously state and nationally endangered.*

Gyrfalcon. *Falco rusticolus*. Casual wintering migrant (mainly central).

FAMILY PHASIANIDAE — PARTRIDGES, GROUSE, AND TURKEYS

Gray Partridge. *Perdix perdix.* Uncommon to rare resident (northeast).[*]
Ring-necked Pheasant. *Phasianus colchicus.* Common resident.[*]
Greater Prairie-chicken. *Tympanuchus cupido.* Regular resident (east and central).[*]
Lesser Prairie-chicken. *Tympanuchus pallidicinctus.* Extirpated.
Sharp-tailed Grouse. *Tympanuchus phasianellus.* Regular resident (west and central).[*]
Wild Turkey. *Meleagris gallopavo.* Locally common resident.[*]

FAMILY ODONTOPHORIDAE — NEW WORLD QUAIL

Northern Bobwhite. *Colinus virginianus.* Common resident.[*]

FAMILY RALLIDAE — RAILS, GALLINULES, AND COOTS

King Rail. *Rallus elegans.* Casual summer resident (east).[*]
Virginia Rail. *Rallus limicola.* Regular summer resident.[*]
Sora. *Porzana carolina.* Common summer resident.[*]
Common Moorhen. *Gallinula chloropus.* Occasional summer resident (east).[*]
American Coot. *Fulica americana.* Common summer resident.[*]

FAMILY GRUIDAE — CRANES

Sandhill Crane. *Grus canadensis.* Common migrant, very rare summer resident.[*]
Whooping Crane. *Grus americana.* Regular migrant (central), nationally endangered.

FAMILY CHARADRIIDAE — PLOVERS

Black-bellied Plover. *Pluvialis squatarola.* Regular migrant.
American Golden-Plover. *Pluvialis dominica.* Regular migrant.
Snowy Plover. *Charadrius alexandrinus.* Casual migrant.
Semipalmated Plover. *Charadrius semipalmatus.* Regular migrant.
Piping Plover. *Charadrius melodus.* Regular summer resident, nationally threatened.[*]
Killdeer. *Charadrius vociferus.* Common summer resident.[*]
Mountain Plover. *Charadrius montanus.* Rare summer resident (southwest), state threatened.[*]

FAMILY RECURVIROSTRIDAE — STILTS AND AVOCETS

Black-necked Stilt. *Himantopus mexicanus.* Local summer resident (west).[*]
American Avocet. *Recurvirostra americana.* Regular summer resident (central and west).[*]

FAMILY SCOLOPACIDAE — SANDPIPERS AND PHALAROPES

Greater Yellowlegs. *Tringa melanoleuca.* Common migrant.
Lesser Yellowlegs. *Tringa flavipes.* Common migrant.

Solitary Sandpiper. *Tringa solitaria.* Regular migrant.

Willet. *Catoptrophorus semipalmatus.* Regular summer resident (central and west).*

Spotted Sandpiper. *Actitis macularia.* Common summer resident.*

Upland Sandpiper. *Bartramia longicauda.* Common summer resident.*

Whimbrel. *Numenius phaeopus.* Casual migrant.

Long-billed Curlew. *Numenius americanus.* Regular summer resident (central and west).*

Eskimo Curlew. *Numenius borealis.* No recent state records, nationally endangered; possibly extinct.

Hudsonian Godwit. *Limosa haemastica.* Regular migrant (east).

Marbled Godwit. *Limosa fedoa.* Regular migrant, possible rare summer resident.

Ruddy Turnstone. *Arenaria interpres.* Regular migrant (east).

Red Knot. *Calidris canutus.* Casual migrant (east).

Sanderling. *Calidris alba.* Regular migrant.

Semipalmated Sandpiper. *Calidris pusilla.* Regular migrant.

Western Sandpiper. *Calidris mauri.* Regular migrant (mainly west).

Least Sandpiper. *Calidris minutilla.* Common migrant.

White-rumped Sandpiper. *Calidris fuscicollis.* Common migrant, mainly in spring.

Baird's Sandpiper. *Calidris bairdii.* Common migrant.

Pectoral Sandpiper. *Calidris melanotos.* Common migrant.

Dunlin. *Calidris alpina.* Regular migrant (east).

Stilt Sandpiper. *Calidris himantopus.* Common migrant.

Buff-breasted Sandpiper. *Tryngites subruficollis.* Regular migrant (east).

Short-billed Dowitcher. *Limnodromus griseus.* Casual migrant (east).

Long-billed Dowitcher. *Limnodromus scolopaceus.* Common migrant.

Common Snipe. *Gallinago gallinago.* Regular but local summer resident.*

American Woodcock. *Scolopax minor.* Regular summer resident (east and central).*

Wilson's Phalarope. *Phalaropus tricolor.* Regular summer resident (west and central).*

Red-necked Phalarope. *Phalaropus lobatus.* Regular migrant.

Red Phalarope. *Phalaropus fulicaria.* Casual migrant.

FAMILY LARIDAE—GULLS AND TERNS

Laughing Gull. *Larus atricilla.* Casual migrant.

Franklin's Gull. *Larus pipixcan.* Common migrant.

Bonaparte's Gull. *Larus philadelphia.* Regular migrant.

Ring-billed Gull. *Larus delawarensis.* Common nonbreeding resident.

California Gull. *Larus californicus.* Regular migrant (west).

Herring Gull. *Larus argentatus.* Regular nonbreeding resident.

Thayer's Gull. *Larus thayeri.* Casual migrant.

Iceland Gull. *Larus glaucoides*. Casual migrant.

Lesser Black-backed Gull. *Larus fuscus*. Casual migrant.

Glaucous Gull. *Larus hyperboreus*. Regular migrant (mainly west).

Great Black-backed Gull. *Larus marinus*. Casual migrant.

Black-legged Kittiwake. *Rissa tridactyla*. Casual migrant.

Caspian Tern. *Sterna caspia*. Regular migrant (east), uncommon summer nonbreeder (west).

Common Tern. *Sterna hirundo*. Regular migrant (east).

Forster's Tern. *Sterna forsteri*. Regular summer resident (west and central).[*]

Least Tern. *Sterna antillarum*. Regular summer resident (east to west-central).[*]

Black Tern. *Chlidonias niger*. Regular summer resident (west and central).[*]

FAMILY COLUMBIDAE — PIGEONS AND DOVES

Rock Dove. *Columba livia*. Common resident.[*]

Eurasian Collared-dove. *Streptopelia decaocto*. Very rare summer resident (central).[*]

Mourning Dove. *Zenaida macroura*. Common summer resident.[*]

FAMILY CUCULIDAE — CUCKOOS AND ANIS

Black-billed Cuckoo. *Coccyzus erythropthalmus*. Regular summer resident.[*]

Yellow-billed Cuckoo. *Coccyzus americanus*. Regular summer resident.[*]

FAMILY TYTONIDAE — BARN OWLS

Barn Owl. *Tyto alba*. Uncommon resident.[*]

FAMILY STRIGIDAE — TYPICAL OWLS

Eastern Screech-owl. *Otus asio*. Common resident.[*]

Great Horned Owl. *Bubo virginianus*. Common resident.[*]

Snowy Owl. *Nyctea scandiaca*. Regular wintering migrant.

Burrowing Owl. *Athene cunicularia*. Uncommon to rare summer resident (west to central).[*]

Barred Owl. *Strix varia*. Uncommon resident (east).[*]

Long-eared Owl. *Asio otus*. Uncommon resident (mainly east).[*]

Short-eared Owl. *Asio flammeus*. Uncommon resident.[*]

Northern Saw-whet Owl. *Aegolius acadicus*. Regular migrant, possible rare summer resident (west).

FAMILY CAPRIMULGIDAE — GOATSUCKERS

Common Nighthawk. *Chordeiles minor*. Common summer resident (east).[*]

Common Poorwill. *Phalaenoptilus nuttallii*. Regular summer resident (west).[*]

Chuck-will's-widow. *Caprimulgus carolinensis*. Regular summer resident (southeast).[*]

Whip-poor-will. *Caprimulgus vociferus.* Common summer resident (east).*

FAMILY APODIDAE — SWIFTS

Chimney Swift. *Chaetura pelagica.* Common summer resident (mainly east).*

White-throated Swift. *Aeronautes saxatalis.* Regular summer resident (west).*

FAMILY TROCHILIDAE — HUMMINGBIRDS

Ruby-throated Hummingbird. *Archilochus colubris.* Regular summer resident (east).*

Calliope Hummingbird. *Stellula calliope.* Very rare migrant (west).

Broad-tailed Hummingbird. *Selasphorus platycercus.* Casual migrant (west).

Rufous Hummingbird. *Selasphorus rufus.* Casual migrant (west).

FAMILY ALCEDINIDAE — KINGFISHERS

Belted Kingfisher. *Ceryle alcyon.* Regular summer resident.*

FAMILY PICIDAE — WOODPECKERS

Lewis' Woodpecker. *Melanerpes lewis.* Regular summer resident (west).*

Red-headed Woodpecker. *Melanerpes erythrocephalus.* Common summer resident.*

Red-bellied Woodpecker. *Melanerpes carolinus.* Common summer resident (east).*

Yellow-bellied Sapsucker. *Sphyrapicus varius.* Regular migrant (east).

Downy Woodpecker. *Picoides pubescens.* Common resident.*

Hairy Woodpecker. *Picoides villosus.* Common resident.*

Northern Flicker. *Colaptes auratus.* Common resident (yellow-shafted east, red-shafted west).*

Pileated Woodpecker. *Dryocopus pileatus.* Rare local resident (east).*

FAMILY TYRANNIDAE — TYRANT FLYCATCHERS

Olive-sided Flycatcher. *Contopus cooperi.* Regular migrant.

Western Wood-pewee. *Contopus sordidulus.* Regular summer resident (west).*

Eastern Wood-pewee. *Contopus virens.* Regular summer resident (east).*

Yellow-bellied Flycatcher. *Empidonax flaviventris.* Uncommon migrant, probable rare summer resident (east).

Acadian Flycatcher. *Empidonax virescens.* Regular migrant, probable rare summer resident (southeast).

Alder Flycatcher. *Empidonax alnorum.* Regular migrant.

Willow Flycatcher. *Empidonax traillii.* Regular summer resident (east).*

Least Flycatcher. *Empidonax minimus.* Regular migrant, rare summer resident (north).*

Cordilleran Flycatcher. *Empidonax occidentalis*. Very local summer resident (northwest).[*]
Eastern Phoebe. *Sayornis phoebe*. Common summer resident (east).[*]
Say's Phoebe. *Sayornis saya*. Regular summer resident (west).[*]
Great Crested Flycatcher. *Myiarchus crinitus*. Regular summer resident (east).[*]
Cassin's Kingbird. *Tyrannus vociferans*. Regular summer resident (west).[*]
Western Kingbird. *Tyrannus verticalis*. Common summer resident.[*]
Eastern Kingbird. *Tyrannus tyrannus*. Common summer resident.[*]
Scissor-tailed Flycatcher. *Tyrannus forficatus*. Casual summer resident (east).[*]

FAMILY LANIIDAE — SHRIKES

Northern Shrike. *Lanius excubitor*. Uncommon wintering migrant.
Loggerhead Shrike. *Lanius ludovicianus*. Common summer resident.[*]

FAMILY VIREONIDAE — VIREOS

White-eyed Vireo. *Vireo griseus*. Casual summer resident (southeast).[*]
Bell's Vireo. *Vireo bellii*. Common summer resident.[*]
Blue-headed Vireo. *Vireo solitarius*. Regular migrant (east).
Plumbeous Vireo. *Vireo plumbeus*. Local summer resident (northwest).[*]
Yellow-throated Vireo. *Vireo flavifrons*. Regular summer resident (east).[*]
Warbling Vireo. *Vireo gilvus*. Common summer resident.[*]
Philadelphia Vireo. *Vireo philadelphicus*. Regular migrant (east).
Red-eyed Vireo. *Vireo olivaceus*. Regular summer resident (east).[*]

FAMILY CORVIDAE — JAYS, MAGPIES, AND CROWS

Blue Jay. *Cyanocitta cristata*. Common resident.[*]
Pinyon Jay. *Gymnorhinus cyanocephalus*. Rare and local resident (west).[*]
Clark's Nutcracker. *Nucifraga columbiana*. Rare resident (west).[*]
Black-billed Magpie. *Pica pica*. Common resident (west).[*]
American Crow. *Corvus brachyrhynchos*. Common resident.[*]

FAMILY ALAUDIDAE — LARKS

Horned Lark. *Eremophila alpestris*. Common resident.[*]

FAMILY HIRUNDINIDAE — SWALLOWS

Purple Martin. *Progne subis*. Common summer resident (mainly east).[*]
Tree Swallow. *Tachycineta bicolor*. Common summer resident.[*]
Violet-green Swallow. *Tachycineta thalassina*. Regular summer resident (west).[*]
Northern Rough-winged Swallow. *Stelgidopteryx serripennis*. Common summer resident.[*]
Bank Swallow. *Riparia riparia*. Common summer resident.[*]
Barn Swallow. *Hirundo rustica*. Common summer resident.[*]
Cliff Swallow. *Petrochelidon pyrrhonota*. Common summer resident.[*]

FAMILY PARIDAE — TITMICE

Black-capped Chickadee. *Poecile atricapillus.* Common resident.*
Tufted Titmouse. *Baeolophus bicolor.* Uncommon resident (east).*

FAMILY SITTIDAE — NUTHATCHES

Red-breasted Nuthatch. *Sitta canadensis.* Common migrant; rare summer resident (north).*
White-breasted Nuthatch. *Sitta carolinensis.* Common resident.*
Pygmy Nuthatch. *Sitta pygmaea.* Uncommon resident (west).*

FAMILY CERTHIIDAE — CREEPERS

Brown Creeper. *Certhia americana.* Uncommon resident.*

FAMILY TROGLODYTIDAE — WRENS

Rock Wren. *Salpinctes obsoletus.* Regular summer resident (west).*
Carolina Wren. *Thryothorus ludovicianus.* Regular resident (east).*
Bewick's Wren. *Thryomanes bewickii.* Rare migrant or summer resident (southeast).*
House Wren. *Troglodytes aedon.* Common summer resident.*
Winter Wren. *Troglodytes troglodytes.* Regular migrant.
Sedge Wren. *Cistothorus platensis.* Regular summer resident (east).*
Marsh Wren. *Cistothorus palustris.* Regular summer resident.*

FAMILY REGULIDAE — KINGLETS

Golden-crowned Kinglet. *Regulus satrapa.* Regular wintering migrant.
Ruby-crowned Kinglet. *Regulus calendula.* Common migrant.

FAMILY SYLVIIDAE — GNATCATCHERS

Blue-gray Gnatcatcher. *Polioptila caerulea.* Uncommon summer resident (east).*

FAMILY TURDIDAE — THRUSHES AND ALLIES

Eastern Bluebird. *Sialia sialis.* Regular summer resident.*
Mountain Bluebird. *Sialia currucoides.* Regular summer resident (west).*
Townsend's Solitaire. *Myadestes townsendi.* Common migrant (west), rare resident.*
Veery. *Catharus fuscescens.* Regular migrant.
Gray-cheeked Thrush. *Catharus minimus.* Regular migrant.
Swainson's Thrush. *Catharus ustulatus.* Common migrant.
Hermit Thrush. *Catharus guttatus.* Regular migrant.
Wood Thrush. *Hylocichla mustelina.* Regular summer resident (east).*
American Robin. *Turdus migratorius.* Common summer resident.*
Varied Thrush. *Ixoreus naevius.* Casual migrant (west).

FAMILY MIMIDAE — MOCKINGBIRDS AND THRASHERS

Gray Catbird. *Dumetella carolinensis.* Common summer resident.*
Northern Mockingbird. *Mimus polyglottos.* Regular summer resident.*
Sage Thrasher. *Oreoscoptes montanus.* Casual summer resident (west).*

Brown Thrasher. *Toxostoma rufum*. Common summer resident.[*]

FAMILY STURNIDAE—STARLINGS

European Starling. *Sturnus vulgaris*. Common resident.[*]

FAMILY MOTACILLIDAE—PIPITS

American Pipit. *Anthus rubescens*. Regular migrant.

Sprague's Pipit. *Anthus spragueii*. Casual migrant.

FAMILY BOMBYCILLIDAE—WAXWINGS

Bohemian Waxwing. *Bombycilla garrulus*. Casual wintering migrant (north).

Cedar Waxwing. *Bombycilla cedrorum*. Regular summer resident.[*]

FAMILY PARULIDAE—WOOD WARBLERS

Blue-winged Warbler. *Vermivora pinus*. Casual migrant (east).

Golden-winged Warbler. *Vermivora chrysoptera*. Regular migrant (east).

Tennessee Warbler. *Vermivora peregrina*. Regular migrant (east).

Orange-crowned Warbler. *Vermivora celata*. Regular migrant.

Nashville Warbler. *Vermivora ruficapilla*. Common migrant.

Northern Parula. *Parula americana*. Regular summer resident (east).[*]

Yellow Warbler. *Dendroica petechia*. Common summer resident.[*]

Chestnut-sided Warbler. *Dendroica pensylvanica*. Regular migrant (east).

Magnolia Warbler. *Dendroica magnolia*. Regular migrant (east).

Cape May Warbler. *Dendroica tigrina*. Casual migrant (east).

Black-throated Blue Warbler. *Dendroica caerulescens*. Casual migrant (east).

Yellow-rumped Warbler. *Dendroica coronata*. Common summer resident (northwest).[*]

Townsend's Warbler. *Dendroica townsendi*. Casual migrant (west).

Black-throated Green Warbler. *Dendroica virens*. Regular migrant (east).

Blackburnian Warbler. *Dendroica fusca*. Regular migrant (east).

Yellow-throated Warbler. *Dendroica dominica*. Local summer resident (southeast).[*]

Pine Warbler. *Dendroica pinus*. Casual migrant (east).

Palm Warbler. *Dendroica palmarum*. Regular migrant (east).

Bay-breasted Warbler. *Dendroica castanea*. Regular migrant (east).

Blackpoll Warbler. *Dendroica striata*. Regular migrant.

Cerulean Warbler. *Dendroica cerulea*. Regular migrant, rare summer resident (east).[*]

Black-and-white Warbler. *Mniotilta varia*. Regular summer resident (north).[*]

American Redstart. *Setophaga ruticilla*. Regular summer resident (north).[*]

Prothonotary Warbler. *Protonotaria citrea*. Local summer resident (southeast).[*]

Worm-eating Warbler. *Helmitheros vermivorus*. Casual migrant (east).

Ovenbird. *Seiurus aurocapillus*. Regular summer resident (east).[*]

Northern Waterthrush. *Seiurus noveboracensis*. Regular migrant (east).

Louisiana Waterthrush. *Seiurus motacilla*. Regular migrant, rare summer resident (east).[*]

Kentucky Warbler. *Oporornis formosus*. Local summer resident (southeast).[*]

Connecticut Warbler. *Oporornis agilis*. Regular migrant (east).

Mourning Warbler. *Oporornis philadelphia*. Regular migrant (east).

MacGillivray's Warbler. *Oporornis tolmiei*. Regular migrant (west).

Common Yellowthroat. *Geothlypis trichas*. Common summer resident.[*]

Hooded Warbler. *Wilsonia citrina*. Casual migrant (east).

Wilson's Warbler. *Wilsonia pusilla*. Regular migrant (east).

Canada Warbler. *Wilsonia canadensis*. Regular migrant (east).

Yellow-breasted Chat. *Icteria virens*. Regular summer resident.[*]

FAMILY THRAUPIDAE — TANAGERS

Summer Tanager. *Piranga rubra*. Regular summer resident (southeast).[*]

Scarlet Tanager. *Piranga olivacea*. Regular summer resident (east).[*]

Western Tanager. *Piranga ludoviciana*. Regular summer resident (west).[*]

FAMILY EMBERIZIDAE — TOWHEES, SPARROWS, AND LONGSPURS

Green-tailed Towhee. *Pipilo chlorurus*. Casual migrant (west).

Eastern Towhee. *Pipilo erythrophthalmus*. Regular summer resident (east and central).[*]

Spotted Towhee. *Pipilo maculatus*. Regular summer resident (west and central).[*]

Cassin's Sparrow. *Aimophila cassinii*. Casual summer resident (west).[*]

American Tree Sparrow. *Spizella arborea*. Common wintering migrant.

Chipping Sparrow. *Spizella passerina*. Common summer resident.[*]

Clay-colored Sparrow. *Spizella pallida*. Regular migrant.

Brewer's Sparrow. *Spizella breweri*. Local summer resident (northwest).[*]

Field Sparrow. *Spizella pusilla*. Common summer resident.[*]

Vesper Sparrow. *Pooecetes gramineus*. Regular summer resident.[*]

Lark Sparrow. *Chondestes grammacus*. Common summer resident.[*]

Lark Bunting. *Calamospiza melanocorys*. Common summer resident (west).[*]

Savannah Sparrow. *Passerculus sandwichensis*. Regular migrant.

Baird's Sparrow. *Ammodramus bairdii*. Rare migrant (mainly west).

Grasshopper Sparrow. *Ammodramus savannarum*. Regular summer resident.[*]

Henslow's Sparrow. *Ammodramus henslowii*. Regular summer resident (southeast).[*]

Le Conte's Sparrow. *Ammodramus leconteii*. Regular migrant.

Nelson's Sharp-tailed Sparrow. *Ammodramus nelsoni*. Casual migrant (east).

Fox Sparrow. *Passerella iliaca*. Regular migrant.

Song Sparrow. *Melospiza melodia*. Common summer resident.*

Lincoln's Sparrow. *Melospiza lincolnii*. Regular migrant.

Swamp Sparrow. *Melospiza georgiana*. Regular summer resident (east and central).*

White-throated Sparrow. *Zonotrichia albicollis*. Common wintering migrant.

Harris's Sparrow. *Zonotrichia querula*. Common wintering migrant.

White-crowned Sparrow. *Zonotrichia leucophrys*. Common wintering migrant.

Dark-eyed Junco. *Junco hyemalis*. Common migrant; local resident (northwest).*

McCown's Longspur. *Calcarius mccownii*. Regular summer resident (west).*

Lapland Longspur. *Calcarius lapponicus*. Regular wintering migrant.

Chestnut-collared Longspur. *Calcarius ornatus*. Regular summer resident (northwest).*

Snow Bunting. *Plectrophenax nivalis*. Regular wintering migrant.

FAMILY CARDINALIDAE—CARDINALS, GROSBEAKS, AND ALLIES

Northern Cardinal. *Cardinalis cardinalis*. Common resident.*

Rose-breasted Grosbeak. *Pheucticus ludovicianus*. Common summer resident (east).*

Black-headed Grosbeak. *Pheucticus melanocephalus*. Common summer resident (west).*

Blue Grosbeak. *Guiraca caerulea*. Regular summer resident.*

Lazuli Bunting. *Passerina amoena*. Regular summer resident (west and central).*

Indigo Bunting. *Passerina cyanea*. Common summer resident (east and central).*

Dickcissel. *Spiza americana*. Common summer resident (mainly east).*

FAMILY ICTERIDAE—MEADOWLARKS, BLACKBIRDS, ORIOLES, AND ALLIES

Bobolink. *Dolichonyx oryzivorus*. Regular summer resident.*

Red-winged Blackbird. *Agelaius phoeniceus*. Common summer resident.*

Eastern Meadowlark. *Sturnella magna*. Regular summer resident (mainly east).*

Western Meadowlark. *Sturnella neglecta*. Common summer resident.*

Yellow-headed Blackbird. *Xanthocephalus xanthocephalus*. Common summer resident.*

Rusty Blackbird. *Euphagus carolinus*. Regular migrant.

Brewer's Blackbird. *Euphagus cyanocephalus.* Regular summer resident (west).[*]

Common Grackle. *Quiscalus quiscula.* Common summer resident.[*]

Great-tailed Grackle. *Quiscalus mexicanus.* Regular summer resident (south).[*]

Brown-headed Cowbird. *Molothrus ater.* Common summer resident.[*]

Orchard Oriole. *Icterus spurius.* Common summer resident.[*]

Baltimore Oriole. *Icterus galbula.* Common summer resident (east).[*]

Bullock's Oriole. *Icterus bullockii.* Common summer resident (west).[*]

FAMILY FRINGILLIDAE — FINCHES

Gray-crowned Rosy-finch. *Leucosticte tephrocotis.* Casual migrant (northwest).

Purple Finch. *Carpodacus purpureus.* Regular wintering migrant.

Cassin's Finch. *Carpodacus cassinii.* Regular wintering migrant.

House Finch. *Carpodacus mexicanus.* Common resident.[*]

Red Crossbill. *Loxia curvirostra.* Regular winter migrant, rare resident (northwest).[*]

White-winged Crossbill. *Loxia leucoptera.* Casual wintering migrant.

Common Redpoll. *Carduelis flammea.* Regular wintering migrant.

Pine Siskin. *Carduelis pinus.* Regularly wintering migrant, irregular resident.[*]

American Goldfinch. *Carduelis tristis.* Common resident.[*]

Evening Grosbeak. *Coccothraustes vespertinus.* Regular wintering migrant (west).

FAMILY PASSERIDAE — OLD WORLD SPARROWS

House Sparrow. *Passer domesticus.* Common widespread resident.[*]

Amphibians and Reptiles

This list is based primarily on Lynch (1985). Extended species descriptions, identification keys, and (often badly outdated) range maps may also be found in Hudson (1958). All Nebraska species of herpetiles are included, and all are illustrated in the field guide by Conant (1998), whose common names are preferentially used here.

ORDER CAUDATA — SALAMANDERS

Small-mouth Salamander. *Ambystoma texanum.* Southeast, rare in ponds.

Tiger Salamander. *Ambystoma tigrinum.* Widespread, common in shallow ponds.

ORDER ANURA — FROGS AND TOADS

Northern Cricket Frog. *Acris crepitans.* East and central, common in streams and ponds.

American Toad. *Bufo americanus.* Missouri Valley, local and probably rare, in grasslands.

Great Plains Toad. *Bufo cognatus.* Widespread, fairly common, in grasslands.

Woodhouse's (Rocky Mountain) Toad. *Bufo woodhousei.* Statewide, common in grasslands and forest edges.

Great Plains Narrowmouth Toad. *Gastrophryne olivacea.* Southeast, local and rare in semiwooded pastures.

Gray Treefrog. *Hyla chrysocelis.* Southeast only, wooded and temporary ponds.

Western (Striped) Chorus Frog. *Pseudacris triseriata.* Widespread, common in ditches and marshes.

Plains Leopard Frog. *Rana blairi.* Eastern half and southwest, in permanent waters.

Bull Frog. *Rana catesbiana.* Widespread in aquatic habitats.

Northern Leopard Frog. *Rana pipiens.* Widespread from Platte north, in sandy streams and marshes.

Plains Spadefoot Toad. *Spea bombifrons.* Widespread, especially in sandy or loess areas.

ORDER CHELONIA — TURTLES

Snapping Turtle. *Chelydra serpentina.* Widespread in permanent water habitats.

Painted Turtle. *Chrysemys picta.* Widespread in all aquatic habitats.

Blanding's Turtle. *Emydoidea blandingi.* From Platte north, mostly limited to Sandhills marshes.

False Map Turtle. *Graptemys pseudogeographica.* Missouri River and adjoining oxbows.

Yellow Mud Turtle. *Kinosternum flavescens.* Republican River and Sandhills ponds and lakes.

Ornate Box Turtle. *Terrapena ornata.* Widespread, especially Sandhills and mixed-grass prairies.

Smooth Softshell. *Trionyx muticus.* Larger eastern rivers.

Spiny Softshell. *Trionyx spiniferus.* Widespread in larger rivers and reservoirs.

ORDER LACERTILIA — LIZARDS

Six-lined Racerunner. *Cnemidophorus sexlineatus.* Widespread, especially in Sandhills.

Five-lined Skink. *Eumeces fasciatus.* Rare and local, southeast, in deciduous woods.

Many-lined Skink. *Eumeces multivirgatus.* Sandhills and Panhandle, Sandhills and sand sagebrush prairies.

Great Plains Skink. *Eumeces obsoletus.* Extreme southern counties along Republican River.

Prairie Skink. *Eumeces septentrionalis.* Eastern third, in tallgrass prairies.

Lesser Earless Lizard. *Holbrookia maculata.* Widespread in sandy areas.

Slender Glass Lizard. *Ophisaurus attenuatus.* Possibly now extirpated, recorded in Franklin and Johnson Counties.

Short-horned Horned Toad. *Phrynosoma douglasii.* Panhandle only, in dry plains.

Sagebrush Lizard. *Sceloperus graciosus.* Morrill County only, along Platte River.

Northern Prairie Lizard. *Sceloperus undulatus.* Mainly in west, common in Sandhills.

ORDER SERPENTES — SNAKES

Copperhead Snake. *Agkristodon contortrix.* Extreme southeast (Gage and Richardson Counties) only; venomous.

Glossy (Faded) Snake. *Arizona elegans.* Dundy County only.

Eastern Worm Snake. *Carphophis amoenus.* Lower Platte and lower Missouri Valleys, in forests.

Blue (Green) Racer. *Coluber constrictor.* Widespread in many habitats.

Timber Rattlesnake. *Crotalus horridus.* Southeast (Gage and Richardson Counties) only, in wooded habitats; venomous.

Prairie Rattlesnake. *Crotalus viridis.* West and central, in rocky grasslands; venomous.

Ringneck Snake. *Diadophis punctatus.* East and central, in deciduous woods.

Corn snake. *Elaphe guttata.* Republican Valley, west to about Harlan County.

Black (Pilot) Rat Snake. *Elaphe obsoleta.* East and southeast, in deciduous forests.

Fox Snake. *Elaphe vulpina.* Northeastern and central eastern, mainly near streams.

Western Hognose Snake. *Heterodon nasicus.* West and central, especially in sandy areas.

Prairie Kingsnake. *Lampropeltis calligaster.* Southeast, from Omaha and Lincoln southward.

Eastern (Common) Kingsnake. *Lampropeltis getulus.* Southeast only (rare), in Lancaster, Gage and Pawnee Counties.

Milk Snake. *Lampropeltis triangulum.* Widespread, in varied habitats.

Western Coachwhip Snake. *Masticophis flagellum.* Southwest only, along Republican drainage.

Common (Northern) Watersnake. *Nerodia sipedon.* Widespread, in marshes, streams, and rivers.

Smooth Green Snake. *Opheodrys vernalis.* Rare; eastern two-thirds in wet meadows; probably near extirpation.

Bullsnake. *Pituophis catenifer.* Statewide, common in varied habitats.

Graham's Watersnake. *Regina grahamii.* Eastern and southeastern; rare in streams and lakes.

Massasauga Rattlesnake. *Sistrurus catenatus*. Rare; now known only from Colfax and Pawnee Counties; venomous; state threatened.

Brown Snake. *Storeria dekayi*. Southeastern, in moist woods.

Red-bellied Snake. *Stoereia occipitomaculata*. Extremely rare; two specimens known from central Missouri River and central Platte River.

Plains Black-headed Snake. *Tantilla nigriceps*. Southwest only, in dry grasslands; now of questionable Nebraska occurrence, perhaps extirpated.

Wandering Garter Snake. *Thamnophis elegans*. Extreme northwest, diverse habitats.

Western Ribbon Snake. *Thamnophis proximus*. East only (Salt Creek, Platte, and Missouri Rivers).

Plains Garter Snake. *Thamnophis radix*. Statewide, very common in diverse habitats.

Common (Red-sided) Garter Snake. *Thamnophis sirtalis parietalis*. Statewide, common along all watercourses.

Lined Snake. *Tropicoclonion lineatum*. East-central and southeast, varied habitats.

Native Fishes

This list is based mainly on Jones (1963) but excludes introductions and some uncertain records. The sequence used here is that of generally accepted taxonomic order. Platte River status in part is based on information from John Lynch (pers. comm.) or Lynch and Roh (1996).

[a] indicates species illustrated in color by Tomelleri and Eberle (1990).

FAMILY PETROMYZONTIDAE — LAMPREYS

Chestnut Lamprey. *Ichthyomyzon castaneus*. Missouri River.[a]

FAMILY ACIPENSERIDAE — STURGEONS

Lake Sturgeon. *Acipenser fulvescens*. Missouri and lower Platte Rivers; state threatened.[a]

Pallid Sturgeon. *Scaphirhynchus albus*. Missouri and lower Platte Rivers; nationally endangered.[a]

Shovelnose Sturgeon. *Scaphirhynchus platyrhynchus*. Missouri and lower Platte Rivers.[a]

FAMILY POLYDONTIDAE — PADDLEFISH

Paddlefish. *Polydon spathula*. Lower Platte (to Fremont) and Missouri Rivers.[a]

FAMILY LEPISOSTEIDAE — GARS

Longnose Gar. *Lepidosteus osseus*. Missouri and lower Platte (west to Kearney) Rivers.[a]

Shortnose Gar. *Lepidosteus playtostomus*. Missouri, lower Platte (west to Columbus), and Elkhorn Rivers.[a]

FAMILY AMIIDAE — BOWFINS
Bowfin. *Amia calva*. Missouri River.[a]

FAMILY CLUPEIDAE — HERRINGS
Gizzard Shad. *Dorosoma cepidanum*. Widespread.[a]

FAMILY SALMONIDAE — TROUT
Cutthroat Trout. *Salmo clarki*. Niobrara tributaries.[a]

FAMILY HIODONTIDAE — MOONEYES
Goldeye. *Hiodon alosoides*. Missouri, lower Platte (west to Kearney), and Elkhorn Rivers.[a]

FAMILY ESOCIDAE — PIKES
Grass Pickerel. *Esox vermiculatus*. Niobrara, northern streams.[a]
Northern Pike. *Esox lucius*. Widespread.[a]

FAMILY CYPRINIDAE — MINNOWS
Stoneroller. *Campostoma anomalum*. Widespread.[a]
Northern Redbelly Dace. *Chrosomus eos*. Rare and local, mainly in Sandhills; state threatened.[a]
Finescale Dace. *Chrosomus neogaeus*. Rare and local, mainly in Sandhills; state threatened.[a]
Brassy Minnow. *Hybognathus hankinsoni*. Widespread.
Western Silvery Minnow. *Hybognathus argyritis*. Widespread, Platte River to North Platte.
Speckled Chub. *Hybopsis aestivalis*. Platte (west to Kearney) and Republican Rivers, smaller eastern streams.
Hornyhead Chub. *Hybopsis biguttatus*. Elkhorn River, Papillion and Lodgepole Creeks
Sturgeon Chub. *Hybopsis gelida*. Platte River west to North Platte, other larger rivers.
Flathead Chub. *Hybopsis gracilis*. Widespread.[a]
Sicklefin Chub. *Hybopsis meeki*. Missouri River.
Silver Chub. *Hybopsis storeriana*. Eastern streams, Platte River west to Grand Island.
Sturgeon Chub. *Macrhybopsis gelada*. Very rare, lower Missouri and lower Platte Rivers; state endangered.
Pearl Dace. *Margariscus margarita*. Rare; recommended for removal from threatened category.
Golden Shiner. *Notemigonus crysoleucus*. Widespread but local.[a]
Emerald Shiner. *Notropis atherinoides*. Eastern streams; Platte River to Grand Island.
River Shiner. *Notropis blennis*. Platte (west to Keystone); Sandhills drainages.
Common Shiner. *Notropis cornutus*. Elkhorn River, also local in western Nebraska.[a]
Bigmouth Shiner. *Notropis dorsalis*. Widespread.

Blacknose Shiner. *Notropis heterolepis*. Niobrara and Sandhills streams (Bush and Holt Creeks); state endangered.

Red Shiner. *Notropis lutrensis*. Widespread.[a]

Plains Shiner. *Notropis percobromus*. Lower Platte and Elkhorn Rivers.

Sand Shiner. *Notropis stramineus*. Widespread.[a]

Topeka Shiner. *Notropis topeka*. Rare and local, eastern streams; nationally endangered.

Silverband Shiner. *Notropis illecebrosus*. Northern and eastern streams.

Suckermouth Shiner. *Phenacobius mirabilis*. Platte River and southward.[a]

Bluntnose Minnow. *Pimephales notatus*. Elkhorn River.[a]

Blacknose Dace. *Rhinichthys atratulus*. Niobrara tributaries.

Longnose Dace. *Rhinichthys cataractae*. North Platte River (east to North Platte), Niobrara River, Sandhills streams.

Pearl Dace. *Semotilus margarita*. Local and rare.

Creek Chub. *Semotilus atromculatus*. Widespread.[a]

FAMILY CATOSTOMIDAE — SUCKERS

River Carpsucker. *Carpoides carpio*. Widespread.[a]

Quillback. *Carpoides cyprinus*. Missouri, Platte, and Republican Rivers.[a]

Plains Carpsucker. *Carpoides forbesi*. Widespread, east and central streams.

Highfin Carpsucker. *Carpoides velifer*. Niobrara, Missouri, Platte, Loup, and Big Nemaha Rivers.[a]

Longnose Sucker. *Catastomus catostomus*. Western streams.[a]

White Sucker. *Catostomus commersoni*. Widespread.[a]

Blue Sucker. *Cycleptus elongatus*. Missouri River.[a]

Smallmouth Buffalo. *Ictiobus bubalus*. Eastern rivers.[a]

Bigmouth Buffalo. *Ictiobus cyrpinellus*. Eastern streams.[a]

Black Buffalo. *Ictiobus niger*. Local, north and east.[a]

Golden Redhorse. *Moxostoma erythurum*. Northern and eastern streams.[a]

Northern (Shorthead) Redhorse. *Moxostoma macrolepidotum*. Widespread.[a]

Mountain Sucker. *Pantosteus platyrhynchus*. Chadron and Hat Creeks.[a]

FAMILY ICTALURIDAE — FRESHWATER CATFISH

Black Bullhead. *Ictalurus melas*. Widespread.[a]

Yellow Bullhead. *Ictalurus natalis*. Eastern streams.[a]

Brown Bullhead. *Ictalurus nebulosus*. Lower Platte River.[a]

Blue Catfish. *Ictalurus furcatus*. Missouri River.[a]

Channel Catfish. *Ictalurus punctatus*. Widespread.[a]

Flathead Catfish. *Pylodictus olivaria*. Eastern streams.[a]

Stonecat. *Noturus flavus*. Widespread.[a]

Tadpole Madtom. *Noturus gyrinus*. Platte River and tributaries.

FAMILY ANGUILLIDAE — FRESHWATER EELS

American Eel. *Anguilla rostrata*. Missouri River.[a]

FAMILY CYPRINODONTIDAE — KILLIFISH
Plains Topminnow. *Fundulus sciadicus.* Widespread from Platte River northward.
Plains Killifish. *Fundulus kansae.* Platte and Republican drainages.[a]

FAMILY GADIDAE — CODFISH
Burbot. *Lota lota.* Niobrara, Missouri, and lower Platte Rivers.[a]

FAMILY GASTEROSTEIDAE — STICKLEBACKS
Brook Stickleback. *Eucalis inconstans.* Sandhills and lower Platte River.[a]

FAMILY PERCOPSIDAE — TROUT-PERCHES
Trout-perch. *Percopsis omiscomaycus.* Missouri River.[a]

FAMILY SERRANIDAE (PERICHTHYIIDAE) — TEMPERATE BASSES
White Bass. *Roccus (Morone) chrysops.* Larger streams, widespread.[a]

FAMILY CENTRARCHIDAE — SUNFISHES
Rock Bass. *Ambloplites rupestris.* Widespread.[a]
Largemouth Bass. *Micropterus salmoides.* Widespread.[a]
Green Sunfish. *Lepomis cyanellus.* Widespread.[a]
Pumpkinseed. *Lepomis gibbosis.* Local and scattered.[a]
Orangespotted Sunfish. *Lepomis humilis.* Widespread.[a]
Bluegill. *Lepomis macrochirus.* Widespread.[a]
White Crappie. *Pomoxis annularis.* Widespread.[a]
Black Crappie. *Pomoxis nigromaculatus.* Widespread.[a]

FAMILY PERCIDAE — PERCHES
Iowa Darter. *Etheostoma exile.* Local and scattered.
Johnny Darter. *Etheostoma nigrum.* Eastern streams.[a]
Orangethroat Darter. *Etheostoma spectabile.* North Platte and Republican Rivers.[a]
Yellow Perch. *Perca flavescens.* Widespread but rather local.[a]
Sauger. *Stizostedion canadensis.* Missouri, Elkhorn, and lower Platte Rivers.[a]
Walleye. *Stizostedion vitreum.* Widespread.[a]

FAMILY SCIAENIDAE — FRESHWATER DRUMS
Freshwater Drum. *Aplodinotus grunniens.* Missouri, Platte, and Republican drainages.[a]

Representative Invertebrates

Tiger Beetles

The grouping that follows is related to occurrence in saline wetlands and distinguishes four species that are confined to such wetlands in Nebraska. All the tiger beetles that are known to occur in Nebraska were described by Carter (1989), who provided identification keys and illustrations for these species. The species sequence within each of the groups is alphabetical by generic and specific names.

FAMILY CICINDELIDAE — TIGER BEETLES

Amblychila cylindriformis. Northern; nocturnal, in ditches and gullies.

Cicindela celepes. Eastern; along rivers.

Cicindela cuprascens. Missouri Valley, eastern and western plains; sandy sites near water.

Cicindela curcumpicta johnsoni. Uncommon on saline flats; eastern and western plains.[a]

Cicindela cursitans. Missouri Valley, eastern and western plains; moist and saline ditches.

Cicindela denverensis. Western plains and Pine Ridge; banks and canyons.

Cicindela duodecimguttata. Widespread but uncommon along creeks.

Cicindela formosa. Widely distributed; eroded bands and dry sand.

Cicindela fulgida. Saline flats and creek banks; Lancaster County, Sandhills, and western plains.

Cicindela hirticollis. Throughout state; moist, sandy areas.

Cicindela lengi. Sandhills and Pine Ridge; sandy and eroded areas.

Cicindela lepida. Missouri Valley; Sandhills; dry, sandy areas.

Cicindela limbalis. Eastern half; loess and clay banks.

Cicindela limmbata. Sandhills; sandy habitats.

Cicindela longilabrus. Lancaster County records only.

Cicindela macra. Missouri Valley, eastern and western plains; sandy sites near water.

Cicindela nebraskana. Northwestern; dirt roads in grasslands.

Cicindela nevadica. Lancaster County and western plains (two subspecies); near water.

Cicindela nevadica lincolniana. Salt Creek Tiger Beetle. Limited to Lancaster County along Salt Creek and Little Salt Creek; nationally endangered subspecies.

Cicindela pulcra. Southwestern; possibly extirpated.

Cicindela punctulata. Most common tiger beetle in Nebraska; all habitats, often in cities.

Cicindela purpurea. Western plains and Pine Ridge; clay banks, dirt roads, and paths.

Cicindela repanda. Very common along creeks and rivers.

Cicindela scutellaris. Widespread in sandy areas; two subspecies.

Cicindela sexguttata. Usually in deciduous woods; eastern and northern.

Cicindela splendida. Fairly common in clay or loess banks; widespread.

Cicindela terricola. Platte Valley, western plains, Pine Ridge; near saline or fresh streams.

Cicindela togata. Common and widespread on saline flats; eastern plains.

Cicindela tranquebarica. Statewide, open sunny areas and sandy or saline sites.

Megacephala virginica. Mainly in southeast; nocturnal.

Carrion Beetles

This list is based on Ratcliffe (1995), who provides color plates of all 18 species.

FAMILY SILPHIDAE — CARRION BEETLES
Subfamily Silphinae—Typical Carrion Beetles
Heterosilpha ramosa. Records from 12 western counties.
Necrodes surinamensis. Records from eastern three-quarters; mainly in wooded areas.
Necrophila americana. Probably statewide, mainly in marshy and forested areas.
Oiceoptoma inequale. Eastern two-thirds, in deciduous forests.
Oicoptoma novaboracense. Statewide except for southwest in prairies and forests.
Thanatophilis lapponicus. Probably statewide, more common in northwest in open areas.
Thanatophilus truncatus. Southern half.
Subfamily Nicrophorinae—Burying Beetles
Nicrophorus americanus. American Burying Beetle. Sandhills and northern part of state, in prairies, forest edge, and scrubland; nationally endangered.
Nicrophorus carolinus. Western half.
Nicrophorus guttula. Mainly in western half.
Nicrophorus hybridus. Northern.
Nicrophorus investigator. Known only from North Platte.
Nicrophorus marginatus. Widespread.
Nicrophorus maxicanus. Known only from Custer County.
Nicrophorus obscurus. Widespread.
Nicrophorus orbicollis. Widespread; state's most abundant burying beetle.
Nicrophorus pustulatus. Eastern two-thirds.
Nicrophorus tomentosus. Widespread.

Common Scarab Beetles

Only the more common species are listed here; see Ratcliffe's (1991) monograph, on which this list is based, for a complete state listing, illustrations, and identification keys.

FAMILY SCARABAEIDAE — SCARAB
OR LAMELLICORN BEETLES
Subfamily Aphodiinae—Aphodiine Dung Beetles
Aphodius distinctus. Widespread.
Aphodius fimetarius. Widespread and abundant.

Aphodius stercorosus. Eastern two-thirds.

Ataenius strigatus. Mainly eastern.

Subfamily Geotrupinae—Earth-boring Dung Beetles

Geotropes splendidus. Eastern half.

Bolboceras filicornis. Probably statewide.

Bolbocerosoma bruneri. Widespread.

Eucanthus lazarus. Widespread.

Subfamily Scarabaeinae—Dung Beetles and Tumblebugs

Canthon ebenus. Tumblebug. Western half.

Canthon pilularis. Tumblebug. Widespread.

Melanocanthon nigricornis. Mainly western half.

Ateuchus histeroides. Eastern and southeastern.

Copris fricator. Widespread but more common in east.

Phanaeus vindex. Widespread.

Onthophagus hecate. Widespread and abundant.

Onthophagus knausi. Northern.

Onthophagus pennsylvanicus. Widespread.

Subfamily Ochodaeinae—Ochodaeine Scarabs

Ochodaeus musculus. Widespread in sandy areas.

Subfamily Troginae—Skin Beetles

Trox scaber. Probably widespread.

Trox unistriatus. Widespread.

Omorgus suberosus. Widespread and abundant.

Subfamily Melolonthinae—May or June Beetles

Serica sericea. Widespread.

Diplotaxis blanchardi. Widespread.

Diplotaxis rudis. Western.

Phyllophaga anxia. Widespread.

Phyllophaga corrosa. Widespread.

Phyllophaga crassimima. Widespread and abundant.

Phyllophaga fusca. Widespread.

Phyllophaga implicita. Widespread.

Phyllophaga lanceolata. Widespread, more common in west.

Phyllophaga prunina. Mainly western.

Phyllophaga rugosa. Widespread.

Polyphylla decemlineata. Ten-lined June beetle. Widespread except in southeast.

Polyphylla hammondi. Widespread.

Dichelonyx subvittata. Eastern half.

Macrodactylus subspinosa. Rose chafer. Widespread.

Hoplia laticollis. Grape-vine beetle. Widespread, more common in east.

Subfamily Rutelinae—Shining Leaf Chafers

Pelidnota punctata. Grapevine beetle. Widespread, more common in east.

Cotalpa lanigera. Gold (Goldsmith) beetle. Widespread.

Anomala binotata. Eastern and central.

Anomala flavipennis. Mainly eastern.

Anomala innuba. Probably mainly eastern.

Strigoderma arboricola. Widespread.

Subfamily Dynastinae—Rhinoceros Beetles

Cyclocephala lurida. Widespread, abundant in east.

Cyclocephala pasadenae. Widespread.

Ligyrus gibbosus. Carrot beetle. Widespread and abundant.

Ligyris relictus. Widespread and common.

Subfamily Cetoniinae—Flower Scarabs

Euphoria fulgida. Green flower beetle. Widespread.

Euphoria inda. Bumblebee flower beetle (Brown fruit chafer). Widespread.

Euphoria kerni. Southwestern.

Subfamily Trichinae—Trichine Scarabs

Osmoderma subplanata. Mainly southeastern.

Trichiotinus piger. Eastern quarter, in wooded areas.

Grasshoppers, Mantids, and Walking-sticks

The grasshopper list is based mainly on Hagen and Rabe (1991) and excludes most rarer species. Hauke (1953) provided a complete listing of the state's approximately 130 species of grasshoppers. Common names of grasshoppers are mostly after Pfadt (1994). Otte (1981, 1984) provides identification criteria, range maps, and color illustrations. The mantid and walking-stick lists are based on Hauke (1949) and should be complete for the state. Illustrations for some of these species may also be found in Milne and Milne (1981). Identification of mantids and walking-sticks that are also found in Colorado is provided by Alexander (1941).

[b] indicates species illustrated in Pfadt (1994).

[c] indicates species illustrated in Otte (1981).

[d] indicates species illustrated in Otte (1984).

[e] indicates species illustrated in Milne and Milne (1981).

FAMILY ACRIDIDAE — SHORT-HORNED GRASSHOPPERS

Acrolophitus hirtipes. Widespread, mainly in west, shortgrass prairies.[c]

Aeolophus turnbulli. Mainly in Panhandle.

Aeropedellus clavatus. Club-horned Grasshopper. Mainly west and high plains and ridges.[cb]

Ageneotettetix deorum. White-whiskered Grasshopper. Widespread, sandy and dry grasslands.[bc]

Amphitornus coloradus. Striped Grasshopper. Mainly west and central, dry grasslands.[b]

Arphia conspersa. Speckle-winged Grasshopper. Mainly west and central, shortgrass prairies.[bc]

Arphia pseudonietana. Widespread, open prairies and brush or forest openings.[d]

Arphia simplex. Mormon cricket. Central and east, local.[bd]

Arphia xanthoptera. East and central, open woodlands, grassy fields, woodland edges.[d]

Aulocara eliotti. Big-headed Grasshopper. Mainly west and central, shorgrass prairies.[bc]

Aulocara femoratum. Mainly west and central, shortgrass prairies.[c]

Boopedon nubilum. West and central, shortgrass prairies.[c]

Brachystola magna. Lubber. Widespread, mainly west and central.[e]

Camnula pellucida. Clear-wing Grasshopper. West and central, grassy pastures.[bd]

Campylacantha olivacea. Mainly Sandhills and sandsage.

Chloealtis conspersa. Widespread, dry upland woods.[c]

Chorthippus curtipennis. Meadow Grasshopper. Widespread; tall, moist grasses.[bc]

Chortophaga viridifasciata. Green-striped Grasshopper. Widespread, shorter grasses.[bd]

Circotettix rabula. Panhandle, rocky hillsides, eroded hills.[d]

Cordillacris occipitalis. Spotted-winged Grasshopper. Mainly west and central, thin grasslands.[bc]

Cratypedes neglectus. Southwestern, sageshrub and dry fields.[d]

Dactylotum bicolor pictum. Painted (Rainbow) Grasshopper. Central and western.[e]

Derotmema haydeni. Hayden Grasshopper. Mainly west and southwest, grasslands with bare ground.[bd]

Dichromorpha viridis. Scattered records, mainly eastern; lower grasses, shady woods.[c]

Dissoteira carolina. Roadside Grasshopper. Widespread and common, open roadsides.[de]

Encoptolophus costalis. Dusky Grasshopper. Widespread but local, prairies and open grassland.[bd]

Eritettix simplex. Velvet-striped Grasshopper. Widespread, mainly west and central; medium to short grasses.[bc]

Hadrotettix trifasciatus. Three-banded Grasshopper. Mainly west and central, gravelly grasslands.[bde]

Hesperotettix speciosus. Sunflower Grasshopper. Common.

Hesperotettix viridis. Snakeweed Grasshopper. Mainly west and central.[b]

Hippiscus oceolote and *H. rugosus*. Widespread, pastures and weedy prairies.[d]

Hypochlora alba. Cudweed Grasshopper. Widespread, mainly west and central.[b]

Melanoplus augustipennis. Narrow-winged Sand Grasshopper. Widespread, mainly central.[b]

Melanoplus bivittatus. Two-striped Grasshopper. Widespread and very common.[b]

Melanoplus bowditchi. Sagebrush Grasshopper. Western half.[b]

Melanoplus confusus. Pasture Grasshopper. Widespread.[b]

Melanoplus dawsoni. Dawson Grasshopper. Widespread.[b]

Melanoplus differentialis. Differential Grasshopper. Widespread and extremely common.[be]

Melanoplus femurrubrum. Red-legged Grasshopper. Widespread and very common.[be]

Melanoplus gladstoni. Gladston Grasshopper. Western half.[b]

Melanoplus infantilis. Little Spur-throated Grasshopper. Western half.[b]

Melanoplus keeleri. Keeler Grasshopper. Widespread.[b]

Melanoplus lakinus. Western half.

Melanoplus mexicanus. Lesser Migratory Grasshopper (Rocky Mountain Locust). Very common.

Melanoplus occidentalis. Flabellate Grasshopper. Panhandle and southwest.[b]

Melanoplus packardi. Packard Grasshopper. Mainly western half.[b]

Melanoplus sanguinipes. Migratory Grasshopper. Widespread.[b]

Mermiria bivittata. Two-striped Slant-faced Grasshopper. Widespread, taller grasses.[bc]

Metator pardalinus. Blue-legged Grasshopper. Western half, shortgrass prairies.[bd]

Opeia obscura. Obscure Grasshopper. Widespread, shortgrass prairies.[bc]

Orphulella speciosa. Widespread and common, especially westwardly; dry grasslands.[c]

Paradalophora haldemai. Haldeman's Grasshopper. Mainly western half, pastures and weedy or sandy prairies.[d]

Paropomela wyomingensis. Western half, shortgrass prairies.[c]

Phlibostroma quadrimaculatum. Four-spotted Grasshopper. Western two-thirds, shortgrass prairies and eroded areas.[bc]

Phoetaliotes nebrascensis. Large-headed Grasshopper. Widespread.[b]

Pseudopomela brachyptera. Widespread but local, taller prairies.[c]

Psoloessa delicatula. Brown-spotted Grasshopper. Panhandle, local eastwardly, shortgrass prairies.[bc]

Psoloessa texana. Mainly western half, bare open ground.[c]

Schistocerca alutacea. Alutacea Bird Grasshopper. Mainly western half.[e]

Shistocerca americana. American Bird Grasshopper. Southeastern quarter.[e]

Spharegemon bolli. Panhandle, open woodlands and sandy fields.[d]

Spharegemon campestris. Panhandle.[d]

Spharagemon collare. Mottled Sand Grasshopper. Widespread, rangeland and sandy fields.[bd]

Spharagemon equale. Orange-legged Grasshopper. Western half, prairies with bare ground.[bd]

Syrbula admirabilis. Mainly eastern half, dry uplands.[c]

Trachyrhachys aspera. Southwestern Panhandle, usually in highlands.[d]

Trachyrhachys kiowa. Kiowa Grasshopper. Widespread, bare gravelly ground.[bd]

Trimerotropis agrrestis. Central and western, in sand dunes.[d]

Trimerotropis fratercula. Panhandle, slopes and eroded banks.[d]

Trimerotropis latifasciata. Mainly Panhandle; open, dry grasslands.[d]

Trimerotropis pallidipennis. Mainly Panhandle, dry grasslands and low prairies.[d]

Trimerotropis pistrinaria. Panhandle; open, eroded, and stony habitats.[d]

Trimerotropis sparsa. Northern Panhandle, alkaline flats and eroded hillsides.[d]

Xanthippus corallipes. Red-shanked Grasshopper. West and central, open dry prairies[bd]

FAMILY MANTIDAE — MANTIDS

Litaneutria minor. Agile ground mantid. Western half.

Oligonicella scudderi. South and east of Lincoln.

Stagmomantis carolina. Carolina Mantid. Mainly south and east of Lincoln.[e]

FAMILY PHASMIDA — WALKING-STICKS

Anisomorpha ferruginea. Southeastern corner.

Diapheromera femorata. Northern Walking-stick. Easternmost edge of state.[e]

Diapheromera velei. Prairie Walking-stick. Common and widespread.

Parabacillus coloradus. Western and northern.

Butterflies

This list is based mainly on that of Dankert et al. (1993). Common English names are those of Pyle (1981). "Records" refers to counties with accepted specimen records. Organized by accepted taxonomic sequence.

[f] indicates species illustrated in color
by Heitzman and Heitzman (1987).

FAMILY HESPERIIDAE — SKIPPERS

Eparygyreus clarus. Silver-spotted Skipper. Widespread.[f]

Achalarus lyciades. Hoary Edge. Southeastern records only.[f]

Thorybes bathyllus. Southern Cloudywing. Mostly eastern records.[f]

Thorybes pylades. Northern Cloudywing. Scattered records across state.[f]

Staphylus hayhurstii. Scalloped Cloudywing. Mostly eastern records.[f]

Erynnis brizo. Sleepy Duskywing. Southeastern records only.[f]

Erynnis juvenalis. Juvenal's Duskywing. Mostly eastern records.[f]

Erynnis horatius. Horace's Duskywing. Eastern records.[f]
Erynnis martialis. Mottled Duskywing. Six scattered records.[f]
Erynnis funeralis. Funereal Duskywing. Mostly eastern records.[f]
Erynnis baptisiae. Wild Indigo Duskywing. Scattered records across state.[f]
Erynnis afranius. Afranius Duskywing. Panhandle records only.
Erynnis persuis. Persius Duskywing. Panhandle records only.[f]
Pyrgus scriptura. Small Checkered Skipper. Panhandle records only.
Pyrgus communis. Common Checkered Skipper. Widespread.[f]
Pholisora catulls. Common Sootywing. Widespread.[f]
Ancyloxypha numitor. Least Skipperling. Probably extends across state.[f]
Oarisma garitra. Garita Skipperling. Panhandle records only.
Thymelicus lineola. European Skipper. One east-central record.[f]
Hylephila phyleus. Fiery Skipper. Mostly eastern records.[f]
Yvretta rhesus. Plains Gray Skipper. Panhandle records only.
Hesperia uncas. Uncas Skipper. Scattered records across state.
Hesperia comma. Common Branded Skipper. Records for western half of state.
Hesperia ottoe. Ottoe Skipper. Probably extends across state.
Hesperia leonardus pawnee. Pawnee Skipper. Probably extends across state.[f]
Hesperia pahaska. Pahaska Skipper. Panhandle records only.
Hesperia metea licinus. Cobweb Skipper. One south-central record.[f]
Hesperia viridus. Green Skipper. Panhandle records only.[f]
Hesperia attalus. Dotted Skipper. One eastern record.[f]
Polites peckius. Yellowpatch Skipper. Widespread.
Polites sabuleti. Sandhill Skipper. Panhandle record only.
Polites themistocles. Tawny-edged Skipper. Widespread.
Polites origenes. Crossline Skipper. Probably extends across state.
Polites mystic. Long Dash. Scattered records across state.
Wallengrenia egeremeti. Northern Broken Dash. Mostly eastern records.[f]
Pompeius verna. Little Glasswing. Mostly eastern records.[f]
Atalopedes campestris. Sachem. Widespread.[f]
Atrytone arogos. Beard-grass (Arogos) Skipper. Probably extends across state.[f]
Atrytone logan. Delaware Skipper. Widespread.[f]
Poanes hobomok. Hobomok Skipper. Probably extends across state.[f]
Poanes zabulon. Zabulon Skipper. Only eastern records.[f]
Poanes taxiles. Golden Skipper. Records for northern counties.
Poanes viator. Broad-winged Skipper. Scattered records across state.[f]
Euphyes dion. Sedge Skipper. Only eastern records.[f]
Euphyes conspicuus. Black Dash. Mostly eastern records.
Euphyes bimacula. Two-spotted Skipper. Scattered records across state.
Euphyes vestris. Dun Skipper. Probably extends across state.
Artytonopsis hianna. Dusted Skipper. Scattered records across state.[f]

Amblyscirtes simius. Orange Roadside Skipper. One Panhandle record.

Amblyscirtes oslari. Oslar's Roadside Skipper. Scattered records across state.

Amblyscrites nysa. Mottled Roadside Skipper. One southeastern record.[f]

Amblyscrites vialis. Roadside Skipper. Probably extends across state.[f]

Lerodea eufala. Eufala Skipper. Mostly eastern records.[f]

Calpodes ethlius. Brazilian Skipper. One eastern record.[f]

Megathymus coloradensis. Colorado Giant Skipper. Only southwestern records.

Megathymus streckeri. Strecker's Giant Skipper. Records for western half of state.

FAMILY PAPILIONIDAE — SWALLOWTAILS

Parnassius phoebus—Phoebus Parnassian. Panhandle records only.

Battus philenor. Pipevine (Blue) Swallowtail. Mostly eastern records.[f]

Eurytides marcellus. Zebra Swallowtail. Mostly eastern records.[f]

Papilio polyxenes asterius. Eastern Black Swallowtail. Probably extends across state.[f]

Papilio bairdii. Western Black Swallowtail. West and southwestern records.

Papilio zelicaon. Anise Swallowtail. Panhandle records only.

Papilio indra. Short-tailed Black Swallowtail. Panhandle records only.

Heraclides cresphontes. Giant Swallowtail. Mostly eastern records.[f]

Pterourus glaucus. Tiger Swallowtail. Widespread.[f]

Pterourus rutulus. Western Tiger Swallowtail. One south-central record.

Pterourus multicaudatus. Two-tailed Tiger Swallowtail. Mainly western half.

Pterourus troilus. Spicebush Swallowtail. Two southeastern records.[f]

Pterourus palamedes. Palamedes Swallowtail. Mostly eastern records.[f]

FAMILY PIERIDAE — SULPHURS, ORANGE-TIPS, AND WHITES

Neophasia menapia. Pine White. Panhandle records only.[f]

Appias drusilla. Florida White. One eastern record.

Pontia sisymbrii. Spring (California) White. Panhandle records only.

Pontia protodice. Checkered White. Widespread.[f]

Pontia occidentalis. Western White. Panhandle records only.

Pieris rapae. Imported Cabbage White. Widespread.[f]

Euchloe ausonides. Creamy Marblewing. Panhandle records only.

Euchloe olympia. Olympia Marblewing. Probably extends across state.[f]

Anthocharis midea. Falcate Orange-tip. Three eastern records.[f]

Colias philodice. Common Sulfur. Widespread.[f]

Colias euytheme. Orange Sulphur (Alfalfa Butterfly). Widespread.[f]

Colias alexandra. Queen Alexandra's Sulphur. Panhandle records only.

Zerena cesonia. Dogface. Probably extends across state.[f]

Anteos maerula. Yellow Angled Sulfur. One eastern record.

Phoebis sennae. Cloudless Giant Sulfur. Mostly eastern records.[f]

Phoebis philea. Orange-barred Giant Sulfur. One eastern record.[f]

Phoebis agarithe. Orange Giant Sulfur. Mostly eastern records.[f]

Kricogonia lyside. Lyside. Five southern records.[f]

Eurema daira. Fairy Yellow. One southeastern record.[f]

Eurema mexicanum. Mexican Yellow. Probably extends across state.[f]

Eurema proterpia. Tailed Orange. One eastern record.

Eurema lisa. Little Yellow. Mostly eastern records.[f]

Eurema nicippe. Sleepy Orange. Southern half.[f]

Nathalis iole. Dwarf Yellow (Dainty Sulfur). Widespread.[f]

FAMILY LYCAENIDAE — BLUES, COPPERS, AND HAIRSTREAKS

Feniseca tarquinius. Harvester. Mostly southeastern records.[f]

Lycaena phlaeas. American Copper. Three eastern records.[f]

Gaeides xanthoides. Great Gray Copper. Widespread.[f]

Hyllolycaena hyllus. Bronze Copper. Widespread.[f]

Chalceria rubida. Ruddy Copper. Records for western half.

Epidemia helloides. Purplish Copper. Probably extends across state.[f]

Harkenclenus titus. Coral Hairstreak. Probably extends across state.[f]

Satyrium acadicum. Acadian Hairstreak. Probably extends across state.[f]

Satyrium edwardsii. Edward's Hairstreak. Four scattered records.[f]

Satyrium calanus. Banded Hairstreak. Scattered records across state.[f]

Satyrium liparops. Striped Hairstreak. Probably extends across state.[f]

Ministrymon leda. Leda Hairstreak. Panhandle record only.

Calycopis cecrops. Red-banded Hairstreak. Two southeastern records.[f]

Callophrys apama. Canyon Green Hairstreak. Panhandle record only.

Mitoura siva. Juniper Hairstreak. Scattered records across state.

Mitoura grynea. Olive Hairstreak. Mostly eastern records.[f]

Incisalis henrici. Henry's Elfin. Southeastern records only.[f]

Incisalia niphon. Eastern Pine Elphin. Southeastern record only.[f]

Incisalia eryphon. Western Pine Elphin. Records for western half.

Strymon melinus. Gray Hairstreak. Widespread.[f]

Brephidium exilis. Western Pygmy Blue. Three southeastern records, one Panhandle record.[f]

Leptotes marina. Marine Blue. Probably extends across state.[f]

Hemiargus isola. Reakirt's Blue. Widespread.[f]

Everes comyntas. Eastern Tailed Blue. Widespread.[f]

Everes amyntula. Western Tailed Blue. Panhandle record only.

Celastrina argiolus. Spring Azure. Widespread.[f]

Eupilotes enoptes. Dotted Blue. Panhandle record only.

Euphilotes rita. Rita Blue. Panhandle record only.

Glaucopsyche piasus. Arrowhead Blue. Panhandle record only.

Glaucopsyche lygdamus. Silvery Blue. Western half.[f]

Plebeius saepolis. Greenish Blue. Panhandle records only.

Icarica icariodes. Common Blue. Panhandle records only.

Lycaeides melissa. Melissa Blue. Probably extends across state.

Icaricia shasta. Shasta Blue. Panhandle records only.

Icaricia acmon. Acmon Blue. Scattered records across state.

FAMILY LIBYTHEIDAE — SNOUT BUTTERFLIES

Lybytheana bachmanii. Snout Butterfly. Records for southern half.[f]

FAMILY NYMPHALIDAE — BRUSH-FOOTED BUTTERFLIES

Agraulis vanillae. Gulf Fritillary. Scattered records across state, mainly east.[f]

Dryas julia. Julia. One eastern record.[f]

Heliconius charitonius. Zebra Butterfly. Four scattered records.[f]

Euptoieta claudia. Variegated Fritillary. Widespread.[f]

Speyaria cybele. Great Spangled Fritillary. Widespread.[f]

Speyaria aphrodite. Aphrodite Fritillary. Probably extends across state.

Speyaria idalia. Regal Fritillary. Widespread.[f]

Speyaria edwardsii. Edward's Fritillary. Panhandle and northern records only.

Speyaria coronis. Coronis Fritillary. Panhandle records only.

Speyaria zerene. Zerene Fritillary. Panhandle records only.

Speyaria callippe. Callippe Fritillary. Panhandle records only.

Speyaria atlantis. Atlantis Fritillary. Panhandle record only.

Speyaria mormonia. Mormon Fritillary. Panhandle record only.

Clossinia selene. Silver-bordered Fritillary. Probably extends across state.[f]

Clossinia bellona. Meadow Fritillary. Mostly eastern records.[f]

Phycoides vesta. Vesta Crescentspot. One eastern record.

Phycoides paon. Phaon Crescentspot. Four scattered records.[f]

Phycoides tharos. Pearly Crescentspot. Widespread.[f]

Phycoides batesii. Tawny Crescentspot. Panhandle records only.

Phycoides pratensis. Field Crescentspot. Panhandle records only.

Phycoides pictus. Painted Crescentspot. Western half.

Phycoides pallidus. Pallid Crescentspot. Panhandle record only.

Euphydryas anicia. Anicia Crescentspot. Panhandle records only.

Euphydryas phaeton. Baltimore Crescentspot. Two scattered records.[f]

Polygonia interogationis. Question Mark. Widespread.[f]

Polygonia comma. Comma (Hop Merchant). Widespread.[f]

Polygonia satyrus. Satyr Angle-wing. Panhandle record only.

Polygonia zephyrus. Zephyrus Anglewing. Four western records.

Polygonia progne. Gray Comma. Probably extends across state.[f]

Nymphalis vau-albrum. Compton Tortoiseshell. Four scattered records.[f]

Nymphalis antiopa. Mourning Cloak. Widespread.[f]

Aglais milberti. Milbert's Tortoiseshell. Probably extends across state.[f]

Vanessa virginiensis. American Painted Lady. Widespread.[f]
Vanessa cardui. Painted Lady. Widespread.[f]
Vanessa annabella. West Coast Lady. Panhandle records only.
Vanessa atalanta. Red Admiral. Widespread.[f]
Junonia coenia. Buckeye. Widespread.[f]
Anartia jatrophe. White Peacock. One southern record.[f]
Basilarchia arthemis astyanax. Red-spotted Purple. Mostly eastern records.[f]
Basilarchia archippus. Eastern Viceroy. Widespread.[f]
Basilarchia weidemeyerii. Weidemeyer's Admiral. Records for western half.
Mestra amymone. Amymone. Three Platte Valley records.[f]
Marpesia petreus. Ruddy Daggerwing. One east-central record.

FAMILY APATURIDAE —
GOATWEED AND HACKBERRY BUTTERFLIES; EMPERORS

Anaea andria. Goatweed Butterfly. Probably extends across state.[f]
Asterocampa celtis. Hackberry Butterfly. Widespread.[f]
Asterocampa clyton. Tawny Emperor. Mostly eastern records.[f]

FAMILY SATYRIDAE —
RINGLETS, SATYRS, AND WOOD NYMPHS

Enodia anthedon. Northern Pearly Eye. Mostly eastern records.[f]
Satyrodes eurydice. Eyed Brown. Probably extends across state.[f]
Megisto cymela. Little Wood Satyr. Widespread.[f]
Coenonympha tullia. Large Heath. Records mostly for western half of state.
Cercyonis pegala. Large Wood Nymph (Grayling). Widespread.[f]
Cercyonis meadii. Red-eyed Wood Nymph. Panhandle records only.
Cercyonis oetus. Dark Wood Nymph. Panhandle records only.
Neominois ridingsii. Riding's Satyr. Panhandle records only.
Oeneis uhleri. Uhler's Arctic. Panhandle records only.

FAMILY DANAIDAE — QUEENS AND MONARCHS

Danaus plexippis. Monarch. Widespread.[f]
Danaus gilippis. Queen. Probably extends across state.[f]

Selected Moth Families

FAMILY SPHINGIDAE — SPHINX MOTHS

This list is based on Messenger (1997), who provides color plates of all the included species (note that in plate 4 *S. vashti* is second from top, not fifth as indicated, and the rest need to be renumbered accordingly). Species-level identification of sphinx moths is sometimes quite difficult and may require microscopic examination of genitalia. Species and genera are organized in accepted taxonomic sequence.

^f indicates species illustrated by Heitzman and Heitzman (1987).

^g indicates species illustrated by Covell (1984).

^e indicates species illustrated by Milne and Milne (1981).

Aellopos titan. Titan Sphinx. Eastern two-thirds.

Agrius cingulatus. Pink-spotted Hawkmoth. Eastern quarter.[fg]

Amphion floridensis. Nessus Sphinx. Northern half.

Ceratomia amyntor. Elm Sphinx. Statewide.[f]

Ceratomia catalpae. Catalpa Sphinx. Eastern third.[g]

Ceratomia hageni. Hagen's Sphinx. Southeastern quarter.[fg]

Ceratomia undulosa. Waved Sphinx. Statewide.[fg]

Darapsa myron. Hog (Grapevine) Sphinx. All but southwestern quarter.[fe]

Deidamia inscripta. Lettered Sphinx. Statewide.

Erynnis obscura. Obscure Sphinx. Eastern quarter (sporadic).

Eumorpha achemon. Achemon Sphinx. Statewide.

Eumorpha pandorus. Pandora's Sphinx. Eastern half.[fg]

Hemaris diffinis. Snowberry Clearwing. Statewide.[f]

Hemaris thysbe. Hummingbird Clearwing. Eastern quarter (probable stray).[fe]

Hyles galii. Gallium Sphinx. Northern half (sporadic).

Hyles lineata. White-lined Sphinx. Statewide.[fe]

Laothoe juglandis. Walnut Sphinx. All but southwestern quarter.[fg]

Manduca quinquemaculata. Five-spotted Hawkmoth (Tomato Hornworm). Statewide.[fge]

Manduca sexta. Carolina Sphinx (Tobacco Hornworm). Probably statewide, mainly south.[fg]

Pachysphinx modesta. Big Poplar (Modest) Sphinx. Statewide.[fge]

Paonias excaecatus. Blinded Sphinx. Statewide.[fg]

Paonias myops. Small-eyed Sphinx. Statewide.[fg]

Paratea plebeja. Plebeian Sphinx. Southeastern quarter.

Proserpinus juanita. Strecker's Day-sphinx. Northwestern quarter.[f]

Smerinthus jamaicensis. Twin-spot Sphinx. Statewide.[fg]

Sphecodina abbotti. Abbott's Sphinx. Southeastern quarter.[f]

Sphinx canadensis. Canadian Sphinx. Northwestern quarter (stray).[g]

Sphinx chersis. Ash Sphinx. Northern three-quarters.[f]

Sphinx drupiferarum. Wild Cherry Sphinx. All but southwestern quarter.[fge]

Sphinx kalmiae. Laurel Sphinx. Northeast quarter (stray).[g]

Sphinx vashti. Vashti Sphinx. Statewide.[f]

Xylophanes tersa. Tersa Sphinx. Statewide (probable stray).[f]

FAMILY SATURNIIDAE—
GIANT SILK MOTHS AND EMPERORS

For a discussion of these species in Nebraska and illustrations of most, see Ratcliffe (1993). All the included species are illustrated in color by Heitzman and Heitzman (1987), Covell (1984), and Milne and Milne (1981).

A few state rarities are excluded from this list. Organized alphabetically by generic and specific names.

Actias luna. Luna Moth. Eastern third.[fg]

Antheraea polyphemus. Polyphemus Moth. Widespread.[fg]

Automeris io. Io Moth. Uncommon.[fg]

Citheronia regalis. Royal Walnut (Regal) Moth. Extreme southeastern; very rare.[fg]

Eacles imperialis. Imperial Moth. Extreme eastern parts of state.[fg]

Hyalaphora cecropia. Cecropia (Robin) Moth. Widespread.[fg]

FAMILY NOCTUIDAE — UNDERWING MOTHS

The species included in this list are those that were photographed in color by John Farrar and described by Jordison (1996). Some of the state's rare species have been excluded. Covell (1984) described and illustrated about 1,300 other moths of eastern North America. Organized alphabetically by specific names.

[g] indicates species illustrated in Covell (1984).

Catocalia aholibah. Aholibah Underwing.

Catocalia amatrix. Sweetheart Underwing.

Catocalia amestris. Three-staff Underwing.[g]

Catocalia amica. Friendly Underwing.

Catocalia andromediae. Gloomy (Andromeda) Underwing.[g]

Catocalia apione. Epione Underwing.

Catocalia briseis. Briseis Underwing.[g]

Catocalia cara. Darling (Bronze) Underwing.[g]

Catocalia coccinata. Scarlet Underwing.[g]

Catocalia concumbens. Sleepy (Pink) Underwing.[g]

Catocalia connubialis. Connubial Underwing.[g]

Catocalia geogama. Bride Underwing.

Catocalia grotelana. Grote's Underwing.

Catocalia grynea. Grynea (Woody) Underwing.[g]

Catocalia ilia. Beloved Underwing.[g]

Catocalia illecta. Magdalen Underwing.[g]

Catocalia innubens. Betrothed Underwing.[g]

Catocalia lacrymosa. Tearful Underwing.[g]

Catocalia luciana. (unnamed underwing).

Catocalia maestosa. Sad Underwing[g]

Catocalia meskei. Meske's Underwing[g]

Catocalia micronympha. Little Nymph Underwing.[g]

Catocalia minuta. Little Underwing.[g]

Catocalia obscura. Obscure Underwing[g]

Catocalia paleogama. Oldwife Underwing.[g]

Catocalia parta. Mother Underwing.[g]

Catocalia relicta. Forsaken Underwing.[g]

Catocalia subnata. Youthful Underwing.[g]
Catocalia vidua. Widow Underwing.[g]
Catocalia whitneyi. Whitney's Underwing

Dragonflies and Damselflies

This list is based largely on R. J. Beckemeyer's website (see references for URL address). It is also based partly on Keech (1934) and specimen data from the University of Nebraska State Museum collection (John Janovy Jr., pers. comm.). Common names of dragonflies are mostly from Beckemeyer and Huggins (1997). A new reference by Dunkle (2000) is the most useful field guide so far available; all Nebraska species are illustrated in color. Other common names and some unpublished Nebraska records were provided by John Sullivan (pers. comm.). Westfall and May (1996) present detailed information on all North American damselflies; most Nebraska Odonata are also described by Walker and Corbet (1953–73).

[h] indicates species illustrated in Beckemeyer and Huggins (1997).

[i] indicates species illustrated by Milne and Milne (1981).

FAMILY AESCHNIDAE—DARNERS OR HAWKERS
Aeshna canadensis. Canada darner. Reported for state by Beckemeyer; widespread in Great Plains.
Aeshna constricta. Lance-tipped darner. Eastern; collected in Buffalo County.[h]
Aeshna interrupta. Variable darner. Probably throughout state; collected at Valentine.
Aeshna multicolor. Blue-eyed darner. Western (Panhandle); collected at Mitchell.
Aeshna palmata. Paddle-tailed darner. Probably mainly in west; collected in Sioux County.
Aeshna umbrosa. Shadow darner. Probably mainly in west; collected at Fort Robinson.
Anax junius. Common green darner. Widely distributed on ponds, sometimes streams.[i]
Boyeria vinosa. Fawn darner. Reported by Beckemeyer; probably mainly eastwardly.[hi]

FAMILY GOMPHIDAE—CLUBTAILS
Arigomphus cornutus. Horned clubtail. Probably mostly eastern, recorded in Cuming County.
Arigomphus submedianus. Jade clubtail. Reported (John Sullivan) for Douglas County.
Gomphurus externus. Plains clubtail. Statewide, especially east.[h]
Gomphus lividus. Ashy clubtail. One record (Sioux County) but probably more common eastwardly.

Gomphus militaris. Sulphur-tipped clubtail. Photographed (John Sullivan) in Lancaster County.

Ophiogomphus severus. Pale snaketail. Western; collected in Sioux County.[h]

Progomphus obscurus. Common sanddraggon. Probably widespread in sandy streams.[h]

Stylurus amnicola. Riverine clubtail. Eastern half.

Stylurus intricatus. Brimstone clubtail. Fairly common.

FAMILY CORDULIDAE — CRUISERS AND EMERALDS

Epitheca cynosura. Common basketail. Probably across eastern half.[h]

Epitheca princeps. Prince basketail. Probably eastern third.[h]

Neurocordulia molesta. Smoky shadowtail. Mainly eastwardly; collected in Boyd County.

Somatochora ensigera. Plains emerald. Probably more common westwardly.

FAMILY LIBELLULIDAE — TYPICAL SKIMMERS

Brachymeasia gravida. Four-spotted pennant. Recorded for Cass County; probably only southeastern.

Celithemis eponina. Halloween pennant (Brown-spotted yellow-wing). Widespread.[hi]

Erythemis simplicicollis. Eastern pondhawk (Green jacket). Very common.[h]

Leucorhinia hudsonica. Hudsonian white-face. Apparently rare; collected at Omaha.

Leucorhinia intacta. Dot-tailed (Johnny) white-face. Widespread.

Libellula flavida. Yellow-sided skimmer. Probably mainly southward; collected at Valentine.

Libellula forensis. Eight-spotted skimmer. Western half; collected east to Mormon Island.

Libellula luctuosa. Widow skimmer. Probably throughout state.[hi]

Libellula lydia. Common whitetail. Probably throughout state.[h]

Libellula pulchella. Twelve-spotted skimmer (Tenspot). Throughout state.[hi]

Libellula quadrimaculata. Four-spot skimmer. Throughout state.[i]

Libellula subornata. Desert whitetail. Western.

Pachidiplax longipennis. Blue dasher (Swift long-winged skimmer). Throughout state.[hi]

Pantala flavescens. Wandering glider (Globe-skimmer). Probably throughout state.

Pantala hymenaea. Spot-winged glider. Mainly southeastern.

Perithemis tenera. Eastern (Low-flying) amber-wing. Probably eastern only.[hi]

Sympetrum ambiguum. Blue-faced meadowhawk. Reported by Becke-meyer.

Sympetrum corruptum. Variegated (Robust pink) meadowhawk. Common.[i]

Sympetrum costiferum. Saffron-winged meadowhawk. Common.

Sympetrum internum. Cherry-faced meadowhawk. Widespread, probably statewide.

Sympetrum madidum. Red-veined meadowhawk. Probably mainly west-wardly.

Sympetrum obtrusum. White-faced meadowhawk. Widespread, more common northwardly.

Sympetrum occidentale. Western Band-winged meadowhawk. Mainly found in west and in Sandhills.[h]

Sympetrum pallipes. Striped meadowhawk. Western.

Sympetrum rubicundulum. Ruby meadowhawk. Widespread, more common northwardly.

Sympetrum semicinctum. Half-banded skimmer. Abundant throughout state.[i]

Sympetrum vicinum. Yellow-legged meadowhawk. Throughout state.

Tramea lacerata. Black (Jagged-wing) saddlebags. Widespread.[hi]

Tramea onusta. Red-mantled saddlebags. Throughout southern part of state.[i]

FAMILY CALOPTYERYGIDAE — BROAD-WINGED DAMSELFLIES

Calopteryx aequabilis. Western, widespread in Great Plains; cold streams.

Calopteryx maculata. Back-winged damselfly. Common throughout, widespread in Great Plains streams.[i]

Genus Hetaerina—Ruby-Spots

Hetaerina americana. American ruby spot. Common throughout, widespread in Great Plains; rapid streams.[i]

Hetaerina titia. Smoky ruby-spot. Eastern, also in Kansas, Iowa, Missouri, and southward.

FAMILY LESTIDAE — SPREAD-WINGED DAMSELFLIES

Archilestes grandis. Great spread-wing. Reported from western part of state, probably widespread in slow streams and ponds.

Lestes congener. Spotted spread-wing (Dark lestes). Reported from Nebraska, abundant and widespread in North American ponds.[i]

Lestes disjunctus. Common spread-wing. Reported from east and central Nebraska, abundant and widespread in North America.

Lestes dryas. Emerald spread-wing (Stocky lestes). Reported from eastern Nebraska, widespread in Great Plains; temporary ponds.[i]

Lestes forcipatus. Sweetflag spread-wing. Reported from eastern Nebraska, widespread in Great Plains; temporary and permanent ponds.

Lestes rectangularis. Slender spread-wing. Reported from northern Nebraska, widespread in Great Plains; shady ponds.

Lestes unguiculatus. Lyre-tipped spread-wing. Widespread in Nebraska and in Great Plains; small ponds and sloughs.

FAMILY COENAGRIONIDAE — NARROW-WINGED DAMSELFLIES

Genus Amphiagrion—Short (Red) Damselflies

Amphiagrion abbreviatum. Western red damselfly (Southwestern short damselfly). Widespread in Nebraska and in Great Plains; marshes and sloughs.[i]

Amphiagrion saucium. Eastern red damselfly (Northeastern short damselfly). Widespread in Nebraska and in Great Plains; ponds and bogs.

Genus Argia—Short-Stalked (Dancer) Damselflies

Argia alberta. Piute dancer. Reported from central Nebraska, widespread in western Great Plains; creeks and springs.

Argia apicallis. Blue-fronted dancer. Reported from central Nebraska, widespread in Great Plains; rivers and small streams.

Argia emma. Emma's dancer. Reported from central Nebraska, widespread in Great Plains; slow and rapid streams.

Argia fumipennis. Variable dancer. Reported from Nebraska, widespread in Great Plains; lakes and ponds.

Argia moesta. Powdered dancer. Reported from central Nebraska, widespread in Great Plains; rocky streams and lakes.

Argia plana. Springwater dancer. Reported from Nebraska, widespread in southern Great Plains.

Argia sedula. Blue-ringed dancer. Reported from central Nebraska, widespread in Great Plains; lakes and gentle streams.

Argia tibialis. Blue-tipped dancer. Reported from south-central Nebraska, widespread in Great Plains; mainly cold streams.

Argia vivida. Vivid dancer. Reported from Nebraska, widespread in Great Plains; spring-fed streams.

Coenagrion resulotum. Tiaga bluet. Reported from Nebraska, widespread in northern Great Plains.

Genus Enallagma—Bluets

Enallagma anna. River bluet. Reported from central and western Nebraska, widespread in western Great Plains; slow streams.

Enallagma antennatum. Rainbow bluet. Reported from central Nebraska, widespread in Great Plains; slow streams and lakes.

Enallagma basidens. Double-striped bluet. Reported from central Nebraska, widespread in Great Plains; ponds.

Enallagma boreale. Boreal bluet. Reported from western Nebraska, widespread in northern Great Plains; slow streams and ponds.

Enallagma carunculatum. Double-striped bluet. Reported from western Nebraska, widespread in Great Plains; slow rivers, ponds, and lakes.

Enallagma civile. Familiar bluet. Reported from eastern Nebraska, widespread and abundant in North America; ponds and slow streams.

Enallagma clausum. Alkali bluet. Reported from western Nebraska, widespread in Great Plains; usually alkaline waters.

Enallagma cyathigerum. Northern bluet. Circumpolar bluet. Reported from western Nebraska, widespread in northern North America; ponds, marshes, and lakes.[i]

Enallagma ebrium. Marsh bluet. Reported from northern Nebraska, widespread in Great Plains; marshes and ponds.

Enallagma exsulans. Stream bluet. Reported from Nebraska, also eastern Great Plains; rivers and larger streams.

Enallagma germinatum. Skimming bluet. Reported from northern Nebraska, also eastern Great Plains; slow streams, ponds, and lakes.

Enallagma hageni. Hagen's bluet. Widespread in Nebraska and in Great Plains; ponds, marshes, and bogs.

Enallagma praevarum. Arroyo bluet. Reported from western Nebraska, widespread in western Great Plains; ponds and streams.

Enallagma signatum. Orange bluet. Reported from Nebraska, also eastern Great Plains; slow streams and lakes.

Genus Ishnura—Forktails

Ishnura barberi. Desert forktail. Reported from Nebraska, also Kansas, Colorado, and the Southwest; pools and ditches.

Ishnura damula. Plains forktail. Reported from Nebraska, also western Great Plains; ponds.

Ishnura perparva. Western forktail. Reported from Nebraska, also western Great Plains; ponds and slow streams.

Ishnura verticalis. Eastern forktail. Widespread in Nebraska and in the Great Plains; ponds and slow streams.

Nehalennia irene. Sedge sprite. Reported from Nebraska, widespread in northern Great Plains; marshes and fens.

Freshwater Mussels

This is not a complete list of Nebraska bivalve mussels and clams. It includes those described and illustrated by Freeman and Perkins (1994) for the Platte River, plus additional species listed for the Platte and its tributaries by Hoke (1995). It also includes Hoke's lists for the Nemaha basin (1996) and for the Elkhorn basin (1994). These surveys collectively suggest that more than 50 species probably occur in Nebraska. Nearly all the Nebraska mussels were included in a survey of Missouri mussels by Oesch (1984). This abbreviated list is arranged alphabetically by genus and species.

indicates species illustrated by Freeman and Perkins (1994).

ᵏ indicates species illustrated by Oesch (1984).

Actinonaias ligamentina carinata. Mucket. Reported for the Elkhorn and Nemaha basins.[k]

Amblema p. plicata. Three-ridge. Reported for the Elkhorn and Nemaha basins.[k]

Anodontia g. grandis. Giant floater. Common in Platte; also in Elkhorn and Nemaha basins.[jk]

Anodontia imbecilis. Paper pond shell. Sandy or muddy lakes and Platte backwaters; also in Elkhorn and Nemaha basins.[jk]

Anodontoides ferrusacianus. Cylindrical paper shell. Common in unpolluted areas of Platte; also in Elkhorn and Nemaha basins.[jk]

Corbicula fluminea. Asiatic clam. Exotic, undesirable species, introduced locally in Platte basin.[jk]

Fusconaia flava. Wabash pig-toe. Reported for the Platte, Elkhorn, and Nemaha basins.[k]

Lampisilis radiata. Fat mucket. Reported for the Platte, Elkhorn, and Nemaha basins.[k]

Lampisilis teres. Sand shell. Reported for the Platte, Elkhorn, and Nemaha basins.[k]

Lampisilis ventricosa. Pocketbook. Reported for the Platte, Elkhorn, and Nemaha basins.[k]

Lasmigona complanata. White heel-splitter. Common in Platte basin; also in Elkhorn and Nemaha basins.[jk]

Lasmagona compresa. Reported for the Elkhorn and Nemaha basins.

Leptodea fragilis. Fragile heel-splitter. Local in Platte River system; also in Elkhorn and Nemaha basins.[jk]

Ligumia recta. Black sand shell. Reported for the Elkhorn and Nemaha basins.[k]

Ligumia subrostrata. Pond mussel. Reported for the Platte, Elkhorn, and Nemaha basins.[k]

Obovaria olivaria. Hickory-nut. Reported for the Elkhorn and Nemaha basins.[k]

Pomatilus alatus. Purple heel-splitter. Apparently rare in Platte basin; also in Elkhorn and Nemaha basins.[jk]

Potamilus ohiensis. Pink heel-splitter. Common and increasing in Platte River system; also in Elkhorn and Nemaha basins.[jk]

Quadula pustulosa. Pimple-back. Reported for the Platte, Elkhorn, and Nemaha basins.[k]

Quadrula quadrula. Maple-leaf mussel. Abundant in Platte basin; also in Elkhorn and Nemaha basins.[jk]

Strophitus u. undulatus. Squaw-foot. Local in Platte basin, abundant in Niobrara River; also in Elkhorn and Nemaha basins.[jk]

Toxolasma parva. Liliput shell. Reported for the Platte, Elkhorn, and Nemaha basins.[k]

Tritogonia verruscosa. Pistol-grip. Reported for the Elkhorn and Nemaha basins.[k]

Truncilla donaciformis. Fawn's foot. Reported for the Elkhorn and Nemaha basins.[k]

Truncilla truncata. Deer toe. Reported for the Elkhorn and Nemaha basins.[k]

Uniomerus tetralasmus. Pond horn shell. Mud-bottom lakes and pools of Platte basin; also in Elkhorn and Nemaha basins.[jk]

Common Vascular Plants

Native Trees

This list is based partly on Pool (1961), who provides identification keys and illustrations. See Barkley (1977) for more recent range maps. Brockman (1986) provides identification and range maps of nearly all species listed here. Species are also discussed by Stephens (1969). Families are arranged alphabetically; lower taxa are arranged alphabetically by generic and specific names. A few nonnative species that are mentioned in the text are also included.

[l] indicates species illustrated by Barkley (1977).
[m] indicates species illustrated by Stephens (1969).
[n] indicates species illustrated by Johnson and Larson (1999).
[o] indicates species illustrated by Larson and Johnson (1999).

ACERACEAE — MAPLE FAMILY

Silver maple. *Acer saccharinum.* Eastern quarter.[lm]
Box elder. *Acer negundo.* Statewide.[lmno]

ANNONACEAE — PAWPAW FAMILY

Pawpaw. *Asimia triloba.*[lm] Southeastern corner.

BETULACEAE — BIRCH FAMILY

Mountain (water) birch. *Betula occidentalis.* Pine Ridge.[o]
Paper birch. *Betula papyrifera.* Central Niobrara Valley.[lo]
Blue beech (Hornbeam). *Carpinus caroliniana.* Southeastern corner.[l]
Hop hornbeam (Ironwood). *Ostrya virginiana.* Niobrara and Missouri Valleys.[lmo]

CAESALPINIACEAE — SENNA FAMILY

Redbud. *Cercis canadensis.* Lower Missouri Valley.[lm]
Honey locust. *Gleditsia triacanthos.* Eastern half.[lm]
Kentucky coffee tree. *Gymnocladus dioca.* Niobrara and Missouri Valleys.[lm]

CELASTRACEAE — BITTERSWEET FAMILY
Wahoo. *Eonymus atropurpurea.* Southeastern (also occurs as a shrub).[lm]

CAPRIFOLIACEAE — HONEYSUCKLE FAMILY
Black haw. *Viburnum lentago.* Southeastern.[l]

CUPRESSACEAE — CYPRESS FAMILY
Western red cedar (Western red juniper). *Juniperus scopulorum.* Western third.[lo]

Eastern red cedar (Red juniper). *Juniperus virginiana.* Eastern two-thirds.[lm]

FAGACEAE — OAK FAMILY
White oak. *Quercus alba.* Extreme southeastern Nebraska.[lm]
Red oak. *Quercus (rubra) borealis.* Eastern quarter.[lm]
Scarlet oak. *Quercus coccinea.* Southeastern corner.[l]
Laurel (Shingle) oak. *Quercus imbricaria.* Southeastern corner.[lm]
Bur oak. *Quercus macrocarpa.* Eastern half.[lmno]
Black jack oak. *Quercus marilandica.* Southeastern corner.[lm]
Chinquapin (Chestnut/Yellow) oak. *Quercus muehlenbergii.* Southeastern corner.[lm]

Dwarf chinquapin (Low Yellow) oak. *Quercus prinoides.* Southeastern corner.[l]

Shumard's red oak. *Quercus shumardii.* Extreme southeastern Nebraska.[lm]

Black oak. *Quercus velutina.* Lower Missouri Valley.[lm]

HIPPOCASTANACEAE — BUCKEYE FAMILY
Ohio buckeye. *Aesculus glabra.* Southeastern corner.[lm]

JUGLANDACEAE — WALNUT FAMILY
Bitternut hickory. *Carya cordiformis.* Southeastern.[lm]
Bignut hickory. *Carya laciniosa.* Southeastern corner.[lm]
Shellbark (Shagbark) hickory. *Carya ovata.* Southeastern.[lm]
Mockernut hickory. *Carya tomentosa.* Southeastern.[lm]
Black walnut. *Juglans nigra.* Eastern half.[lm]
Butternut. *Juglans cinerea.* Southeastern corner.[l]

MORACEAE — MULBERRY FAMILY
Osage orange. *Maclura pomifera.* Widespread.[lm]
Mulberry. *Morus rubra.* Eastern quarter.[lm]

OLEACEAE — ASH FAMILY
White ash. *Fraxinus americana.* Lower Missouri Valley.[lm]
Green (and Red) ash. *Fraxinus pennsylvanicua* (including *subintegerrima*). Statewide.[lmno]

PINACEAE — PINE FAMILY
Limber pine. *Pinus flexilis.* Limited to Kimball County.[o]

Lodgepole pine. *Pinus contorta*. Not native to state.[o]

Western yellow (Ponderosa) pine. *Pinus ponderosa*. Panhandle and Niobrara Valley.[o]

PLANTANACEAE — SYCAMORE FAMILY

Sycamore. *Plantanus occidentalis*. Lower Missouri Valley.[lm]

ROSACEAE — PLUM FAMILY

Juneberry. *Amelanchier canadensis*. Lower Missouri Valley (often occurs as a shrub).

Hawthorn. *Crataegus* spp. Widespread.[lmo]

Wild plum. *Prunus americana*. Statewide (often occurs as a shrub).[lmno]

Wild black cherry. *Prunus serotina*. Southern and southeastern.[lm]

Eastern chokecherry. *Prunus virginiana*. Statewide (often occurs as a shrub).[lmno]

Prairie apple. *Pyrus iowensis*. Eastern.[m]

SALICACEAE — WILLOW FAMILY

Narrow-leaf cottonwood (poplar). *Populus angustifolia*. Panhandle.[lo]

Balsam poplar. *Populus balsamifera*. Pine Ridge.[lo]

Cottonwood. *Populus deltoides*. Statewide.[lmno]

Bigtooth poplar. *Populus grandidentata*. Hybrid population with *tremuloides* in the Niobrara Valley; also in western Pine Ridge.

Quaking aspen. *Populus tremuloides*. Panhandle.[lo]

Peach-leafed willow. *Salix amygdaloides*. Statewide, usually riparian (often shrubby).[lmno]

Sand-bar willow. *Salix interior*. Widespread in riparian sites (often occurs as a shrub).[lm]

Black willow. *Salix nigra*. Statewide.[lm]

TILIACEAE — BASSWOOD FAMILY

Basswood (Linden). *Tilia americana*. Eastern half.[lm]

ULMACEAE — ELM FAMILY

Hackberry. *Celtis occidentalis*. Statewide.[lm]

White (American) elm. *Ulmus americana*. Statewide.[lmn]

Red (Slippery) elm. *Ulmus rubra*. Eastern half.[lm]

Cork elm. *Ulmus thomasi*. Northern and eastern.[l]

Common Shrubs and Woody Vines

Families, genera, and specific names are organized alphabetically. Some species appearing in the tree list may also occur as shrubs.

[p] indicates mostly eastern species described by Petrides (1958).

[q] indicates predominantly western species described and illustrated by Stubbendieck et al. (1997).

[o] indicates species described and illustrated by Stephens (1969).

[m] indicates species described and
illustrated by Johnson and Larson (1999).
[n] indicates species described and
illustrated by Larson and Johnson (1999).

AGAVACEAE — AGAVE FAMILY
Great Plains yucca (Small soapweed). *Yucca glauca.* Widespread in west.[mon]

ANACARDIACEAE — CASHEW FAMILY
Aromatic sumac. *Rhus aromatica.* Widespread.[qpomn]
Smooth sumac. *Rhus glabra.* Widespread.[pomn]
Poison ivy. *Toxicodendron (Rhus) radicans.* and *T. rydbergii.* Statewide (often vinelike or partially shrubby); poisonous to touch, causing severe dermatitis.[pomn]

ASTERACEAE — ASTER FAMILY
Silver sagebrush. *Artemisia cana.* Northwestern corner.[qmn]
Sand sagebrush. *Artemisia filifolia.* Western and northern.[qom]
Fringed sagebrush. *Artemisia ludoviciana.* Western half.[qm]
Big sagebrush. *Artemisia tridentata.* Panhandle.[qmn]
Rabbitbrush. *Chrysothamnus naseosus.* Western half.[omn]

BERBERIDACEAE — BARBERRY FAMILY
Oregon grape. *Berberis repens.* Northwestern Panhandle.[n]

BETULACEAE — BIRCH FAMILY
Paper birch. *Betula papyrifera.* Niobrara Valley.[n]
American hazelnut. *Corylus americana.* Widespread in riparian areas.[p o]
Ironwood (hop-hornbeam). *Ostrya virginiana.* Eastern and northern.[n]

CAPRIFOLIACEAE — HONEYSUCKLE FAMILY
Elderberry. *Sambucus canadensis.* Mainly eastern.[po]
Snowberry. *Symphoricarpos alba.* Northwestern.[pn]
Western snowberry. *Symphoricarpos occidentalis.* Statewide.[pqomn]
Buckbrush (Coralberry). *Symphoricarpus orbiculatus.* Mainly southeastern.[pqo]

CELASTRACEAE — STAFF-TREE FAMILY
Climbing Bittersweet. *Celastrus scandens.* Forest, mainly eastern (woody vine).[pon]
Wahoo. *Euonymus atropurpureus.* Eastern half.[po]

CHENOPODIACEAE — GOOSEFOOT FAMILY
Fourwing saltbush. *Atriplex canescens.* Western.[om]
Winterfat. *Ceratoides lanata.* Western.[mn]
Black greasewood. *Sarcobatus vermiculatus.* Western.[m]

CORNACEAE — DOGWOOD FAMILY
Pale (Silky) dogwood. *Cornus amonium.* Eastern third.[po]

Rough-leaved dogwood. *Cornus drummondi*. Eastern half.[po]
Red oiser dogwood. *Cornus stolonifera*. Widespread, riparian.[pn]

ELAEAGINACEAE — RUSSIAN OLIVE FAMILY

Buffaloberry. *Shepherdia argentea*. Widespread.[pmn]

FABACEAE — BEAN FAMILY

Leadplant. *Amorpha canescens*. Statewide in prairies.[pqomn]
False indigo. *Amorpha fructicosa*. Statewide.[pon]
Shrubby (Bushy) cinquefoil. *Potentilla paradoxa*. Widespread.

GROSSULARIACEAE — CURRANT FAMILY

Black currant. *Ribes americana*. Widespread, moist soils.[n]
Buffalo (golden)currant. *Ribes odoratum*. Sandy soils, widespread.[n]
Northern gooseberry. *Ribes oxycanthoides*. Varied habitats, northern.[n]

LILIACEAE — LILY FAMILY

Bristly greenbrier. *Smilax hispida*. Eastern third.[p]

RANUNCULACEAE — BUTTERCUP FAMILY

Western clematis (Virgin's bower). *Clematis ligusticifolia*. Western half.

RHAMNACEAE — BUCKTHORN FAMILY

New Jersey tea. *Ceanothus americanus*. Eastern third.[po]
Lance-leaved buckthorn. *Rhamnus lanceolata*. Eastern half.[po]

ROSACEAE — ROSE FAMILY

Serviceberry. *Amelanchier alnifolia*. Northern and northwestern.[pqn]
Juneberry. *Amelanchier canadensis*. Northern.[p]
True mountain mahogany. *Cercocarpus montanus*. Panhandle; may be poisonous.[qn]
Hawthorn. *Crataegus* spp. Widespread.[qpm]
Wild plum. *Prunus americana*. Statewide.[qpom]
Sand cherry. *Prunus besseyi*. Widespread.[o]
Eastern chokecherry. *Prunus virginiana*. Statewide.[qpomn]
Prairie wild rose. *Rosa arkansana*. Widespread, woodland edges.[mn]
Western wild rose. *Rosa woodsi*. Western two-thirds.[qon]
Black raspberry. *Rubus occidentalis*. Eastern, riparian.[po]
Mountain ash. *Sorbus scopulina*. Widespread.[n]

RUTACEAE — RUE FAMILY

Prickly ash. *Zanthoxylum americanum*. Eastern third.[po]

SALICACEAE — WILLOW FAMILY

Peach-leaved willow. *Salix amygdaloides*. Widespread.[omn]
Bebb's (Long-beaked) willow. *Salix bebbiana*. Panhandle.[pqn]
Sandbar (Coyote) willow. *Salix exugua*. Widespread in riparian areas.[qmn]
Shining willow. *Salix lucida*. Panhandle.[p]
Diamond willow. *Salix rigida*. Statewide, riparian areas.[po]

SAXFRAGINACEAE — SAXIFRAGE FAMILY
Missouri gooseberry. *Ribes missouriensis.* Widespread, moist woods.[pom]
Buffalo current. *Ribes odoratum.* Widespread.[pom]

STAPHYLEACEAE — BLADDERNUT FAMILY
Bladdernut. *Staphylea trifolia.* Missouri Valley.[p]

VITACEAE — GRAPE FAMILY
Woodbine. *Parthenocissus vitacea.* Statewide (woody vine).[o]
River-bank grape. *Vitis riparia.* Statewide (woody vine).[po]

Common Terrestrial Forbs

The following list of more than 400 herbaceous species (or about a fourth of the state's known total) is mostly limited to those described and illustrated in regional identification guides or to other species mentioned in the text. Nomenclature follows the Great Plains Flora Association monograph (Barkley 1986), which also provides keys to the species. Stubbendieck et al. (1995), Farrar (1990a), and Lommasson (1973) describe and illustrate these species. Johnson and Larson (1999) is an excellent regional guide, as is Larson and Johnson (1999), which is especially good for western Nebraska. Common names preferentially follow Farrar, which are primarily based on those of Barkley (1977). Familial, generic, and specific names are organized alphabetically. A few species that are variably woody are also included in the list of shrubs and woody vines; some others are likewise included in the list of shoreline and aquatic plants. The minimum species numbers for each family are based on range maps in Barkley (1977) and include any woody or variably aquatic species. The species numbers are based on records more than two decades old and thus are minimum estimates. See Blackwell (1990) and Emboden (1979) for information on poisonous and narcotic plants.

[r] indicates invasive or "weedy" species described and illustrated by Stubbendieck et al. (1995).

[s] indicates generally less invasive wildflowers described and illustrated by Farrar (1990).

[t] indicates wildflowers described and illustrated by Lommasson (1973).

[o] indicates species described and illustrated by Johnson and Larson (1999).

ACANTHUS FAMILY — ACANTHACEAE
(At least 2 species in Nebraska)
Fringeleaf ruellia. *Ruellia humilis.* Southeastern corner, dry prairies, rocky banks, open woods.[s]

PIGWEED FAMILY — AMARANTHACEAE
(At least 9 species in Nebraska)
Tumble pigweed. *Amaranthus albus.* Widespread, weedy.[r]

Prostrate pigweed. *Amaranthus graecizans*. Widespread, weedy.[r]
Redroot pigweed. *Amaranthus retroflexus*. Widespread, weedy.[r]
Common water hemp. *Amaranthus rudis*. Mainly eastern, weedy.[r]
Field snakecotton. *Froelichia floridana*. Widespread, sand dunes, rocky open woods.[rs]
Slender froelichia. *Froelichia gracilis*. Mainly eastern, sandy areas, rocky open woods.[t]

CASHEW FAMILY — ANACARDIACEAE
(At least 6 species in Nebraska)
Poison ivy. *Toxicodendron* (*Rhus*) spp. Statewide; also a variably woody shrub or vine, in shaded woods; extremely poisonous, avoid touching.[o]

PARSLEY FAMILY — APIACEAE (UMBELLIFEREAE)
(At least 30 species in Nebraska)
Spotted waterhemlock. *Cicuta maculata*. Widespread, near streams.[rsto]
Poison hemlock. *Conium maculatum*. Widespread, weedy; poisonous.[rsto]
Queen Anne's lace. *Daucus carota*. Mainly eastern, weedy.[st]
Cow parsnip. *Heracleum sphondylium*. Eastern and northern, shaded woods.[st]
Wild parsley (desert biscuitroot). *Lomatium foeniculaceum*. Eastern and Panhandle, dry prairies.[sto]
Leafy musineon. *Musineon divaricatum*. Western half, rocky prairies, open woods.[so]
Narrow-leaved musineon. *Musineon tenuifolium*. Panhandle, dry prairies, open woods.[s]
Sweet cicely. *Osmorhiza claytonii*. Missouri Valley, wooded hillsides.[s]
Prairie parsley. *Polytaenia nuttallii*. Eastern third; low, moist prairies.[s]
Heartleaf Alexander. *Zizia aptera*. Moist meadows.[o]
Golden alexander. *Zizia aurea*. Missouri Valley, low prairies, ditch margins.[r]

DOGBANE FAMILY — APOCYNACEAE
(At least 4 species in Nebraska)
Hemp (Prairie) dogbane. *Apocynum cannabinum*. Widespread, prairies, open woods, weedy; poisonous.[trs]

ARUM FAMILY — ARACEAE
(At least 3 species in Nebraska)
Jack-in-the-pulpit. *Arisaema triphyllum*. Eastern third; moist, humid areas; poisonous.[st]

GINSENG FAMILY — ARALACEAE
(At least 3 species in Nebraska)
Wild sarsaparilla. *Aralia nudicaulis*. Northern counties.
Spiknard. *Aralia racemosa*. Niobrara and Missouri Valleys.

Ginseng. *Panax quinquefolium.* Missouri Valley; shaded, moist woods; state threatened.

MILKWEED FAMILY — ASCLEPIADACEAE
(At least 17 species in Nebraska)
Sand milkweed. *Asclepias arenaria.* Widespread, sandy upland prairies.[st]
Swamp milkweed. *Asclepias incarnata.* Widespread, wet prairies, moist banks.[rst]
Wooly milkweed. *Asclepias lanuginosa.* Mainly central, sandy prairies.[t]
Plains milkweed. *Asclepias pumila.* Widespread.[s]
Showy milkweed. *Asclepias speciosa.* Mainly western, moist prairies, near water.[sto]
Narrow-leaved milkweed. *Asclepias stenophylla.* Widespread, sandy or rocky prairies.[st]
Smooth milkweed. *Asclepias sullivantii.* Eastern third; sandy, loamy, or rocky prairies.[st]
Common milkweed. *Asclepias syriaca.* Mainly eastern, banks, floodplains, waste areas.[rsto]
Butterfly milkweed. *Asclepias tuberosa.* Eastern half, sandy or rocky prairies.[so]
Whorled milkweed. *Asclepias verticillata.* Widespread, sandy or rocky prairies.[rsto]
Green milkweed. *Asclepias viridiflora.* Widespread, sandy or rocky prairies.[so]
Spider milkweed. *Asclepias viridis.* Southeastern corner, sandy or rocky prairies.[st]
Honeyvine (Sand vine) milkweed. *Cynanchum laeve.* Eastern, weedy and spreading.[r]

SUNFLOWER FAMILY — ASTERACEAE (COMPOSITAE)
(At least 230 species in Nebraska)
Common yarrow. *Achillea millefolium.* Widespread, grasslands, open woods, weedy.[rsto]
Common ragweed. *Ambrosia artemisiifolia.* Mainly eastern, disturbed sites, weedy.[ro]
Woollyleaf bursage. *Ambrosia grayii.* Southwestern, mildly saline sites.[r]
Western ragweed. *Ambrosia psilostachya.* Widespread, open prairies, waste sites.[ro]
Pussy-toes. *Antennaria neglecta.* Widespread, prairies, open woodlands, pastures.[to]
Rocky Mountain pussy-toes. *Antennaria parviflora.* Mainly western, prairies, open woods, roadsides.[so]
Plainleaf pussy-toes. *Antennaria plantaginifolia.* Mainly southeastern, woods, thickets.[so]
Common burdock. *Arctium minus.* Eastern half, weedy.[rs]

Absinth wormwood. *Artemisia absinthium.* Weedy, induced.[o]

Biennial wormwood. *Artemisia biennis.* Widespread; damp, sandy soil; streambanks.[r]

Western sagewort. *Artemisia campestris.* Widespread, upland grasslands.[o]

Silky wormwood (Tarragon). *Artemisia dracunculus.* Widespread; dry, open sites.[ro]

Fringed sagewort. *Artemisia frigida.* Western half, shortgrass prairies.[o]

Cudweed sagewort. *Artemisia ludoviciana.* Widespread, dry plains.[s]

White (Heath) aster. *Aster ericoides.* Widespread; open, upland prairies.[rsto]

White prairie aster. *Aster falcatus.* Mainly western, dry plains.[s]

Fendler's aster. *Aster fendleri.* Local along Kansas border in central Nebraska.

New England aster. *Aster novae-angliae.* Eastern half; moist, sandy areas.[sto]

Aromatic aster. *Aster oblongifolius.* Widespread, rocky or sandy open sites.[s]

Azure aster. *Aster oolentangiensis* (*azureus*). Missouri Valley, prairies, open woods.[s]

Willowleaf aster. *Aster praealtus.* Eastern half, damp or drying sites.[s]

Silky aster. *Aster sericeous.* Eastern third; dry, upland sites; open woods.[t]

Slender aster. *Aster subulatus.* Southeastern; damp, saline sites; weedy.[r]

Spanish needles. *Bidens bipinnata.* Eastern half; damp, disturbed sites; weedy.[r]

Nodding beggarticks. *Bidens cernua.* Widespread; muddy, disturbed sites; weedy.[rst]

Tickseed sunflower. *Bidens coronata.* Eastern half; damp, drying sandy sites.[st]

Devils beggarticks. *Bidens frondosa.* Widespread; moist, wooded sites.[r]

Tuberous Indian plantain. *Cacalia plantaginea* (*tuberosa*). Widespread; damp, rocky prairies.[ts]

Musk thistle. *Carduus nutans.* Widespread, pastures, prairies, weedy.[rsto]

Russian knapweed. *Centaurea repens.* Scattered records, introduced weed.[o]

Oxeye daisy. *Chrysanthemum leucantheremum.* Widespread, fields, waste places, weedy.[t]

Golden aster. *Chrysopsis villosa.* Widespread; open, sandy uplands.[sto]

Chicory. *Cichorium intybus.* Widespread, weedy, introduced species.[rst]

Tall thistle. *Cirsium altissimum.* Mainly eastern, waste sites, weedy.[rt]

Canada thistle. *Cirsium arvense.* Widespread, waste sites, weedy.[rsto]

Platte thistle. *Cirsium canescens.* Mainly western, sandy upland prairies.[rst]

Flodman's thistle. *Cirsium flodmanii.* Widespread; moist, open pastures; weedy.[rso]

Yellowspine thistle. *Cirsium ochrocentrum.* Mainly southwestern; dry, sandy prairies.[r]

Wavyleaf thistle. *Cirsium undulatum*. Widespread, dry prairies, weedy.[ro]

Bull thistle. *Cirsium vulgare*. Widespread, waste sites, weedy, introduced species.[rsto]

Horseweed. *Conyza canadensis*. Widespread, disturbed sites, weedy.[ro]

Plains coreopsis. *Coreopsis tinctoria*. Widespread, sandy ground, disturbed sites.[rsto]

Hawk's-beard. *Crepis runcinata*. Mainly western; open, often damp meadows.[to]

Fetid marigold. *Dyssodia papposa*. Widespread, open fields, disturbed sites.[ro]

Purple coneflower. *Echinacea angustifolia*. Widespread; open, rocky prairies.[sto]

Western fleabane. *Erigeron bellidiastrum*. Mainly western; open, damp, sandy sites.[so]

Low fleabane. *Erigeron pumilus*. Western half; open, dry prairies.[st]

Daisy (Rough) fleabane. *Erigeron strigosus*. Widespread; moist, damp prairies.[rsto]

Tall Joe-pye weed. *Eupatorium altissimum*. Southeastern, pastures, disturbed sites.[r]

Spotted Joe-pye weed. *Eupatorium maculatum*. Mainly northern; moist, wooded sites.[s]

Boneset. *Eupatorium perfoliatum*. Widespread; damp, low ground.[st]

White snakeroot. *Eupatorium rugosum*. Eastern half, open woods, disturbed sites.[rs]

Blanket flower. *Gaillardia aristata*. Western half (local).[o]

Fragrant cudweed. *Gnaphalium obtusifolium*. Southeastern, prairies, open woods.[s]

Curly-top gumweed. *Grindelia squarrosa*. Widespread, waste places, weedy.[rsto]

Broom snakeweed. *Gutierrezia sarothrae*. Mainly western; dry, open plains.[rso]

Cutleaf ironplant. *Haplopappus spinulosus*. Widespread, open prairies.[rsto]

Common sneezeweed. *Helenium autumnale*. Widespread; moist, open sites.[rsto]

Common sunflower. *Helianthus annuus*. Widespread, open sites.[rsto]

Sawtooth sunflower. *Helianthus grosseserratus*. Mainly eastern, damp prairies, open bottomlands.[rsto]

Maximilian sunflower. *Helianthus maxmiliani*. Widespread, dry or damp prairies, sandy areas.[sto]

Nuttall's sunflower. *Helianthus nuttallii*. Scattered records.[o]

Plains sunflower. *Helianthus petiolaris*. Widespread; open, sandy sites.[sto]

Stiff sunflower. *Helianthus rigidus*. Eastern half, dry or damp prairies.[to]

Jerusalem artichoke. *Helianthus tuberosus*. Widespread, open or shaded, moist sites.[rst]

False sunflower (Oxeye). *Heliopsis helianthoides.* Mainly eastern; dry open woods; weedy.[rsto]

Camphorweed. *Heterotheca latifolia.* Extreme southern; open, sandy, disturbed sites.[r]

Stemless hymenoxys. *Hymenoxys acaulis.* Western third, rocky breaks, calcareous soils.[sto]

Poverty sumpweed. *Iva axillaris.* Western half; dry, often alkaline soils; prairies.[r]

Marshelder. *Iva xanthifolia.* Widespread, borders of streams and local drying sites.[r]

Falseboneset. *Kuhnia eupatorioides.* Widespread, open prairies.[rsto]

Blue lettuce. *Lactuca oblongifolia.* Widespread; low, moist meadows.[rst]

Prickly lettuce. *Lactuca serriola.* Widespread, disturbed sites, weedy.[s]

Rough gayfeather. *Liatris aspera.* Eastern half, open woods on sandy sites.[sto]

Blazing star (scaly gayfeather). *Liatris glabrata.* Widespread, especially Sandhills.[o]

Dotted gayfeather. *Liatris punctata.* Widespread; dry, sandy upland prairies.[sto]

Thick-spike gayfeather. *Liatris pycnostachya.* Eastern half; open, damp prairies.[st]

Skeletonweed. *Lygodesmia juncea.* Widespread; open, high prairies.[rsto]

Viscid aster. *Machaeranthera linearis.* Western half; open, dry, sandy sites.[st]

Pineappleweed. *Matricaria matricarioides.* Eastern and northern, waste sites, weedy.[r]

False dandelion. *Microseris cuspidata.* Widespread, dry or drying open prairies.[st]

Snakeroot. *Prenanthes racemosa.* Scattered records, varied habitats.[s]

Prairie coneflower. *Ratibida columnifera.* Widespread, disturbed sites, weedy.[so]

Gray-headed coneflower. *Ratibida pinnata.* Eastern quarter, disturbed prairies, open woods.[to]

Black-eyed susan. *Rudbeckia hirta.* Widespread, disturbed prairies, waste sites.[st]

Golden glow. *Rudbeckia laciniata.* Eastern half, moist places.[s]

Gray ragwort. *Senecio canus.* Western half.[o]

Lambstongue groundsel. *Senecio integerrinus.* Mainly northern; sometimes poisonous.[o]

Prairie ragwort. *Senecio plattensis.* Widespread, open prairies; may be poisonous.[sto]

Riddle groundsel. *Senecio riddellii.* Western half, open sites, sandy areas; poisonous to livestock.[ro]

Threetooth ragwort. *Senecio tridenticulatus.* Mainly western, sandy plains.[s]

Rosinweed. *Silphium integrifolium.* Eastern third; open, disturbed sites.[st]

Compass plant. *Silphium laciniatum.* Eastern third, open prairies.[s]

Cup plant. *Silphium perfoliatum.* Eastern third; moist, low ground.[so]

Canada goldenrod. *Solidago canadensis.* Widespread, dry or drying open sites.[so]

Giant goldenrod. *Solidago gigantea.* Widespread, damp soils.[s]

Prairie goldenrod. *Solidago missouriensis.* Widespread, open prairies, sparse woods.[rso]

Soft goldenrod. *Solidago mollis.* Widespread, dry plains.[o]

Rigid goldenrod. *Solidago rigida.* Widespread, sandy or rocky prairies, drying sites.[sto]

Showy-wand goldenrod. *Solidago speciosa.* Eastern and northern; prairies; dry, open woods.[s]

Field sow thistle. *Sonchus arvensis.* Eastern edge, disturbed sites, weedy, introduced species.[ro]

Common dandelion. *Taraxacum officinale.* Widespread, introduced weed.[o]

Greenthread. *Thelesperma filifolium.* Widespread; open, weedy sites.[st]

Large-flowered Townsendia. *Townsendia grandiflora.* Panhandle, dry plains and hillsides.[st]

Goat's beard (Western salsify). *Tragopogon dubius.* Widespread, disturbed sites, introduced weed.[rsto]

Baldwin's ironweed. *Vernonia baldwinii.* Southeastern; open, dry pastures.[rs]

Western ironweed. *Vernonia fasciculata.* Eastern half, damp prairies.[sto]

Common cocklebur. *Xanthium strumarium.* Widespread, waste sites, weedy.[ro]

TOUCH-ME-NOT FAMILY — BALSAMINACEAE
(At least 2 species in Nebraska)
Spotted touch-me-not. *Impatiens capensis.* Widespread, shady woods.[st]

BARBERRY FAMILY — BERBERIDACEAE
(At least 3 species in Nebraska)
May-apple. *Podophyllum peltatum.* Southeastern corner, shady woods; poisonous.[st]

BORAGE FAMILY — BORAGINACEAE
(At least 23 species in Nebraska)
Butte candle. *Cryptantha celosoides.* Panhandle, dry hillsides.[o]

Miner's candle. *Cryptantha thyrsiflora.* Western third, rocky outcrops, open pine forests.[st]

Blueweed. *Echium vulgare.* Southeastern, waste sites, weedy.[t]

Large-flowered tickseed. *Hackelia floribunda.* Panhandle, moist creek-banks, open woods.[s]

Western sticktight. *Lappula redoweskii.* Widespread; open, often sandy waste sites.[r]

Hoary puccoon. *Lithospermum canescens.* Eastern third, dry prairies, open woods.[so]

Hairy puccoon. *Lithospermum carolinense.* Widespread, sandy prairies, open woods.[sto]

Narrow-leaved puccoon. *Lithospermum incisum.* Widespread, dry prairies, open woods.[st]

Lanceleaf bluebells. *Mertensia lanceolata.* Panhandle, brushy prairies.[o]

False gromwell. *Onosmodium molle.* Widespread, prairies, meadows, open woods.[sto]

MUSTARD FAMILY — BRASSICACEAE
(At least 64 species in Nebraska)

Hoary false alyssum. *Berteroa incana.* Northern half, waste sites, weedy.[st]

Indian mustard. *Brassica juncea.* Eastern and northern, waste sites, introduced weed.[t]

Hoary cress. *Cardaria draba.* Widespread, introduced weed.[o]

Toothwort. *Cardamine (Dentaria) concatenata.* Southeastern, moist woods.[t]

Tansy-mustard. *Descurainia pinnata.* Widespread, dry prairies, open woods.[r]

Western wallflower. *Erysimum aspersum.* Widespread, prairies, sandhills, open woods.[sto]

Bushy wallflower. *Erysimum repandum.* Widespread, waste places, weedy.[r]

Dame's rocket. *Hesperis matronalis.* Widespread, roadsides, waste sites, introduced weed.[t]

Greenflower pepperweed. *Lepidium densiflorum.* Widespread, waste places, weedy.[r]

Silvery bladderpod. *Lesquerella ludoviciana.* Mainly western, sandy and gravelly soils.[t]

Spreading yellowcress. *Rorippa sinuata.* Widespread, dry and wet sites, ditches.[r]

Tall hedge mustard. *Sisymbrium loeselii.* Widespread, waste sites, introduced weed.[st]

Prince's plume. *Stanleya pinnata.* Western half, selenium indicator species.[o]

Pennycress. *Thlaspi arvense.* Widespread, waste sites, weedy.[to]

CACTUS FAMILY — CACTACEAE
(At least 6 species in Nebraska)

Missouri pincushion. *Corypantha missouriensis.* Scattered records, dry soils.[o]

Pincushion cactus. *Corypantha vivipara*. Western half; dry, sandy, or rocky prairies.[sto]

Little (Brittle) prickly pear. *Opuntia fragilis*. Western half, sandy or rocky prairies.[rsto]

Bigroot prickly pear. *Opuntia macrohiza*. Widespread; sandy, gravelly, or rocky prairies.[rso]

Plains prickly pear. *Opuntia polyacantha*. Mainly western; dry, sandy prairies.[sto]

CAESALPINIA FAMILY — CAESALPINACEAE
(At least 6 species in Nebraska)

Partridge-pea. *Cassia fasciculata*. Eastern half, rocky or sandy prairies.[srt]

BELLFLOWER FAMILY — CAMPANULACEAE
(At least 12 species in Nebraska)

American bellflower. *Campanula americana*. Eastern third, open woods, wet sites.[st]

Harebell. *Campanula rotundifolia*. Northwestern, dry woods, meadows.[st]

Blue lobelia. *Lobelia siphilitica*. Widespread, moist soil, woods and meadows.[st]

Palespike lobelia. *Lobelia spicata*. Widespread, prairies, meadows, open woods.[s]

Western looking-glass. *Triodanis leptocarpa*. Southeastern, prairies, pastures, disturbed sites.[t]

Venus' looking glass. *Triodanis perfoliata*. Widespread, sandy to gravelly prairies, disturbed sites.[rst]

CAPER FAMILY — CAPPARACEAE
(At least 4 species in Nebraska)

Rocky Mountain beeplant. *Cleome serrulata*. Widespread, prairies, open woods.[rsto]

PINK FAMILY — CARYOPHYLLACEAE
(At least 30 species in Nebraska)

Sandwort. *Arenaria hookeri*. Panhandle, sandy to rocky hillsides, ledges.[st]

Grove sandwort. *Arenaria lateriflora*. Northern.

Rock sandwort. *Arenaria stricta*. Southwestern, gravelly to rocky sites.[s]

Prairie chickweed. *Cerastium arvense*. Western half, disturbed areas.[s]

Deptford pink. *Dianthus armeria*. Southeastern, disturbed sites, introduced weed.[t]

Bouncing bet. *Saponaria officinalis*. Widespread, disturbed sites, introduced weed.[st]

Catchfly. *Silene noctiflora*. Widespread, disturbed sites, introduced weed.[t]

Chickweed. *Stellaria media*. Mainly eastern, disturbed sites, introduced weed.[t]

GOOSEFOOT FAMILY — CHENOPODIACEAE
(At least 32 species in Nebraska)

Silverscale saltbush. *Atriplex argentea.* Mainly western, alkaline soils.[o]
Shadscale. *Atriplex canescens.* Northern panhandle.[o]
Mapleleaf goosefoot. *Chenopodium gigantospermum.* Widespread, disturbed sites.[r]
Winged pigweed. *Cycloloma atriplicifolium.* Widespread, sandy sites, weedy.[t]
Kochia. *Kochia scoparia.* Widespread, introduced weed.[o]
Nuttall povertyweed. *Monolepis nuttalliana.* Widespread, disturbed sites, weedy.[r]
Saltwort. *Salicornia rubra.* Very rare, Lancaster County saltflats; state endangered.

ST. JOHN'S WORT FAMILY—CLUSIACEAE (HYPERICACEAE)
(At least 4 species in Nebraska)
Common St. John's wort. *Hypericum perforatum.* Widespread, open sites, introduced weed.[rs]

SPIDERWORT FAMILY—COMMELINACEAE
(At least 5 species in Nebraska)
Erect dayflower. *Commelina erecta.* Scattered records, sandy or rocky soils.[st]
Long-bracted spiderwort. *Tradescantia bracteata.* Widespread, prairies, disturbed sites.[sto]
Prairie spiderwort. *Tradescantia occidentalis.* Widespread, prairies, disturbed sites.[so]
Ohio spiderwort. *Tradescantia ohiensis.* Southeastern, prairies, disturbed sites.[s]

MORNING-GLORY FAMILY—CONVOLVULACEAE
(At least 7 species in Nebraska)
Field bindweed. *Convolvulus arvensis.* Widespread, introduced weed.[sto]
Ivyleaf morning-glory. *Ipomoea hederacea.* Southeastern, open ground, weedy.[r]
Bush morning-glory. *Ipomoea leptophylla.* Widespread, plains and prairies.[sto]

STONECROP FAMILY—CRASSULACEAE
(At least 2 species in Nebraska)
Virginia stonecrop. *Penthorum sedoides.* Mainly eastern, ditches, streambanks.[t]
Stonecrop. *Sedum lanceolatum.* Panhandle; open, rocky sites.[s]

CUCUMBER FAMILY—CUCURBITACEAE
(At least 4 species in Nebraska)
Buffalo gourd. *Cucurbita foetidissima.* Southern half, sandy waste sites.[r]
Wild cucumber. *Echinocystis lobata.* Widespread; moist, open woods.[rt]

Bur cucumber. *Sicyos angulatus.* Southeastern, damp river soils, waste sites.[r]

SPURGE FAMILY — CUSCUTACEAE
(At least 10 species in Nebraska)
Field dodder. *Cuscuta pentagona.* Scattered records, parasitic.[r]

HORSETAIL FAMILY — EQUISETACEAE
(At least 5 species in Nebraska)
Field horsetail. *Equisetum arvense.* Widespread, disturbed sites.[r]

SPURGE FAMILY — EUPHORBIACEAE
(At least 26 species in Nebraska)
Texas croton. *Croton texensis.* Western half, sandy soils, weedy.[r]
Flowering spurge. *Euphorbia corollata.* Eastern quarter, rocky prairies, waste sites.[s]
Toothed spurge. *Euphorbia dentata.* Widespread, prairies, waste sites.[r]
Leafy spurge. *Euphorbia esula.* Introduced noxious weed.[o]
Spotted spurge. *Euphorbia maculata.* Eastern half, prairies, waste sites.[r]
Snow-on-the-mountain. *Euphorbia marginata.* Widespread, prairies, waste sites; poisonous.[rsto]

BEAN FAMILY — FABACEAE (LEGUMINACEAE)
(At least 90 species in Nebraska)
Standing milk vetch. *Astragalus adsurgens.* Mainly western, dry prairies, open woods.[sto]
Barr's milk vetch. *Astragalus barrii.* Panhandle, barren knolls.[o]
Two-grooved milk vetch. *Astragalus bisulcatus.* Northwestern Panhandle; poisonous.[o]
Canada milk vetch. *Astragalus canadensis.* Widespread, moist prairies, open woods.[st]
Painted milk vetch. *Astragalus ceramicus.* Mainly western, sandy prairies.[st]
Ground-plum. *Astragalus crassicarpus.* Widespread, prairies, rocky soils.[sto]
Drummond milk vetch. *Astragalus drummondii.* Western.
Plains milk vetch. *Astragalus gilviflorus.* Panhandle.[o]
Lotus milk vetch. *Astragalus lotiflorus.* Widespread.
Missouri milk vetch. *Astragalus missouriensis.* Mainly western, prairies, bluffs, ravines.[ro]
Woolly locoweed. *Astragalus mollissimus.* Mainly western; poisonous to livestock.[rto]
Alkali (creamy) milk vetch. *Astragalus racemosus.* Scattered records; poisonous to livestock.[sto]
Draba milk vetch. *Astragalus spatulatus.* Panhandle, rocky hills, prairies.[so]
Pulse (loose-flowered) milk vetch. *Astragalus tenellus.* Northern Panhandle.

Large wild indigo. *Baptisia lactea.* Southeastern, rocky prairies, hillsides; poisonous.[t]

Plains wild indigo. *Baptisia (leucophea) bracteata.* Southeastern, prairies, pastures.[ts]

Golden prairie clover. *Dalea aurea.* Mainly west, loamy prairies.[o]

Slender (nineanther) dalea. *Dalea enneandra.* Widespread.[o]

Canada tickclover. *Desmodium canadense.* Widespread, rocky or sandy prairies.[s]

Tick trefoil. *Desmodium illinoensis.* Widespread, prairie ravines, hillsides.[t]

Wild licorice. *Glycyrrhiza lepidota.* Widespread, prairie ravines, moist areas.[rsto]

Hoary vetchling. *Lathyrus polymorphus.* Widespread; dry, sandy to rocky prairies; woods.[sto]

Bird's-foot trefoil. *Lotus corniculatus.* Scattered records, waste sites, introduced weed.[sto]

American deervetch. *Lotus purshianus.* Widespread.[o]

Silvery lupine. *Lupinus argenteus.* Western half, rocky prairies, open woods.[s]

Low lupine. *Lupinus pusillus.* Western half, sandy prairies.[r]

Alfalfa. *Medicago falcata.* Introduced forage crop.[o]

Sweet-clover. *Melilotis officinalis* and *M. albas.* Widespread, waste places, introduced weed and forage crop.[to]

Purple (Lambert) locoweed. *Oxytropis lambertii.* Widespread; poisonous to livestock.[rso]

White locoweed. *Oxytropis sericea.* Panhandle, rocky prairies; poisonous.[o]

White prairie-clover. *Petalostemon (Dalea) candida.* Widespread, waste sites, weedy.[sto]

Round-headed prairie-clover. *Petalostemon (Dalea) multiflora.* Southeastern corner.

Purple prairie-clover. *Petalostemon (Dalea) purpurea.* Widespread, rocky prairies, open woods.[sto]

Silky prairie-clover. *Petalostemon (Dalea) villosa.* Widespread, sandy prairies, open woods.[so]

Silver-leaf scurf-pea. *Psoralea argophylla.* Widespread, prairies, open woods.[sto]

Tall breadroot scurf-pea. *Psoralea cuspidata.* Widespread.[o]

Palmleaf scurf-pea. *Psoralea digitata.* Widespread.[o]

Broad-leaf scurf-pea (Prairie-turnip). *Psorales esculenta.* Widespread, prairies, open woods.[so]

Wild alfalfa. *Psoralea tenuiflora.* Widespread, prairies, roadsides.[sto]

Prairie buck-bean. *Thermopsis rhombifolia.* Western half, prairies, open woods.[sto]

Clovers. *Trifolium* spp. Mostly introduced and cultivated forage plants.[o]

American vetch. *Vicia americana*. Widespread, mostly sandy soils.[o]

FUMITORY FAMILY — FUMARIACEAE
(At least 4 species in Nebraska)
Golden corydalis. *Corydalis aurea*. Panhandle, prairies, open wood.[t]
Dutchman's breeches. *Dicentra cucullaria*. Missouri Valley, moist woods.[st]

GENTIAN FAMILY — GENTIANACEAE
(At least 4 species in Nebraska)
Prairie gentian. *Eustoma grandiflorum*. Mainly western, moist meadows and prairies.[st]
Closed gentian. *Gentiana andrewsii*. Northeastern, wet meadows, prairies, or woods.[sto]
Downy gentian. *Gentiana puberulenta*. Eastern half, dry woods and prairies.[sto]

GERANIUM FAMILY — GERANIACEAE
(At least 4 species in Nebraska)
Carolina geranium. *Geranium carolinianum*. Southeastern, open woods, prairie ravines.[t]
Wild cranesbill. *Geranium maculatum*. Southeastern, rich or rocky woods.[s]

WATERLEAF FAMILY — HYDROPHYLLACEAE
(At least 4 species in Nebraska)
Waterpod. *Ellisia nyctelea*. Widespread, sandy prairies, open woods.[r]
Waterleaf. *Hydrophyllum virginianum*. Missouri Valley, moist woods.[st]
Scorpionweed. *Phacelia hastata*. Panhandle, sandy to rocky soils, disturbed sites.[s]

IRIS FAMILY — IRIDACEAE
(At least 7 species in Nebraska)
White-eyed grass. *Sisyrinchium campestre*. Eastern half, prairies, open woods.[s]
Blue-eyed grass. *Sisyrhynchium montanum*. Western half, prairies, open woods.[sto]

ARROWGRASS FAMILY — JUNCAGINACEAE
(At least 13 species in Nebraska)
Arrowgrass. *Triglochin maritimum*. Western half, moist, alkaline sites.[r]

MINT FAMILY — LAMIACEAE (LABITAE)
(At least 41 species in Nebraska)
Rough false pennyroyal. *Hedeoma hispidum*. Widespread, waste sites, open ground.[r]
Motherwort. *Leonurus cardiaca*. Mainly eastern, waste sites, weedy.[t]
American bugleweed. *Lycopus americanus*. Widespread; moist, exposed sites.[r]

Field mint. *Mentha arvensis*. Widespread, moist sites.[r]

Wild bergamot. *Monarda fistulosa*. Widespread, prairies, open woods.[rsto]

Plains (Spotted) beebalm. *Monarda pectinata*. Mainly western, upland prairies.[rst]

Catnip. *Nepeta cataria*. Widespread, waste sites, introduced weed.[st]

Virginia mountain mint. *Pycnanthemum virginianum*. Eastern half, moist woods, wetlands.[s]

Healall. *Prunella vulgaris*. Widespread, waste sites, streambanks.[rst]

Pitcher's (Blue) sage. *Salvia azurea (pitcheri)*.[st] Southeastern, rocky to sandy prairies.

Lanceleaf sage. *Salvia reflexa*. Widespread, disturbed sites.[r]

Marsh scullcap. *Scutellaria galericulata*. Mainly western, wet sites.[s]

Leonard small scullcap. *Scutellaria parvula*. Eastern, upland prairies, open woods.[s]

Marsh hedge-nettle. *Stachys palustris*. Mainly eastern, dry to wet prairies.[s]

American germander. *Teucrium canadense*. Widespread, streambanks, pastures.[rst]

LILY FAMILY — LILIACEAE

(At least 29 species in Nebraska)

Wild onion. *Allium canadense*. Widespread, prairies, open woods.[rst]

Pink wild onion. *Allium stellatum*. Eastern half.[o]

Wild white onion. *Allium textile*. Western half, prairies, coniferous woods.[so]

Sego lily. *Calochortus gunnisonii*. Panhandle, dry prairies, open coniferous woods.[st]

Mariposa lily. *Calochortus nuttallii*. Pine Ridge, open coniferous woods.[o]

White fawn lily. *Erythronium albidum*. Missouri Valley, moist woods.[st]

Prairie fawn lily. *Erythronium mesochoreum*. Eastern, prairies, open woods.[s]

Yellow stargrass. *Hypoxis hirsuta*. Mainly eastern, prairies, open woods.[st]

Mountain lily. *Leucocrinum montanum*. Western half, shortgrass prairies, coniferous woods.[sto]

Turk's cap (Canada) lily. *Lilium canadense*. Missouri Valley, moist prairies and woods.[s]

Western red lily. *Lilium philadelphicum*. Northern half, open woods, prairies.[sto]

Solomon's seal. *Polygonatum biflorum*. Widespread, moist deciduous woods.[rst]

False Solomon's seal. *Smilacina stellata*. Widespread, moist to dry woods.[st]

Death Camass. *Zigadenus venenosus*. Panhandle, dry prairies, open woods; poisonous.[rsto]

FLAX FAMILY — LINACEAE

(At least 6 species in Nebraska)

Stiffstem flax. *Linum rigidum.* Widespread, sandy prairies and hillsides.
Grooved (prairie) flax. *Linum sulcatum.* Eastern half, prairies, open woods.[st]

STICKLEAF FAMILY — LOASACEAE
(At least 4 species in Nebraska)
Ten-petal stickleaf (mentzelia). *Mentzelia decapetala.* Western half, waste sites.[rso]

LOOSESTRIFE FAMILY — LYTHRACEAE
(At least 6 species in Nebraska)
Winged loosestrife. *Lythrum alatum* (*decotanum*). Widespread, wet soils.[st]
Purple loosestrife. *Lythrum salicaria.* Widespread, moist sites, introduced weed.[rso]

MALLOW FAMILY — MALVACEAE
(At least 12 species in Nebraska)
Pink poppy mallow. *Callirhoe alcaeoides.* Eastern half; dry, sandy prairies.[s]
Purple poppy mallow. *Callirhoe involucrata.* Widespread; dry, sandy prairies.[rst]
Flower-of-an-hour. *Hibiscus* (*Malviscus*) *trionum.* Eastern half, waste sites, introduced weed.[st]
Running mallow. *Malva rotundifolia.* Widespread, waste sites, introduced weed.[t]
Red false (Scarlet globe) mallow. *Sphaeralcea coccinea.* Widespread, dry prairies, hillsides.[rsto]

MIMOSA FAMILY — MIMOSACEAE
(At least 2 species in Nebraska)
Illinois bundleflower. *Desmanthus illinoensis.* Widespread, rocky or sandy prairies.[s]
Sensitive brier. *Schrankia nuttallii.* Mainly eastern, rocky or sandy soils.[so]

INDIAN PIPE FAMILY — MONOTROPACEAE
(At least 2 species in Nebraska)
Indian pipe. *Monotropa uniflora.* Southeastern, saprophytic (myco-trophic) on organic matter.[t]

FOUR-O'CLOCK FAMILY — NYCTAGINACEAE
(At least 6 species in Nebraska)
Sweet sand verbena. *Abronia fragrans.* Panhandle, sandy prairies, waste sites.[st]
Hairy four-o'clock. *Mirabilis hirsuta.* Widespread, prairies, open woods.[st]
Wild four-o'clock. *Mirabilis nyctaginea.* Widespread, waste places, weedy.[rst]

EVENING PRIMROSE FAMILY — ONAGRACEAE
(At least 29 species in Nebraska)

Lavender evening primrose. *Calyphus hartwegii*, var. *lavandulifolius*. Panhandle, dry prairies.[s]

Plains yellow evening primrose. *Calyphus serrulatus*. Widespread, dry prairies, open woods.[so]

Fireweed. *Epilobium* spp. Widespread, disturbed sites, often appears following fire.[t]

Scarlet gaura. *Gaura coccinea*. Widespread, dry prairies, open woods.[rsto]

Large-flowered gaura. *Gaura longiflora*. Southeastern, rocky prairies, open woods.[s]

Colorado butterfly plant. *Gaura neomexicana coloradensis*. Rare along Lodgepole Creek; state endangered.

Common evening primrose. *Oenothera biennis*. Widespread, streambanks, open woods.[rst]

Gumbo evening primrose. *Oenothera caespitosa*. Panhandle, dry prairies, open woods.[so]

Comb-leaf evening primrose. *Oenothera coronopifolia*. Panhandle, sandy to rocky prairies, woods.[s]

Fremont's evening primrose. *Oenothera macrocarpa* var. *fremontii*. Local along Kansas border in central and eastern Nebraska.

White-stemmed evening primrose. *Oenothera nuttallii*. Western half, dry prairies, open woods.[so]

Fourpoint evening primrose. *Oenothera rhombipetala*. Widespread, sand dunes, sandy prairies.[st]

ORCHID FAMILY — ORCHIDACEAE

(At least 17 species in Nebraska)

Spotted coral-root. *Corallorhiza maculata*. Panhandle, dry coniferous woods, largely mycotrophic.[t]

Late coral-root. *Corallorhiza odontorhiza*. Southeastern corner, Missouri Valley.

Wister's coral-root. *Corallorhiza wisteriana*. Southeastern corner, Missouri Valley.

Small white lady's-slipper. *Cypripedium candidum*. Moist soils, now known only from Howard, Pierce, Platte, and Sherman Counties; state threatened.[s]

Large yellow lady-slipper. *Cypripedium calceolus*. Lower Missouri Valley, moist soils, prairies or woods.[st]

Showy orchis. *Galearis (Orchis) spectabilis*. Missouri Valley; moist, upland woods.[st]

Northern green orchid. *Habenaria hyperborea*. Northern Panhandle and Niobrara Valley.

Prairie fringed orchid. *Platanthera ("Habenaria") praeclara*. Scattered records in east; nationally threatened.[s]

Nodding lady's-tresses. *Spiranthes cernua*. Eastern half, prairies, open woods.[st]

Ute lady's-tresses. *Spiranthes diluvialis*. Rare, western (alkaline meadows, Niobrara Valley, Sioux County); nationally threatened.

BROOM-RAPE FAMILY — OROBRANCHACEAE
(At least 3 species in Nebraska)
Cancer-root. *Orobanche fasciculata*. Scattered records, dry prairies, sandy soils.[t]

WOODSORREL FAMILY — OXALIDACEAE
(At least 3 species in Nebraska)
Gray-green wood sorrel. *Oxalis dillenii*. Mainly eastern, open woods, waste sites, weedy.[r]

Yellow wood sorrel. *Oxalis stricta*. Mainly eastern, open woods, waste sites, weedy.[st]

Violet wood sorrel. *Oxalis violacea*. Mainly eastern, open woods, waste places, weedy.[sto]

POPPY FAMILY — PAPAVERACEAE
(At least 2 species in Nebraska)
Annual pricklypoppy. *Argemone polyanthemos*. Widespread, sandy soils, waste sites.[rso]

Bloodroot. *Sanguinaria canadensis*. Missouri Valley, moist woods; poisonous.[s]

UNICORN-PLANT FAMILY — PEDALIACEAE
(At least 1 species in Nebraska)
Devil's claw. *Proboscidea louisianica*. Southern counties, sandy pastures, waste sites, weedy.[r]

POKEWEED FAMILY — PHYTOLACCACEAE
(At least 1 species in Nebraska)
Common pokeweed. *Phytolacca americana*. Southeastern corner, waste sites, weedy; poisonous.[r]

PLANTAIN FAMILY — PLANTAGINACEAE
(At least 9 species in Nebraska)
Buckhorn plantain. *Plantago lanceolata*. Scattered records, waste sites, introduced weed.[r]

Woolly plantain (Indianwheat). *Plantago patagonica*. Widespread, waste sites, weedy.[ro]

Blackseed plantain. *Plantago rugelii*. Mainly eastern, waste sites, shady places, weedy.[r]

PHLOX (POLEMONIUM) FAMILY — POLEMONIACEAE
(At least 12 species in Nebraska)
Slenderleaf collomia. *Collimia linearis*. Mainly northern, native prairies.[o]

Gilia. *Ipomopsis longifolia.* Western half; dry, sandy soil.[st]
Plains phlox. *Phlox andicola.* Western half; dry, sandy prairies.[st]
Blue phlox. *Phlox divaricata.* Eastern quarter, open woods, rocky slopes.[st]
Hood's phlox. *Phlox hoodi.* Panhandle, rocky soils.[o]
Prairie phlox. *Phlox pilosa.* Eastern quarter, open woods, meadows.[sto]

MILKWORT FAMILY — POLYGALACEAE

(At least 4 species in Nebraska)
White milkwort. *Polygala alba.* Widespread, rocky prairies hillsides.[sto]

BUCKWHEAT FAMILY — POLYGONACEAE

(At least 36 species in Nebraska)
Umbrella plant. *Eriogonum annuum.* Widespread; dry, open grasslands.[ro]
Yellow wild buckwheat. *Eriogonum flavum.* Panhandle, dry plains and ridges.[so]
Littleleaf eriogonum. *Eriogonum pauciflorum.* Western panhandle.[o]
Common knotweed. *Polygonum arenostrum.* Widespread, waste sites, introduced weed.[r]
Pink smartweed. *Polygonum bicorne.* Mainly eastern, wet sites.[s]
Pale smartweed. *Polygonum lapathifolium.* Widespread, damp soils.[r]
Pennsylvania smartweed. *Polygonum pensylvanicum.* Widespread, disturbed sites, weedy.[r]
Bushy knotweed. *Polygonum ramosissimum.* Widespread; damp, brackish soils.[r]
Climbing false buckwheat. *Polygonum scandens.* Widespread, waste sites, introduced weed.[t]
Wild begonia (veiny dock). *Rumex venosus.* Widespread, sandy dunes and riverbanks.[sto]

PURSLANE FAMILY — PORTULACACEAE

(At least 4 species in Nebraska)
Virginia spring beauty. *Claytonia virginica.* Lower Missouri Valley, moist woods.[st]

PRIMROSE FAMILY — PRIMULACEAE

(At least 10 species in Nebraska)
Shooting star. *Dodecatheon pulchellum.* Scattered records, moist woods and prairies.[st]
Fringed loosetrife. *Lysimachia ciliata.* Widespread weed, moist woods and wetter sites.[st]
Moneywort. *Lysimachia nummularia.* Southeastern, moist sites, introduced.[t]
Tufted loosestrife. *Lysimachia thyrsiflora.* Widespread, moist to wet sites.[st]

BUTTERCUP FAMILY — RANUNCULACEAE

(At least 37 species in Nebraska)
Meadow anemone. *Anemone canadensis.* Mainly eastern, wet prairies, wet woods.[sto]

Candle anemone. *Anemone cylindrica.* Northern and eastern, open prairies and pastures.[st]

Pasque flower. *Anemone patens.* Mainly northern, open prairies, often rocky soil.[sto]

Wild columbine. *Aquilegia canadensis.* Northern and eastern, moist woods.[st]

Fremont's clematis. *Clematis fremontii.* Southern counties, rocky prairie hillsides.[s]

Virgin's bower. *Clematis virginiana.* Eastern half, partly woody; climbing vine.[t]

Blue larkspur. *Delphinium nuttallianum.* Panhandle, varied habitats.

Prairie larkspur. *Delphinium virescens.* Widespread, prairies and pastures.[rsto]

Early wood (Smallflower) buttercup. *Ranunculus abortivus.* Widespread, moist woods.[r]

Threadleaf buttercup. *Ranunculus flabellaris.* Scattered records, moist and wet sites.[s]

Macoun's buttercup. *Ranunculus macouni.* Western half, streambanks, wet meadows.[s]

Purple meadow rue. *Thalictrum dasycarpum.* Widespread, moist habitats.[so]

ROSE FAMILY — ROSACEAE
(At least 42 species in Nebraska)

Wild strawberry. *Fragaria virginiana.* Mainly eastern, moist soils, prairies, open woods.[st]

Silverweed. *Potentilla anserina.* Scattered records, streambanks, saline soils.[t]

Tall cinquefoil. *Potentilla arguta.* Widespread, prairies, open woods.[sto]

Sulphur cinquefoil. *Potentilla recta.* Mainly eastern, waste sites, prairies.[st]

MADDER FAMILY — RUBIACEAE
(At least 10 species in Nebraska)

Catchweed bedstraw. *Galium aparine.* Widespread, woods, prairies, waste sites.[r]

Northern bedstraw. *Galium boreale.* Northern half, rocky prairies, woods, roadsides.[to]

Narrow-leafed bluets. *Hedyotis nigricans.* Southeastern, rocky prairies, woods.[t]

SANDALWOOD FAMILY — SANTALACEAE
(At least 2 species in Nebraska)

Bastard toad-flax. *Comandra umbellata.* Eastern quarter; dry, sandy to rocky soils; open woods.[sto]

FIGWORT FAMILY — SCROPHULARIACEAE
(At least 46 species in Nebraska)

Rough purple gerardia. *Agalinis aspersa.* Widespread, dry prairies, open woods.[s]

Slender gerardia. *Agalinis tenuifolia.* Widespread, moist woods and prairies.[st]

Butter-and-eggs. *Linaria vulgaris.* Widespread, waste sites, weedy.[t]

Allegheny monkey-flower. *Mimulus ringens.* Eastern half, wet streamsides, sometimes emergent.[s]

White beardtongue. *Penstemon albidus.* Widespread, sandy to gravelly soils, open prairies.[sto]

Narrow beardtongue. *Penstemon angustifolius.* Mainly western, sandhills and sandy prairies.[sto]

Cobea penstemon. *Penstemon cobaea.* Southeastern, open prairies, pastures, weedy.[st]

Crested beardtongue. *Penstemon eriantherus.* Western half, sandy to gravelly soils, dry prairies.[s]

Sawsepal penstemon. *Penstemon glaber.* Panhandle, sandy to gravelly soils, prairies.[so]

Slender penstemon. *Penstemon gracilis.* Widespread, sandy to gravel soils, dry prairies.[sto]

Shell-leaf penstemon. *Penstemon grandiflorus.* Widespread, prairies with sandy to loamy soils.[to]

Hayden's (Blowout) penstemon. *Penstemon haydenii.* Sandhills, bare dunes; nationally endangered.[s]

Common mullein. *Verbascum thapsus.* Widespread, waste sites, introduced weed.[rsto]

Brooklime speedwell. *Veronica americana.* Widespread, emergent in aquatic sites.[s]

Water speedwell. *Veronica anagallis-aquatica.* Widespread, emergent in aquatic sites.[s]

NIGHTSHADE FAMILY — SOLANACEAE

(At least 20 species in Nebraska)

Jimsonweed. *Datura stramonium.* Eastern half, waste sites; poisonous.[rt]

Clammy groundcherry. *Physalis heterophylla.* Widespread, prairies, open woods.[rst]

Virginia groundcherry. *Physalis virginiana.* Widespread, open woods, waste sites.[r]

Carolina horse-nettle. *Solanum carolinense.* Southeastern, waste sites, open woods; poisonous.[rs]

Buffalobur. *Solanum rostratum.* Widespread, waste sites; poisonous.[rsto]

NETTLE FAMILY — URTICACEAE

(At least 5 species in Nebraska)

Wood nettle. *Laportea canadensis.* Missouri Valley, moist woods; skin irritant.[t]

Pennsylvania pellitory. *Parietaria pensylvanica.* Widespread, shaded woods.[r]

Stinging nettle. *Urtica dioica.* Widespread, moist woods, streambanks; skin irritant.[r]

VERVAIN FAMILY — VERBENACEAE
(At least 9 species in Nebraska)
Fog-fruit. *Lippia (Phyla) lanceolata.* Eastern and southern, prairies, ditches, waste sites.[t]

Dakota vervain. *Verbena bipinnatifida.* Widespread, dry plains and prairies.[s]

Prostrate vervain. *Verbena bracteata.* Widespread, waste sites, prairies.[r]

Blue vervain. *Verbena hastata.* Widespread, moist meadows, woods, seepage areas.[s]

Hoary (woolly) vervain. *Verbena stricta.* Widespread, pastures, prairies, waste sites.[rso]

VIOLET FAMILY — VIOLACEAE
(At least 10 species in Nebraska)
Canada violet. *Viola canadensis.* Northern and eastern, shaded woods.[s]

Nuttall's violet. *Viola nuttallii.* Western half, dry prairies, bluffs.[so]

Prairie violet. *Viola pedatifida.* Eastern half, prairies, open woodlands.[sto]

Blue prairie violet. *Viola pratincola.* Widespread, open woods, prairie hillsides.[s]

Downy yellow violet. *Viola pubescens.* Northern and eastern, woods, thickets.[s]

Downy blue violet. *Viola sororia.* Northern and eastern, woods, streamsides.[s]

CALTROP FAMILY — ZYGOPHYLLACEAE
(At least 1 species in Nebraska)
Puncture vine. *Tribulus terrestris.* Widespread, waste places, weedy.[st]

Common Prairie and Range Grasses and Sedges

This list includes nearly 100 of Nebraska's more widespread grasses and sedges, or about a third of the state's reported total. The sequence is alphabetical by genera and species. Johnson and Larson (1999), Larson and Johnson (1999), Stubbendieck et al. (1997), and Brown (1979) describe and illustrate these species. In addition, all of Nebraska's grasses were illustrated and described by Hitchcock (1937).

> [r] indicates grass and sedge species described and illustrated by Johnson and Larson (1999).
>
> [o] indicates "weedy," often introduced, species illustrated by Stubbendieck et al. (1995).

^p indicates species illustrated and
described by Stubbendieck et al. (1997).

^u indicates species described and illustrated by Brown (1979).

Jointed goatgrass. *Aegilops cylindrica.* Scattered records, introduced weed.^o

Slender wheatgrass. *Agropyron caninum (trachycaulum).* Widespread.^{rp}

Crested wheatgrass. *Agropyron cristatum.* Widespread.^{rp}

Thickspike wheatgrass. *Agropyron dasystachium.* Northwestern Panhandle.^r

Intermediate wheatgrass. *Agropyron intermedium.* Scattered records, introduced.^{rp}

Quackgrass. *Agropyron repens.* Introduced weed.^{oru}

Western wheatgrass. *Agropyron smithii.* Widespread.^{rp}

Bluebunch wheatgrass. *Agropyron spicatum.* Northern Panhandle.^{rp}

Redtop bent. *Agrostis stolonifera.* Widespread, introduced weed.^{orp}

Big bluestem. *Andropogon gerardii.* Mainly eastern.^{rpu}

Sandhills bluestem. *Andropogon hallii.* Sandhills.^{rp}

Red threeawn. *Aristida longiseta.* Widespread.^{rp}

Prairie threeawn. *Aristida oligantha.* Mainly eastern.^{opu}

Purple threeawn. *Aristida purpurea.* Panhandle, weedy.^{orp}

Wild oats. *Avena fatua.* Widespread, introduced weed.^{op}

American sloughgrass. *Beckmannia syzigachne.* Northwestern, wetlands.^r

Side-oats grama. *Bouteloua curtipendula.* Widespread.^{rpu}

Blue grama. *Bouteloua gracilis.* Widespread.^{rp}

Hairy grama. *Bouteloua hirsuta.* Mainly western.^{rp}

Smooth brome. *Bromus inermis.* Widespread, introduced weed.^{orp}

Japanese brome. *Bromus japonica.* Widespread, introduced weed.^r

Downy brome. *Bromus tectorum.* Widespread, introduced weed.^{orp}

Buffalo grass. *Buchloe dactyloides.* Mainly western.^{rp}

Bluejoint. *Calamagrostis canadensis.* Widespread.^{pu}

Northern reedgrass. *Calamogrostis inexpansa.* Widespread.

Prairie sandreed. *Calamovilfa longifolia.* Widespread.^{rp}

Fescue sedge. *Carex brevior.* Widespread.^r

Needleleaf sedge. *Carex eleocharis.* Widespread.^r

Threadleaf sedge. *Carex filifolia.* Western half, upland prairies.^{rp}

Sun sedge. *Carex heliophila.* Widespread.^r

Woolly sedge. *Carex languinosa.* Widespread, low prairies.^{ru}

Nebraska sedge. *Carex nebraskensis.* Mainly western, swamps, wet meadows.^{rp}

Fox sedge. *Carex vulpinoides.* Widespread.^{ru}

Longspine sandbur. *Cenchrus longispinus.* Widespread, weedy.^{rou}

Tumble windmillgrass. *Chloris verticillata.* Mainly southern, weedy.^o

Schweinitz flatsedge. *Cyperus schweinitzii.* Widespread.

Orchardgrass. *Dactylis glomerata*. Widespread, introduced weed.[orpu]
Smooth crabgrass. *Digitaria ischaemum*. Eastern half, introduced weed.[o]
Large crabgrass. *Digitaria sanguinalis*. Widespread, introduced weed.[ou]
Saltgrass. *Distichlis spicata*. Widespread, saline soils, weedy.[orpu]
Barnyardgrass. *Echinochloa crusgalli*. Widespread, introduced weed.[oru]
Goosegrass. *Eleusine indica*. Southeastern, introduced weed.[ou]
Canada wildrye. *Elymus canadensis*. Widespread.[rpu]
Stinkgrass. *Eragrostis cilianensis*. Weedy annual.[r]
Purple lovegrass. *Eragrostis spectabilis*. Eastern half, weedy.[oru]
Sand lovegrass. *Eragrostis trichodes*. Widespread.[rp]
Sixweeks fescue. *Festuca (Vulpia) octoflora*. Widespread, weedy.[orp]
Fowl mannagrass. *Glyceria striata*. Widespread, moist soils.[ru]
Foxtail barley. *Hordeum jubatum*. Widespread, weedy.[orpu]
Little barley. *Hordeum pusillum*. Widespread, weedy.[orpu]
Prairie junegrass. *Koeleria pyramidata*. Widespread.[rp]
Bearded sprangletop. *Leptochloa fascicularis*. Widespread, weedy.[o]
Plains muhly. *Muhlenbergia cuspidata*. Widespread, dry soils.[r]
Marsh muhly. *Muhlenbergia racemosa*. Widespread, weedy.[or]
Nimblewill. *Muhlenbergia schreiberi*. Eastern third, weedy.[ou]
False buffalo grass. *Munroa squarrosa*. Western half, dry plains.[r]
Indian ricegrass. *Oryzopsis hymenoides*. Western half.[rp]
Witchgrass. *Panicum capillare*. Widespread, weedy.[oru]
Fall panicum. *Panicum dichotomiflorum*. Eastern half, weedy.[ou]
Proso millet. *Panicum miliaceum*. Scattered records, introduced weed.[o]
Small (Scribner) panicgrass. *Panicum oligosanthes*. Widespread, mainly east.[r]
Switchgrass. *Panicum virgatum*. Widespread, tallgrass prairies.[rpu]
Reed canarygrass. *Phalaris arundinacea*. Widespread.[rpu]
Timothy. *Phleum pratense*. Widespread.[rpu]
Common reed. *Phragmites australis*. Widespread, wet soils.[ru]
Annual bluegrass. *Poa annua*. Eastern quarter, introduced weed.[ou]
Canada bluegrass. *Poa compressa*. Widespread, introduced weed.[ru]
Kentucky bluegrass. *Poa pratensis*. Widespread, introduced weed.[orpu]
Blowoutgrass. *Redfieldia flexuosa*. Sandhills dunes.[rp]
Tumblegrass. *Schedonnardus paniculatus*. Widespread, weedy.[orp]
Little bluestem. *Schizachyrium (Andropogon) scoparium*. Widespread.[rpu]
Spangletop (whitetop). *Scolochloa festucacea*. Scattered records, northern.[u]
Giant foxtail. *Setaria faberi*. Eastern half, introduced weed.[o]
Yellow foxtail. *Setaria glauca*. Widespread, introduced weed.[oru]
Bristly foxtail. *Setaria verticillata*. Widespread, introduced weed.[o]
Green foxtail. *Setaria viridis*. Widespread, introduced weed.[oru]
Squirreltail. *Sitanion hystrix*. Southwestern.[rp]
Indiangrass. *Sorghastrum nutans*. Widespread.[rpu]

Johnsongrass. *Sorghum halepense.* Widespread, weedy.[o]

Alkali cordgrass. *Spartina gracilis.* Western half.[rp]

Prairie cordgrass. *Spartina pectinata.* Widespread.[rp]

Alkali sacaton. *Sporobolus airoides.* Mainly western.[rp]

Tall dropseed. *Sporobolus asper.* Widespread.[rpu]

Sand dropseed. *Sporobolus cryptandrus.* Sandhills, also widespread, weedy.[orp]

Prairie dropseed. *Sporobolus heterolepis.* Widespread, native prairies.[ru]

Poverty dropseed. *Sporobolus vaginiflorus.* Eastern half, weedy.[ou]

Needle-and-thread. *Stipa comata.* Mainly central and western.[rp]

Porcupine grass. *Stipa spartea.* Widespread, native prairies.[r]

Green needlegrass. *Stipa viridula.* Mainly western.[r]

Eastern gamagrass. *Tripsacum dactyloides.* Southeastern.[pu]

Common Shoreline and Aquatic Plants

This list is based largely on Whitley et al. (1990), who illustrate nearly all the species listed here.

SHORELINE SPECIES

Calamus (Sweet flag). *Acorus calamus.* Probably introduced; southeastern corner.

Water plantain. *Alisma gramineum.* North-central.

Buttonbush. *Cephalanthus occidentalis.* Lower Missouri Valley.

Horsetails. *Equisetum* spp. Widespread.

Burhead. *Ecinodorus rostratus.* Missouri Valley.

Mud Plantain. *Heteranthera limosa.* Southern half.

Rose mallow. *Hibiscus militaris.* Missouri Valley.

Wild irises. *Iris* spp. (especially Southern blue flag, *Iris virginica*). Several species.

Purple loosestrife. *Lythrum salicaria.* Introduced, widespread invasive species.

Water pepper. *Polygonum hydropiper.* Eastern half.

Mild water pepper. *Polygonum hydropiperoides.* Southern half.

Threadleaf buttercup. *Ranunculus flabellaris.* Scattered records.

Water parsnip. *Sium suave.* Widespread.

EMERGENT SPECIES

Spikerushes. *Eleocharis* spp. Widespread.

Hedge hyssop. *Gratiola neglecta.* Eastern.

Rush. *Juncus* spp. Widespread, many species in Nebraska.

American lotus. *Nelumbo lutea.* Southeastern.

Spatterdock. *Nuphar luteum.* Widespread.

Water cress. *Rorippa* spp. Widespread.

Arrowheads. *Sagittaria* spp., especially *S. latifolia.* Widespread.

Giant bur-weed. *Sparganium eurycarpum.* Widespread.

Bullrush. *Scirpus* spp. Widespread, many species in Nebraska.

Narrow-leaved cattail. *Typha angustifolia.* Widespread.

Common cattail. *Typha latifolia.* Widespread.

Brooklime speedwell. *Veronica americana.* Widespread.

Water speedwell. *Veronica anagallis-aquatica.* Widespread.

FLOATING-LEAF SPECIES

Water fern. *Azolla mexicana.* Scattered records.

Lesser duckweed. *Lemna minor.* Widespread.

Yellow water lily. *Nuphar luteum.* Widespread but local.

American (Fragrant) water lily. *Nymphea odorata.* Widespread but local.

Floating-leaved pondweeds. *Potamogeton* spp., especially *P. nodosus.* Widespread.

Big duckweed (Duck's meat). *Spirodela polyrhiza.* Widespread.

SUBMERGED SPECIES

Starworts. *Callitriche* spp. Local.

Coontail. *Ceratophyllum demersum.* Widespread.

Elodea. *Elodea canadensis.* Widespread.

Water stargrass. *Heteranthera dubia.* Eastern.

Water purslane. *Ludwigia palustris.* Eastern.

Water milfoil. *Myriophyllum heterophyllum.* Widespread.

Eurasian water milfoil. *Myriophyllum spicatum.* Introduced, invasive species.

Naiads. *Najas* spp. Widespread.

Submerged pondweeds. *Potamogeton* spp., especially *P. pectinatus* (sago pondweed), *P. foliosus* (leafy pondweed), and *P. pusillus* (small pondweed). Widespread.

White water crowfoot. *Ranunculus longirostris.* Widespread.

Bladderwort. *Utricularia vulgaris.* Widespread.

A Guide
to Nebraska's
Natural Areas
and Preserves

This guide represents an effort to identify and describe nearly all of the 400 public-access sites in Nebraska of special interest to naturalists. It includes all of Nebraska's federally owned national wildlife refuges, national monuments, and national grasslands. It also includes nearly all the state parks, the larger state wildlife management areas, the waterfowl production areas, and those state recreation areas having significant natural habitats. Municipal and county parks are not included unless they are of particular biological interest.

This appendix is divided into four general regions: Far Western, West-central, East-central, and Eastern. Within each of these regions, individual counties are discussed in a generally west-to-east and north-to-south sequence. For each Nebraska county, federally owned and state-owned public-access locations are described, followed by any additional sites under other kinds of ownership or control, such as by municipalities, private landowners, the National Audubon Society, or the Nature Conservancy. Most federally owned areas in Nebraska consist of national historic sites, national wildlife refuges, and national monuments. Federally owned areas also include waterfowl production areas (WPAS). State-owned sites include state parks, state recreation areas (SRAS), and wildlife management areas (WMAS). Typically no permit is needed to enter WPAS or WMAS, but annual (or daily) state park entry permits are needed for all SRAS, state parks, and state historical parks. Nature Conservancy sites may require entry permission from their local field office or their state headquarters in Omaha. One national wildlife refuge (DeSoto) also now charges a daily entry fee. All state wildlife management areas offer free, unrestricted birding or other nature study opportunities. They usually provide only primitive camping facilities, and most are open to seasonal hunting and fishing. State recreation areas typically offer more highly developed recreational facilities and modern camping. Annual state park/recreation area entry permits or more information can be obtained from the Parks Division, Game and Parks Commission in Lincoln. Entry permits are also sold at the larger parks and also at many local businesses that sell sporting goods. District Game and Parks offices are also located in Alliance, Bassett, Norfolk, and North Platte. They sell park permits and can supply regional information. The Game and Parks homepage is *www.ngpc.state.ne.us/homepage.html*.

Nebraska's rivers are publicly owned, but the adjoining shorelines and many islands are usually in private ownership. Birding from a canoe is possible on several rivers (Niobrara, Dismal, Calamus, Missouri, Platte, and Republican), but access points are often limited. The longest stretch of river ideal for canoe-based birding is the 76-mile section of the Niobrara designated as a National Scenic River. The Platte is too shallow over much of its length for good canoeing. Watts, Duck, West Long, Hackberry, Dewey, and Clear Lakes at Valentine National Wildlife Refuge also offer wonderful birding and canoeing opportunities, but access may be limited or prohibited during the nesting season.

The locations of most of the sites mentioned are described, which should suffice for finding them with the use of a state highway map. The *Nebraska Sportsman's Atlas* contains maps of all of Nebraska's 93 counties—with descriptions of hunting, fishing, camping, and related outdoor attractions, including nearly all of those listed here. It is, available from Sportsman's Atlas Company in Lytton, Iowa. Sets of individual county road maps (available from the Nebraska Roads Department in Lincoln are also extremely useful when exploring back-country areas. An atlas of 79 topographic maps of the entire state (is available in the *Nebraska Atlas and Gazetteer*, published by DeLorme, Freeport, Maine. This atlas also shows state parks and recreation areas, national lands, campgrounds, wildlife viewing areas, and other outdoor attractions but not state-owned wildlife management areas.

A 96-page booklet by Joseph Krue (1997), titled *Nebraskaland Magazine Wildlife Viewing Guide*, published jointly by the Nebraska Game and Parks Commission and Falcon Press, includes descriptions of 68 public-access sites in the state that offer wildlife viewing opportunities. The Nebraska Division of Travel and Tourism in Lincoln can provide free information on general tourist attractions. Tourism information and free state highway maps are also available at most interstate rest areas.

The most useful field identification guide to the commoner wildflowers of Nebraska is by John Farrar, but it is now sadly out of print. However, a recent (1999) and inexpensive guide (*Grassland Plants of South Dakota and the Northern Great Plains*) produced by the South Dakota Agricultural Experiment Station is even more complete and additionally includes Nebraska's more common grasses, sedges, shrubs, and trees. Most of Nebraska's butterflies and moths are illustrated in *Butterflies and Moths of Missouri*, published by the Missouri Department of Conservation. Similar books published by the same department on the mammals, fish, freshwater mollusks, and aquatic plants of Missouri are equally valuable to Nebraska naturalists. There are many fine bird field guides to choose from for use in Nebraska, but beginners usually find the Golden Press's *Birds of North America* the easiest to use. Golden Press has also produced several other excellent and easy-to-use field guides to North American trees, reptiles, wildflowers, and other groups, all of which are helpful to have readily available when exploring Nebraska's natural areas. Many other references to regional taxonomic or identification guides may be found in the citations section.

The Far Western Region

Counties

Sioux, Dawes, Box Butte, Sheridan, Scotts Bluff, Banner, Kimball, Morrill, Cheyenne, Garden, and Deuel

The beautiful Far Western region of Nebraska, the state's geographic panhandle, is largely a ridge-and-canyon region interspersed with High Plains topography and steppe vegetation. Crazy Horse died trying to protect this land for his people, the Oglala Sioux or Lakota, and it is laced with the bitter history of these people and the Cheyenne as they vainly fought to maintain their sacred lands. The pine-covered hills and escarpments remind one of the Black Hills, and about 3.5 percent of the region's land is covered by woodlands. Several pine-adapted species that are common in the Black Hills occur only in the northwestern corner of Nebraska, such as Lewis' woodpecker (now rare or perhaps even extirpated), pinyon jay, dark-eyed junco, western tanager, yellow-rumped warbler, Swainson's thrush, plumbeous (previously "solitary") vireo, red-breasted nuthatch, and red crossbill. Some of these same species as well as the violet-green swallow, white-throated swift, and pygmy nuthatch also occur in the pine forests of the Scottsbluff area and the Wildcat Hills. However, the canyon-adapted cordilleran (previously "western") flycatcher is seemingly found only in Sowbelly Canyon and nearby Monroe Canyon, Sioux County.

The Panhandle region also has more than 5.6 million acres of grasslands, including the vast Oglala National Grassland, which supports a few quite localized shortgrass or arid plains species such as the McCown's longspur, chestnut-collared longspur, Brewer's sparrow, and mountain plover. Several sage-adapted species, including the sage thrasher and perhaps even the apparently extirpated sage grouse, are rather remote possibilities. It is also a land rich in the fossil remains of early Cenozoic mammals as well as an 8-million-year-old fossil bird bone that appears to be identical to that of a modern sandhill crane. This would make the sandhill crane the most archaic of all known extant birds and provides another reason for considering it a very special species. Tens of thousands of sandhill cranes still pass through this region each spring and fall, but their major migratory pathway lies to the east in the central Platte Valley.

The major birding attractions in the Panhandle include the biologically diverse and scenic Pine Ridge area. A bird checklist encompassing four northwestern counties, based on a study by Richard Rosche, reports 324 species. The area around Lake McConaughy has one of the very few local bird lists exceeding 300 reported species for any site north of Mexico. Crescent Lake National Wildlife Refuge, a wilderness refuge in the western Sandhills, has the second-largest local bird list for the state, with 273 species. North of Crescent Lake, in northern Garden County and southern Sheridan County, are hundreds of relatively saline and still pristine Sandhills marshes that abound with waterfowl and marshland birds.

Sioux County

Sioux County is in the heart of the Pine Ridge region, an area of ridge-and-canyon topography that is a southern outlier of the Black Hills of South Dakota, and includes a north-facing escarpment largely covered by ponderosa pine forest and streamside deciduous forests, totaling some 68,000 acres. As such, it has several species that occur rarely if at all elsewhere in Nebraska, such as the cordilleran flycatcher and plumbeous vireo. There are also more than 1 million acres of shortgrass plains, much of which is included in the Oglala National Grasslands and which support a typical High Plains avifauna. A major part of the population of the rare lady's-tresses orchid occurs in a Niobrara Valley alkaline meadow in Sioux County. There are tourist accommodations at Harrison.

FEDERAL AREAS
Oglala National Grasslands. Area 93,344 acres. The area around Sugarloaf Butte offers Brewer's sparrows, long-billed curlews, Swainson's and ferruginous hawks, chestnut-collared longspurs, and perhaps sage thrashers. Horned larks, western meadowlarks, and lark buntings are common breeders in this vast region, which extends into Dawes County. Buteo hawks are common here (red-tailed, Swainson's, and ferruginous), and golden eagles are also frequent. The rare swift fox also occurs here, and pronghorns are common. For information contact the Forest Service office in Chadron.

Soldier Creek Wilderness. Area 9,600 acres. This is a large roadless area that has an extensive hiking trail network as well as bridle trails. Water must be carried in, and facilities are lacking. Much of the area was burned in a 1989 fire. An 8-mile loop trail over ridges and canyons has its trailhead at the picnic area. For information contact the Forest Service office in Chadron.

Agate Fossil Beds National Monument. Includes nearly 3,000 acres of shortgrass plains with rich Cenozoic fossil deposits dating from about 22 million years ago. Includes the Niobrara River floodplain and channel (a few feet wide), rocky buttes, woodlands, and cottonwood groves. No official bird checklist is yet available, but 156 species have been reported for the site, including many western forms such as ferruginous hawk, mountain plover, burrowing owl, white-throated swift, Cassin's kingbird, pinyon jay, Townsend's warbler, western tanager, black-headed grosbeak, and lazuli bunting. Three species of longspurs occur, including both McCown's and chestnut-collared as breeders. Prairie rattlesnakes are common. Many rock-adapted plants such as Hood's phlox also occur here. There is a 4.5-mile trail. For information contact the National Park office at in Gering.

Toadstool Geologic Park. This area of barren badlands (about 300 acres) supports rock wrens, Say's phoebes, golden eagles, and prairie falcons and sometimes also gray-crowned rosy finches during winter. A 1-mile loop trail through part of the park that begins at the picnic area should turn up rock wrens and other topography-dependent birds. Water is at a premium here, and a canteen should be carried in hot weather. Fossils in the park are 28–34 million years old. A small campground is present.

McKelvie Division, Nebraska National Forest. Area comprises about 51,000 acres, including holdings in Dawes County. This area is much like the Pine Ridge Unit, and most or all of the same birding opportunities should exist (see Dawes County).

STATE AREAS

Fort Robinson State Park. Area 22,000 acres. Although still providing good pine habitat, a forest fire in 1989 destroyed much of the best sections of the park, which does offer lodging and eating facilities. A nesting area for white-throated swifts occurs 6 miles west of the headquarters (Pettingill 1981). Released bighorn sheep have become established here, and burned areas offer a slight chance for Lewis' woodpeckers. There are two hiking trails, totaling 67 miles.

Gilbert-Baker WMA. Area 2,457 acres. This is a region of ridges and valleys covered with ponderosa pines, with scattered areas of grasslands at the forest fringes. Monroe Creek traverses the area, and the cordilleran flycatcher has been seen in its canyon. Located 3 miles north of Harrison via an oil-surfaced road. A gravel road going south along the Wyoming border (turn 8 miles west of Harrison) crosses the Niobrara River and passes into High

Plains topography that supports McCown's longspurs, Say's phoebes, rock wrens, Brewer's sparrows, ferruginous hawks, long-billed curlews, and occasional chestnut-collared longspurs. At about 8 miles south of the turn a road goes east and connects to Highway Hiking trails penetrate the entire region.

Peterson WMA. Area 2,460 acres. This area consists of habitats alternating between mature ponderosa pine forests and grasslands in typical ridge-and-canyon topography. Two streams bisect the area. There are no camping facilities.

Bordeaux Creek WMA. A new area of 1,857 acres, 3 miles east of Chadron on Highway 20, with 1,100 acres of mixed grasslands and timberlands, 704 acres of pasture, 70 acres of croplands, and 3 acres of wetlands. Includes habitat for elk, turkey, and deer.

OTHER AREAS

Sowbelly Canyon. Although privately owned, a county road northeast from Harrison passes through this pine-clad canyon to a creek bottom public picnic area where birding can be done. Many distinctly western species occur here, including not only the very local cordilleran flycatcher but also white-throated swift, violet-green swallow, common poorwill, Say's phoebe, plumbeus vireo, western tanager, Bullock's oriole, and many other western species, plus some eastern forms such as redstart and rose-breasted grosbeak. The nearby Monroe Canyon has many of these same species.

Sioux County Ranch and Guadalcanal Memorial Prairie. A working ranch near Harrison with 5,000 acres of mostly shortgrass prairie, one of several prairie sites that are owned and managed by the Plains Prairie Resource Institute, Aurora.

Dawes County

Dawes County is one of Nebraska's most scenic regions, with nearly 100,000 acres of woodlands and almost 600,000 acres of grasslands within its boundaries. There are tourist accommodations at Chadron and Crawford.

FEDERAL AREAS

Pine Ridge National Recreation Area and Nebraska National Forest, Pine Ridge Unit. Recreation area 50,803 acres, mostly of ponderosa pine forest and intervening grasslands. The topography of this area is often rugged, and the roads may not be in good condition, so it is well to check with the ranger office on Highway 385 before venturing far from the main road. Cattle grazing is permitted here, so attention to gates is needed. There is a 4-mile fairly difficult trail starting at the Iron Horse Road meadow and a less difficult 3-mile hiking trail with its trailhead at East Ash Road. A fairly difficult 8-mile trail leading to Chadron State Park begins at a gravel road off Highway 85. The Pine Ridge Trail totals 52 miles. Bighorn sheep and elk have been established here, and western mammals such as least chipmunks and fringe-tailed bats also occur. Many arid-adapted and cushion-shaped plants, growing low among the rocks, are located here. For more information contact the forest supervisor in Chadron.

Oglala National Grasslands. Area 94,394 acres. See Sioux County.

STATE AREAS

Fort Robinson State Park. Area 20,000 acres. See Sioux County.

Chadron State Park. Area 801 acres. This is the best place in the region to see Lewis' woodpecker, and it also supports pygmy nuthatches, western tanagers, and common poorwills. On the way to the Black Hills overlook watch for Lewis' woodpeckers perched on the tops of snags. At the lookout one might see pinyon jays, yellow-rumped warblers, western tanagers, mountain bluebirds, and several western raptors. Pettingill (1981) has described birding opportunities in this park and provides a list of nesting species. The Spotted Tail hiking trail extends for 8 miles from the park boundary through the Nebraska National Forest, and the Black Hills Overlook trail extends for 4 miles from the park campground. The Cowboy Recreation and Nature Trail, now under development, begins in Chadron and extends east 321 miles to Norfolk and the Elkhorn River. It will generally parallel Highway 20 and the old Chicago Northwestern Railroad right-of-way, crossing 221 bridges and eight counties.

Ponderosa WMA. Area 3,659 acres. Located southeast of Crawford, this WMA is largely covered by ponderosa pine forests, with grasslands on level areas and also some deciduous trees lining Squaw Creek. There is a hiking trail starting at Parking Area 5 that provides an excellent panorama and may offer views of such raptors as prairie falcons. National Forest land adjoins the area to the south and southwest. About 10 miles south of Crawford along Highway 2 are ridgetop pine woodlands where Cassin's kingbirds are rather easily seen, especially during September.

Box Butte SRA. Area 612 acres, 20-acre reservoir. This is an outstanding birding area in the Panhandle; Richard Rosche (pers. comm.) has observed over 200 species in a 20-year span. Rock wrens, Say's phoebes, and ferruginous hawks are among the more interesting western species, and probable eastern breeders include eastern bluebird, eastern wood-pewee, and wood thrush. Small passerines such as warblers and vireos are abundant during migration.

Whitney Lake WMA. Area 900 acres. Located 2 miles northwest of Whitney.

Box Butte County

Box Butte County is only slightly forested (about 5,000 wooded acres) but has over 300,000 acres of remaining grasslands. There are tourist accommodations at Alliance.

STATE AREAS
None

OTHER AREAS

Kilpatrick Lake. Located about 20 miles west of Alliance (west on 10th St. for 15 miles, then south 1 mile, then west for 5 more miles. At a sign indicating the Snake Creek Ranch, a trail goes left and leads to the dam. This small reservoir is a major stopover point for snow geese and a few Ross' geese in spring. The wet meadows to the south of the dam around Snake Creek support willets, long-billed curlews, common snipes, eastern meadowlarks, and savannah sparrows, and the drier areas should be scanned for Cassin's sparrows, which are rare in Nebraska. These birds inhabit sandsage grasslands, and their vocalizations help locate them.

Sheridan County

Sheridan County has about 50,000 acres of woodlands and nearly 1.2 million acres of grasslands. It also has over 20,000 acres of surface wetlands, much of which consists of Sandhills marshes. There are tourist accommodations at Gordon and Rushville.

FEDERAL AREAS
None

STATE AREAS

Metcalf WMA. Area 3,068 acres. Located about 10 miles north of Hay Springs. It has typical Pine Ridge habitat, which is mostly pine covered but with some open grasslands. There are no camping facilities.

Smith Lake WMA. Area 640 acres. Located 20 miles south of Rushville. The area has a 222-acre lake, surrounding marshes and grasslands, and some woodlands. There are primitive camping facilities and toilets. Between Smith Lake and Lakeside, 30 miles apart, excellent birding opportunities exist, especially for viewing water birds. Nesting records for the long-eared owl, black-necked stilt, piping plover, and even the northern parula have been obtained.

Walgren Lake WMA. Area 130 acres. Located near Hay Springs. There are primitive camping facilities. A great variety of migrant species are attracted to this lake, including such rarities as Sabine's and black-headed gulls and Townsend's warbler. One mile south of Walgren Lake is a prairie dog town with nesting burrowing owls and occasional chestnut-collared longspurs. The latter are more common along the first road going east to the north of the colony.

OTHER AREAS

Sandhills marshes near Lakeside. This area, extending west and east from Lakeside on Highway 2 and north on Highway 250, provides views of many highly alkaline marshes that attract waterfowl, such as trumpeter swans, and many shorebirds, such as breeding black-necked stilts, eared grebes, American avocets, willets, and Wilson's phalaropes. The gravel road south from Lakeside takes one (28 miles, no gas stations or facilities) through Sandhills country to Crescent Lake National Wildlife Refuge (see Garden County) and past many wet meadows and very saline marshes that are highly attractive to shorebirds and waterfowl. Sandhills roads are narrow, hilly, and slippery after rains; careful driving is mandatory. Highway 250 north from Lakeside is even more productive and is paved. About 20 miles north of Lakeside an unimproved road going east connects with Highway 27 and can return one south to Highway 2 at Ellsworth. For the less adventurous it might be better to backtrack from Smith Lake to Lakeside and make a similar two-way run north from Ellsworth for about 15 miles, where the marshy wetlands become rare.

Scotts Bluff County

Scotts Bluff County has over 25,000 acres of woodlands and more than 180,000 acres of grasslands. There are tourist accommodations at Gering and Scottsbluff.

FEDERAL AREAS

North Platte National Wildlife Refuge, including Lake Minatare State Recreation Area. Area 5,047 acres. The best part of this refuge is the 500-acre Winters Creek Lake Unit northwest of Lake Minatare, where a marshy lake attracts a large number of migratory and breeding water birds, including western grebes. There is a bird list for the entire refuge. The checklist includes 181 species, with 32 known nesters and 20 additional possible breeders. For information contact the local U.S. Fish and Wildlife Service office.

Scotts Bluff National Monument. Area 3,000 acres. This famous bluff along the Oregon Trail is capped by ponderosa pine woodlands and has steep sides that are used as nesting sites by white-throated swifts. At least 100 species have been reported for the area, including prairie falcon, burrowing owl, common poorwill, pinyon jay, yellow-billed and black-billed cuckoos, rock wren, yellow-rumped warbler, Baltimore and Bullock's orioles, blue and black-headed grosbeaks, green-tailed and spotted towhees, three races of dark-eyed juncos, and lazuli bunting (unpublished staff records). There is a 3-mile nature trail leading from the summit parking lot to the visitor center. Watch for prairie rattlesnakes on the trail. For information contact the superintendent in Gering.

STATE AREAS

Nine-Mile Creek Special Use Area. Area 178 acres. Located north and east of Minatare, it consists of grasslands and a trout stream.

Wildcat Hills SRA *and Buffalo Creek* WMA. Total area 3,935 acres. Buffalo Creek WMA is a few miles east of Wildcat Hills SRA and consists of typical Wildcat Hills ridge-and-canyon habitats covered by pines and junipers. It is nearly all wooded but has a 7-acre pond. Primitive camping facilities are present. Pygmy nuthatches nest in the vicinity, and violet-green swallows are fairly common. Several raptors, such as golden eagles, prairie falcons, and several buteos, are good possibilities. Pettingill (1981) listed 11 species that should potentially be seen here, including common poorwill and white-throated swift. A new nature center is present at the SRA with a 2-mile nature trail, bird-viewing windows, and many ecological exhibits. Red crossbills are regular there. Confined bison and elk are also present on a 310-acre enclosure. There is a challenging 3.5-mile trail starting at the Buffalo Creek parking lot.

Banner County

Banner County is a High Plains county that is slightly wooded (under 25,000 acres of woodlands) and has about 235,000 acres of grassland habitats. Very little surface water is present.

FEDERAL AREAS
None

STATE AREAS
Buffalo Creek WMA. See Scotts Bluff County.
Wildcat Hills SRA. See Scotts Bluff County.

Kimball County

Kimball County is a High Plains county with only about 500 acres of woodlands, about 500 acres of surface water, and about 180,000 acres of grasslands. There are tourist accommodations at Kimball.

FEDERAL AREAS
None

STATE AREAS
Oliver Reservoir SRA. Area 1,187 acres. This reservoir is an excellent birding location, attracting many migrant passerines during spring and fall, especially warblers. Common snipe have been reported to nest here (west-end marshes), and a population of song sparrows is the only one known for western Nebraska.

OTHER AREAS
Tri-state corner (highest point in Nebraska). This remote area can be reached by driving south from Bushnell (south 12.5 miles, west 4.2 miles, south 1 mile, west 2 miles, and south 2 miles nearly to the Kansas line). In this area there are lark buntings, horned larks, McCown's longspurs, and occasional mountain plovers.

Lodgepole Creek. This creek should attract warblers and other passerines. It is the only known location for the Colorado butterfly plant in Nebraska.

Morrill County

Morrill County has nearly 30,000 acres of woodlands, some 660,000 acres of grasslands, and almost 5,000 acres of surface water. There are tourist accommodations at Bayard and Bridgeport.

FEDERAL AREAS

Chimney Rock National Historic Site. Area 83 acres. Chimney Rock is located near Bayard and is worth investigating for nesting golden eagles (either on the column itself or on the eroding escarpment to the north). An old cemetery lies to the northwest of Chimney Rock, and burrowing owls are often found in a nearby prairie dog colony. Lazuli buntings are common in brushy areas.

STATE AREAS

Bridgeport SRA. Area 128 acres. In Bridgeport.
Facus Springs WMA. This recently established WMA preserves one of the best saline marshes in the North Platte Valley. It is a major stopover point for migrant shorebirds and attracts ducks during migration. Some shorebirds such as American avocets and Wilson's phalaropes also nest, and the cinnamon teal, rare in Nebraska, has nested here.
Bridgeport SRA. Area 126 acres.

OTHER AREAS

Saline marsh near Bridgeport. Like Facus Springs, this saline marsh near Bridgeport attracts great numbers of shorebirds during migration. In wet years other marshes may occur here.
Courthouse Rock and Jail Rock. These famous Oregon Trail landmarks often have nesting golden eagles (on Jail Rock), and breeding rock wrens are common. By driving west on Highway 88 to Redington one can take the Redington Gap road to Facus Springs and Bridgeport or go south from Redington (see below).
Redington Gap road and road south of Redington. By driving south past Facus Springs one passes over a long, eroded line of hills (Redington Gap), and many western species typical of the High Plains may be seen. Within 1 mile of turning south off Highways 26/92, one passes a meadow that supports a good population of savannah sparrows (rare in Nebraska). Go to Redington and continue south from there for 4.5 miles, then take a left fork, and go three more miles until pines appear on a north-facing slope. This area supports a good population of Cassin's kingbirds the easternmost known pinyon jay population, and such western birds as western wood-pewees and common poorwills.

Cheyenne County has almost no surface water, about 7,000 acres of woodlands, and over 210,000 acres of grasslands. There are tourist accommodations at Lodgepole and Sidney.

FEDERAL AREAS
None

STATE AREAS
None

Garden County

Garden County has over 22,000 acres of surface water, about 500 acres of woodlands, and nearly 900,000 acres of grasslands. There are tourist accommodations at Oshkosh.

FEDERAL AREAS
Crescent Lake National Wildlife Refuge. 40,900 acres. One of the great wilderness refuges in America, this refuge supports a greater bird diversity than any other Nebraska site except the Lake McConaughy area. However, it is about 30 miles from the nearest source of gas, food, and lodging, and one must plan accordingly, taking a tow rope if possible and never parking on bare sand. Rather, park or turn around on level, grassy meadows if possible. Water and a toilet are available at the refuge headquarters. Goose Lake near the headquarters is excellent for eared grebes, and both Crescent Lake and Smith Lake have good populations of western grebes (plus some Clark's grebes). Rush Lake (not within the refuge) has breeding ruddy ducks, canvasbacks, redheads, and black-crowned night herons. The area near Border Lake is best for avocets, black-necked stilts, cinnamon teal, Wilson's phalaropes, and other shorebirds attracted to saline water conditions; Border Lake marks the boundary of such hypersaline conditions. On most visits no other people will be seen, but the birding will be spectacular and well worth the long ride over sand roads. Pettingill (1981) provides a good description. Pronghorns are sometime seen, as are both species of jack rabbits. For maps, a bird checklist (279 species, 86 nesters), and other information contact the refuge office in Ellsworth.

STATE AREAS
Ash Hollow State Historical Park. This historical park has a wide variety of habitats, from exposed rocky bluffs that are used by great horned owls, American kestrels, and sometimes prairie falcons, through grassy wet meadows where bobolinks and eastern meadowlarks are present, to riparian woodlands used by warbling vireos and other woodland songbirds. There is also upland grasslands, with blue grosbeaks and spotted towhees in shrubby areas, and scattered yuccas, where field and grasshopper sparrows sometimes perch. An air-conditioned interpretive center provides

welcome relief from oppressive summer temperatures. The nearby Highway 26 bridge across the North Platte offers views of many marshland species, including least bitterns on rare occasions. A 1-mile trail leads from the parking lot off Highway 26 to Windlass Hill, where ancient Conestoga wagon ruts are still easily visible.

Clear Creek Waterfowl Management Area. See Keith County.

OTHER AREAS

Oshkosh Sewage Lagoons. These lagoons are reached by driving south on Highway 27 for 0.5 mile from Oshkosh, turning east, and driving until the lagoons appear on the south side of the road. Three lagoons are accessible by walking. They attract a surprising array of waterfowl, including breeding wood ducks and even nesting ruddy ducks.

Deuel County

Deuel County has more than 1,000 acres each of woodlands and surface water and nearly 64,000 acres of grasslands. There are tourist accommodations at Big Springs and Chappell.

FEDERAL AREAS

None

STATE AREAS

Bittersweet WMA. Area 76 acres. Consists of river frontage along the South Platte.
Goldeneye WMA. Area 25 acres, 11 wetland acres.
Goldenrod WMA. Area 97 acres, upland habitat.

The West-Central Region

Counties

Cherry, Keya Paha, Brown, Rock, Grant, Hooker, Thomas, Blaine, Loup, Arthur, McPherson, Logan, Custer, Keith, Perkins, Lincoln, Dawson, Chase, Hayes, Frontier, Gosper, Phelps, Dundy, Hitchcock, Red Willow, Furnas, and Harlan

The West-Central Region of the state includes two of the very best bird-finding localities in Nebraska, namely Valentine and Fort Niobrara National Wildlife Refuges. These two locations have bird lists that are among the largest in the state. Additionally, it includes those portions of the Niobrara and Platte Valleys that lie in the middle of the transition zone between the Rocky Mountain coniferous forest and eastern deciduous forest biogeographic regions. These transition, or suture, zones include areas of hybridization between several species or nascent species pairs of birds that are now in secondary contact after having been isolated geographically for much or all of the Pleistocene period. This transition zone is very wide in the Platte Valley but is compressed to a width of under 100 miles in the Niobrara Valley, most of which is now included within the boundaries of the Fort Niobrara National Wildlife Refuge and the Nature Conservancy's wonderful Niobrara Valley Preserve.

South of these refuges, the region mostly consists of the Nebraska Sandhills. The Sandhills represent the largest natural ecosystem in Nebraska, covering nearly 19,000 square miles, or almost a quarter of the state. It is also the largest remaining grassland ecosystem in the country that is still virtually intact both faunistically and floristically. It is a land with more cattle than people. There are few roads, and tourist facilities and accommodations are almost nonexistent. The little-traveled roads that do exist often consist of only slightly improved sandy trails leading to ranches. But the region is filled with breathtaking vistas, spectacular bird populations in the hundreds of lakes and marshes, and a pioneer spirit that requires everyone to help a neighbor or indeed any stranger who happens to fall afoul of trouble while on the road. It is a land designed for naturalists who would like to study virtually unaltered prairie ecosystems and who are prepared to deal with nature on its own terms. A summary of the natural history of the Nebraska Sandhills, along with an annotated bird checklist (277 species) and lists of its fishes, amphibians, reptiles, mammals, and nearly all of its vascular plants is in Johnsgard (1995).

Cherry County

Cherry County is by far the largest county in the state and consists mostly of Sandhills habitat, with 3.7 million acres of grasslands, about 17,000 acres of woodlands, and 41,000 acres of surface wetlands. There are tourist accommodations at Merriman and Valentine.

FEDERAL AREAS
Fort Niobrara National Wildlife Refuge. Area 19,122 acres. This refuge, originally established to protect bison and other large game animals, lies on the western edge of the east-west ecological transition zone between forest types and thus has a fine mixture of eastern and western avifauna. Western-eastern species pairs that occur and may hybridize include western and eastern wood-pewees, black-headed and rose-breasted grosbeaks, eastern and spotted towhees, and Bullock's and Baltimore orioles. A checklist of the refuge's bird species (excepting accidentals) is available upon request. A total of 201 species (76 breeders) have been reported here. About two-thirds of the refuge consists of Sandhills prairie, and the rest is mostly mixed riparian hardwoods. There is a good population of sharp-tailed grouse, and wild turkey viewing blinds are available. Breeding burrowing owls (and a prairie dog colony), yellow-breasted chats, American redstarts, grasshopper and savannah sparrows, and eastern and west-

ern meadowlarks can be found here. There are also confined elk and bison herds.

Valentine National Wildlife Refuge. Area 71,516 acres. This is Nebraska's largest national wildlife refuge and one that rivals Crescent Lake in its bird diversity, with 221 species (93 breeders) reported. A checklist (excluding accidentals) is available upon request. Most of the refuge consists of Sandhills prairie, with dunes from 40 to 200 feet high and intervening interdune depressions that often contain shallow, marshy lakes. Driving on the sandy trails requires care; a supply of water and a tow rope are recommended. Several prairie-chicken and sharp-tailed grouse leks are present in the refuge, and the numerous marshes and shallow lakes provide breeding habitat for eared, western, and pied-billed grebes; a dozen species of waterfowl; and shorebirds such as soras, common snipes, and American avocets. The higher grasslands offer views of long-billed curlews, upland sandpipers, and Swainson's hawks.

Samuel R. McKelvie District, Nebraska National Forest. Area 115,703 acres. This section of forest is similar to that of the Bessey Division but is not so rich in migrants. The adjoining Merritt Reservoir might also be visited. A sharp-tailed grouse blind available in spring accommodates four people on a first-come basis.

STATE AREAS

Schlegel Creek WMA. Area 600 acres. Consists of Sandhills grasslands including 2 miles of Schlegel Creek. There are no camping facilities.

Big Alkali WMA. Area 889 acres. Consists of 47 lakeside acres plus an 842-acre Sandhills lake. Campground present.

Ballard's Marsh WMA. Area 1,561 acres. Includes a large marsh and adjoining Sandhills grasslands. Campground present.

Smith Falls State Park. Area 244 acres. Located 3 miles west and 4 miles south of Sparks. The river must be crossed (by a new bridge) for a view of the falls.

Merritt Reservoir WMA. Area 2,906-acre reservoir, 350 upland Sandhills acres. Located 26 miles southwest of Valentine. This area abuts national forest land to the north. The reservoir attracts migrant waterfowl, pelicans, western grebes, and other species.

Rat and Beaver Lake WMA. Consists of 240 acres of Sandhills grasslands and parts of two marshy lakes. Located 29 miles south of Valentine on Highway 20, 6 miles west on Sandhills trail.

Willow Lake WMA. Area 440 acres, with 240 acres of marsh, bordered by Valentine National Wildlife Refuge. Located 22 miles south of Valentine and 2 miles west on Sandhills trail.

Shell Lake WMA. Consists of 640 acres of Sandhills grasslands and a 162-acre lake. Located 14 miles northeast of Gordon.

Bowman Bridge WMA. Area 159 acres of bottomland along the Niobrara River in transition east-west forest. Located 1.5 miles southeast of Valentine.

Anderson Bridge WMA. Area 137 acres of river valley bottomland forest and 1 mile of Niobrara River frontage. Located 10.5 miles south and 2 miles east of Kilgore. Includes deciduous floodplain, coniferous upland woods, and wetland habitats.

Cottonwood Lake SRA. Area 180 land acres, 60-acre lake. Located 1 mile southeast of Merriman. Camping facilities are present. Canada geese breed here. Trumpeter swans have nested in a marsh 5 miles east of Merriman, off the north side of Highway 20.

Bowring Ranch State Historical Park. Located 1 mile north of Merriman. Trumpeter swans forage on a marsh just north of this park, which is a working cattle ranch. Park entry permit required.

Keya Paha County

Keya Paha County has over 37,000 acres of woodlands, over 400,000 acres of grasslands, and about 1,300 acres of surface water. There are tourist accommodations at Springview.

FEDERAL AREAS
None

STATE AREAS
Cub Creek Recreation Area. Area 300 acres.
Thomas Creek WMA. Area 692 acres. Two miles east and 3 miles south of Springview. Steep topography around Thomas Creek, with grasslands on the hills and wooded creek bottom. A rugged area with about 90 percent woods, consisting of ponderosa pine on the canyon slopes and bur oak in the ravines. The uplands are native grasslands.

OTHER AREAS
Niobrara Valley Preserve. See Holt County.

Brown County

Brown County has over 21,000 acres of woodlands, nearly 700,000 acres of grasslands, and about 8,000 acres of surface water. There are tourist accommodations at Ainsworth.

FEDERAL AREAS
None

STATE AREAS
Bobcat WMA. Area 893 acres. Nearly 90 percent of this area consists of steep pine- and cedar-covered canyons. Plum Creek passes through. The remainder is Sandhills grasslands.

School Land WPA *and Keller Park* SRA. Area 836 acres (WMA 640 acres, SRA 196 acres). These areas consist of native prairie, wooded canyons, Bone Creek, and five small fishing ponds stocked with trout and other game fish. The ponds attract ducks, eagles, and other water birds; the prairies support grassland sparrows; and the mixed woodlands have a variety of both coniferous and deciduous forest birds including wild turkeys, scarlet tanagers, and American redstarts.

Pine Glen WMA. Area 960 acres. Located 7 miles west and 6.5 miles north of Bassett. It consists of canyons, a trout stream, and mixed grasslands and woodlands. No facilities are present.

Long Pine WMA. Area 160 acres. Consists of about 85 percent pine and red cedar woodlands, and the rest is native Sandhills grasslands, bisected by Long Pine Creek. The terrain is steep, and camping facilities are primitive. Located just off Highway 20 near the town of Long Pine.

South Twin Lake WMA. Area 160 acres. Consists of a 60-acre lake and Sandhills grasslands. This is similar to the next three WMAs and should offer birders some excellent views of Sandhills wildlife.

American Game Marsh WMA. Area 160 acres. Consists of a large Sandhills marsh and surrounding grasslands. No facilities are present.

Long Lake WMA. Area 30 upland acres, 50-acre Sandhills lake.

Willow Lake WMA. Area 511 acres. A Sandhills lake and surrounding grasslands.

OTHER AREAS

Niobrara Valley Preserve. Area about 61,000 acres, including about 40 miles of the Niobrara River, in the heart of the transition zone between western coniferous and eastern deciduous forest types. Ducey (1989) has published an annotated bird checklist for the entire Niobrara Valley, including a list of 125 breeding species. Among the breeding birds of special interest are the eastern and western forms that hybridize here, such as the Baltimore and Bullock's orioles, the lazuli and indigo buntings, and the rose-breasted and black-headed grosbeaks. The eastern and western wood-pewees may also hybridize here. Seventy species of butterflies have been documented. Two trails radiate out from the headquarters that pass through several forest types and the Sandhills prairie vegetation on the uplands. Each trail has a short loop and a long loop; the northern, 3-mile trail is somewhat longer and steeper. Be on the lookout for prairie rattlesnakes. This preserve supports a relict Pleistocene population of white birch and a similar isolated population of eastern woodrats. The preserve lies within the Niobrara National Scenic River District, which extends for 76 miles and is a popular canoeing destination.

Rock County

Rock County is, despite its name, mostly comprised of Sandhills habitat, with about 11,000 acres of woodlands, over 600,000 acres of grasslands, and about 11,000 acres of surface water. There are tourist accommodations at Bassett.

FEDERAL AREAS
None

STATE AREAS
Twin Lakes WMA. Includes 113 acres of surface water (two lakes) and 30 acres of grasslands. It is located 18 miles south and 2 miles east of Bassett.

Grant County

Grant County is a sparsely populated Sandhills county with only about 400 acres of woodlands, about 3,500 acres of surface water, and over 460,000 acres of grasslands.

FEDERAL AREAS
None

STATE AREAS
None

Hooker County

Hooker County is a sparsely populated Sandhills county with 1,800 acres of woodlands, about 400 acres of surface water, and over 450,000 acres of grasslands. There are very limited tourist accommodations at Mullen.

FEDERAL AREAS
None

STATE AREAS
None

Thomas County

Thomas County is in the heart of the Sandhills but has over 16,000 acres of woodlands, most of which lie in the planted pine "forest" near Halsey. There are about 1,500 acres of surface water (mostly Middle Loup and Dismal Rivers) and almost 380,000 acres of grasslands. There are tourist accommodations at Halsey and Thedford.

FEDERAL AREAS
Bessey Division, Nebraska National Forest. Area 90,445 acres. Grasslands in and around this planted "forest" support greater prairie-chickens, sharp-tailed grouse, upland sandpipers, horned larks, and western meadowlarks. The conifers provide habitat for great horned owls, black-capped chickadees, and sometimes red crossbills. Brushy and riparian thicket areas attract several woodpeckers, brown thrashers, towhees, chipping sparrows, and Baltimore orioles. At least six warbler species nest here, including yellow, black-and-white, American redstart, ovenbird, common

yellowthroat, and yellow-breasted chat. Three vireos (Bell's, warbling, and red-eyed) also nest here. There is a bird checklist available at the headquarters, where information on the grouse blinds is also available. A fire in the 1960s burned much of the forest, but 80 percent of 25,000 acres still survive. There is an 8-mile hiking trail (Dismal River Tail) that begins at the parking lot off Highway 2 and a 3-mile trail leading to a lookout tower. For information contact the Forest Service office in Halsey.

STATE AREAS

None

Blaine County

Blaine County is a Sandhills county with 1,600 acres of woodlands, about 1,000 acres of surface water, and nearly 440,000 acres of grasslands.

FEDERAL AREAS

None

STATE AREAS

Bessey Division, Nebraska National Forest. See Thomas County.
Calamus Reservoir SRA/ WMA. Area 10,312 acres total; 5,124-acre reservoir. Developed for fishing with no hiking trails. No information on the birds is available, but eagles, ospreys, and waterfowl are likely.
Milburn Dam WMA. Consists of 672 acres of Middle Loup River valley, with extensive mud flats present around the reservoir. Located 14 miles southeast of Brewster.

Loup County

Loup County is a Sandhills county with about 7,000 acres of surface water, nearly 5,000 acres of woodlands, and over 325,000 acres of grasslands. There are tourist accommodations at Taylor.

FEDERAL AREAS

None

STATE AREAS

Calamus Reservoir SRA/ WMA. Area 10,312 acres total; 5,124-acre reservoir. Located near county boundary. See Blaine County.

Arthur County

Arthur County is a sparsely populated Sandhills county with almost 3,000 acres of surface water, about 200 acres of woodlands, and 427,000 acres of grasslands.

FEDERAL AREAS

None

STATE AREAS
None

OTHER AREAS
Marshes near McPherson County border. These Sandhills marshes and creeks are often used by trumpeter swans.

McPherson County

McPherson County is a very sparsely populated Sandhills county with about 600 acres of surface water, 1,000 acres of woodlands, and over 520,000 acres of grasslands.

FEDERAL AREAS
None

STATE AREAS
None

OTHER AREAS
Marshes near Arthur County border. See Arthur County.

Logan County

Logan County is a Sandhills county with only 250 acres of surface water, about 300 acres of woodlands, and over 520,000 acres of grasslands.

FEDERAL AREAS
None

STATE AREAS
None

Custer County

Custer County is a mostly Sandhills county with about 2,500 acres of surface water, over 10,000 acres of woodlands, and about 1.1 million acres of grasslands. There are tourist accommodations at Arnold, Broken Bow, Callaway, and Sargent.

FEDERAL AREAS
None

STATE AREAS
Victoria Springs SRA. Area 60 acres.
Pressey WMA. Area 1,640 acres. Located 5 miles north of Oconto. Consists of South Loup Valley lands, hills, and steep canyons mostly covered by grasslands. Toilets, a campground, and hiking trails are present. There are sharp-tailed grouse in the area and a great blue heron rookery that can be seen easily.
Arcadia Diversion Dam SRA. Area 925 acres. This area consists of the Middle Loup River valley, mostly of grasslands and tree plantings but with

deciduous woodlands lining the river. There are some campgrounds on both sides of the river. Located 8.5 miles northwest of Arcadia.

Keith County

Keith County is notable in having over 37,000 acres of surface water, nearly 6,000 acres of woodlands, and over 420,000 acres of mainly Sandhills grasslands. There are tourist accommodations at Keystone, Lemoyne, Ogallala, and Paxton.

FEDERAL AREAS
None

STATE AREAS

Clear Creek WMA. Area 5,709 acres. Partly developed as Clear Creek Refuge (2,500 acres, west half) and also as a controlled hunting area. The latter includes the west end of Lake McConaughy and the Platte River inflow area. The low meadows support nesting bobolinks and probably breeding common snipes, and the tall tree groves hold many breeding passerines. White pelicans are common, and least bitterns have been sighted. One of the state's best birding areas, but mosquitoes can be a problem during summer. Barn-owl nest cavities usually can be seen in the cutbanks at the turnoff from the main highway; nests in this part of the state are typically in such excavated sites rather than in old buildings. This is the state's only known nesting area for Clark's grebe.

Lake McConaughy SRA. Area 6,492 acres. Occupies much of the north side of Lake McConaughy, the largest body of water in Nebraska. A small area on the south side is also included. This area has the largest bird list of any location in the state, including more than 340 species, with 104 known breeders, 17 additional possible breeders, and 184 transients. The large water area attracts vast numbers of migrant waterfowl, grebes (especially western grebes), gulls (including many rarities), and shorebirds. A good spotting scope is needed to cover this vast reservoir, but many of the waterfowl congregate near the spillway during winter or toward the western end of the lake in the summer (see Clear Creek WMA). Large numbers of bald eagles also build up in winter, attracted by dead fish and the wintering duck and goose populations. Well over 100 miles of shoreline are present along the lake, with the southern shoreline rocky and steep and the northern shore sandy. The north shoreline supports nesting piping plovers and least terns.

Kingsley Dam and Lake Ogallala SRA. Area 339 acres. Kingsley Dam offers a good vantage point for viewing birds both on the deeper end of Lake McConaughy and on the shallower and much smaller Lake Ogallala located at the base of the dam. Lake Ogallala (and its eastern end, often called Lake Keystone) receives the spillway water from Lake McConaughy, and its level fluctuates greatly. However, it is very attractive to migrant ducks, ospreys, Caspian terns, cliff swallows, gulls, American white pelicans, double-crested cormorants, and other summering species. It is used by waterfowl and gulls and by numerous bald eagles in winter. An eagle-watching blind is available during peak winter periods, when 200–300 eagles are sometimes present. Information on the area's bird life is available at the office on the south side of the dam. Camping is possible along the western and northern shorelines of Lake Ogallala, where the deciduous woodlands offer a rich array of nesting passerines, but lake fluctuations limit nesting for aquatic species. Nearby Cedar Point Biological Station is an extension of the University of Nebraska and a summer field station and thus is not open to the public; ornithological research here has made its avifauna the best known of any area in the state.

Ogallala Strip WMA. Area 453 acres; includes 2.5 miles of river frontage. This stretch of riparian woodlands supports many of the same species found around Lake Ogallala, such as house wren, yellow warbler, common yellowthroat, eastern and western kingbirds, and killdeer. Mississippi kites now breed in nearby Ogallala.

Lakeview Road. The road leading down the canyon to Lakeview and a similar road leading Eagle Canyon 6 miles farther west may offer views of rock wrens, rough-winged swallows, and, with luck, occasional prairie falcons or ferruginous hawks. Turkey vultures nest along the south side of the reservoir, usually in eroded crevices or recesses well out of view. These roads are often in poor condition, and caution must be exercised when driving over them.

Perkins County

Perkins County is a High Plains county with only about 200 acres of surface water, about 1,000 acres of woodlands, and 125,000 acres of grasslands.

FEDERAL AREAS
None

STATE AREAS
None

Lincoln County

Lincoln County straddles the Platte Valley with nearly 10,000 acres of surface water, about 36,000 acres of woodlands, and about 1.2 million acres of grasslands. There are tourist accommodations at North Platte and Sutherland.

FEDERAL AREAS
None

Sutherland Reservoir SRA. Area 3,020-acre reservoir, 37 upland acres. This site has been called the gull capital of western Nebraska, with at least ten species having been observed. These include such rarities as Thayer's, glaucous, great and lesser black-backed, and even Ross' gull. There are often large flocks of wintering grebes, diving ducks, double-crested cormorants, and American white pelicans during mild winters. During spring large flocks of snow, greater white-fronted, and occasional Ross' geese stop here.

Malony Reservoir SRA. Area 1,600-acre reservoir, 1,732 upland acres. This lake is used during spring by American white pelicans and double-crested cormorants and by many shorebirds when the water levels subside (Pettingill 1981).

Jeffrey Canyon WMA *and Reservoir.* Area 900-acre reservoir, 35 upland acres. This area consists of canyon-and-upland topography, with grasses and scattered deciduous trees and cedars.

North River Wildlife WMA. Area 681 acres, 2 miles of river frontage. There are woods along the river and grasslands beyond that is used by sandhill cranes. This is one of the westernmost crane roosting sites; the birds use the southeastern part of the area in less-than-ideal roosting habitat.

Muskrat Run WMA. Area 224 acres. Mostly riparian woodlands and marshy areas.

East Sutherland WMA. Area 27 upland acres, 8-acre lake.

Hershey WMA. Area 53 upland acres, 80-acre lake.

East Hershey WMA. Area 20 upland acres, 20-acre lake.

Birdwood Lake WMA. Area 20 upland acres, 13-acre lake.

Fremont Slough WMA. Area 30 upland acres, 11-acre lake.

Platte WMA. Area 242 upland acres, 0.5 mile of river frontage. Mostly riparian woodlands.

Fort McPherson WMA. Area 30 acres, with pond.

West Brady WMA. Area 10 upland acres, 6-acre lake.

Chester Island WMA. Area 69 acres. Includes ponds and 0.3 mile of river frontage.

Box Elder Canyon WMA. Area 20 acres. Located 3 miles south and 2.5 miles west of Maxwell. Consists of native grasslands and deciduous woodlands along the Tri-County Supply Canal.

Cottonwood Canyon WMA. This small (15.4-acre) site is much like the nearby Cottonwood Canyon WMA and is 4.5 miles south of Maxwell.

Wellfleet WMA. Comprising only 65 acres along Medicine Creek, it provides a diversity of habitats that usually attracts a wide variety of small passerines and water birds. This area is just west of the village of Wellfleet or 20 miles south of North Platte.

OTHER AREAS

North Platte Sewage lagoons. These sites are reached by leaving I-80 at exit 179 and going north on spur road L56G. Cross the South Platte River and turn east on a dead-end gravel road that will take you to the lagoons. These lagoons attract many water birds during migration.

Dawson County

Dawson County is another Platte Valley county with about 8,000 acres of surface water, over 17,000 acres of woodlands, and over 250,000 acres of grasslands.

FEDERAL AREAS
None

STATE AREAS
Willow Island WMA. Area 45 upland acres, 35-acre lake, riparian woodlands.
East Willow Island WMA. Area 16 upland acres, 21 wetland acres. Includes 0.3 mile of river frontage; mostly riparian woodlands.
West Cozad WMA. Area 19 upland acres, 29-acre lake.
Cozad WMA. Area 182 upland acres, 16 wetland acres, 0.5 mile of river frontage.
East Cozad WMA. Area 18 upland acres.
Darr Strip WMA. Area 976 acres; 767 upland acres, 2.5 miles of river frontage.
Dogwood WMA. Area 402 acres, 10-acre lake, 1.5 miles of river frontage.
Midway Lake WMA. A reservoir near the Tri-County Canal; at its upper (southern) end is Midway Canyon, an eroded area of loess hills.
Gallagher Canyon SRA. Area 400-acre reservoir, 424 upland acres. Another canyon in the loess hills and associated reservoir.
Plum Creek WMA. Area 152 acres, 320-acre reservoir.
Johnson Lake SRA. Area 2,061-acre reservoir, 81 upland acres. This lake is rather highly developed, which might reduce its attractiveness to birds somewhat, but it should attract bald eagles during winter. Elwood Reservoir (1,330 acres) is nearby.
Bittern's Call WMA. Consists of 80 acres of mixed upland and wetland habitat. Located about 10 miles north of Lexington on Highway 21.

Chase County

Chase County is a High Plains county with about 2,200 acres of surface water (nearly all reservoir), about 1,400 acres of woodlands, and 290,000 acres of grasslands. There are tourist accommodations at Imperial.

FEDERAL AREAS
None

STATE AREAS
Enders Reservoir SRA. Area 3,643 upland acres, 2,146-acre reservoir. Nearly all open sandsage grasslands with rolling to rugged topography.

Developed facilities. This large reservoir attracts large numbers of mallards and Canada geese; most of the western half of the reservoir and surrounding land is a wildlife refuge. Prairie rattlesnakes are fairly common in rocky areas; ornate box turtles also occur; and Cassin's sparrows may breed in sandsage habitat.

Enders Reservoir WMA. Area 3,643 acres. The area to the west of the reservoir is managed for big game and upland game hunting.

Wannamaker WMA. Area 160 acres. Located about 1 mile west of Imperial, it consists mostly of planted grasslands and shelterbelts.

Champion Lake SRA. Area 13 acres.

Hayes County

Hayes County is a High Plains county with about 800 acres of surface water, about 2,000 acres of woodlands, and 255,000 acres of grasslands. There are tourist accommodations at Benkelman.

FEDERAL AREAS

None

STATE AREAS

Hayes Center WMA. Area 78 acres. Located 12 miles northeast of Hayes Center. It consists of native High Plains grasslands, scattered woodlands, and a 40-acre reservoir. There is a campground, well, and toilets. This area is a migration trap for waterfowl, and wood ducks are notably common in spite of its western location. The shrubby riparian vegetation attracts many passerines, and some eastern species such as eastern phoebes, red-bellied woodpeckers, and northern bobwhites breed here. To the north along Highway 25, western birds such as Say's phoebe, rock wren, and ferruginous hawk may at times be seen.

Frontier County

Frontier County is a High Plains county with about 3,500 acres of surface water (nearly all reservoir), 1,300 acres of woodlands, and almost 330,000 acres of grasslands.

FEDERAL AREAS

None

STATE AREAS

Red Willow Reservoir SRA/WMA. Area 4,320 upland acres, 1,628-acre reservoir. Modern camping facilities are present. This reservoir in a water-poor region attracts good numbers of migratory water birds, including many geese and ducks. Burrowing owls should be searched for in the prairie dog town near Spring Creek. See also Red Willow County.

Medicine Creek Reservoir and Medicine Creek SRA/WMA. Area of WMA 6,726 acres, SRA area 1,768-acre reservoir and 1,200 upland acres. There are 17 hiking trails and both primitive and modern camping facilities present. Wood ducks are surprisingly common, and barn-owls breed in the area. About 5,000 acres are managed for upland habitat, and over 50,000 trees and shrubs have been planted

Gosper County

Gosper County is a mostly High Plains county with about 4,000 acres of surface water, nearly 1,700 acres of woodlands, and nearly 140,000 acres of grasslands. There are tourist accommodations at Elwood.

FEDERAL AREAS
Victor Lake Federal Waterfowl Area. Area 174 wetland acres, 64 upland acres.
Elley Lagoon Federal Waterfowl Area. Area 33 wetland acres, 29 upland acres.
Peterson Basin Federal Waterfowl Area. Area 527 wetland acres, 627 upland acres.

STATE AREAS
Johnson Lake SRA. Area 2,061-acre reservoir, 81 upland acres. See also Dawson County.
Elwood Reservoir WMA. Consists of a 1,330-acre reservoir and 900 adjacent acres of grasslands and some wooded sites. Located 2 miles north of Elwood. There are no camping facilities.

Phelps County

Phelps County is a Platte Valley county at the western edge of the Rainwater Basin, with about 200 acres of permanent surface water (plus temporary wetlands), 3,800 acres of woodlands, and over 72,000 acres of grasslands. There are tourist accommodations at Holdrege.

FEDERAL AREAS
Cottonwood Basin WPA. Area 79 wetland acres, 161 upland acres.
Linder WPA. Area 2 wetland acres, 79 upland acres.
Johnson Lagoon WPA. Area 252 wetland acres, 326 upland acres.
Funk Lagoon WPA. Area 1,163 wetland acres, 826 upland acres. This is the largest and perhaps the best of the lagoons in the western Rainwater Basin. During spring it supports amazing numbers of geese (especially greater white-fronted) and some 20 species of ducks. Breeding birds include great-tailed grackles, yellow-headed blackbirds, and eared and pied-billed grebes. The main parking area has an information kiosk and a nearby observation blind looking out over the marsh. A variety of herons, egrets, and white-faced ibis visit the area in spring and fall.
Atlanta Marsh WPA. Area 453 wetland acres, 659 upland acres.
Jones Marsh WPA. Area 90 wetland acres, 76 upland acres.

West Sacramento WMA. Area 200 wetland acres, 188 upland acres.
Sacramento-Wilcox WMA. Area 1,050 wetland acres, 1263 upland acres.

Dundy County

Dundy County is a High Plains county with about 500 acres of surface water, about 5,200 acres of woodlands, and 384,000 acres of grasslands.

FEDERAL AREAS

None

STATE AREAS

Rock Creek Lake SRA. Area 165 acres; 54-acre reservoir. Located 10 miles west and 4 miles north of Benkelman, 1 mile south of fish hatchery. According to Richard Rosche (pers. comm.), this may be the best birding area in southwestern Nebraska. This reservoir is one of the few locations in the region where migrating water birds can settle; thus it attracts ducks, shorebirds, and other water birds during both spring and fall. It also attracts many passerine migrants, especially in autumn. The nearby fish hatchery often attracts ospreys.

Hitchcock County

Hitchcock County is a Republican Valley county with over 5,600 acres of surface water (mostly reservoir), about 2,000 acres of woodlands, and over 200,000 acres of grasslands. There are tourist accommodations at Trenton and Culbertson.

FEDERAL AREAS

None

STATE AREAS

Swanson Reservoir WMA. Area 1,157 upland acres, 4,973-acre reservoir. Primitive and modern camping facilities are present, and there are 13 hiking trails. Swanson Reservoir attracts many migrant water birds, some of which might overwinter. The wet meadows south of Stratton also attract many water birds during migration, including sandhill cranes and white-faced ibis. About 3,000 acres are open to hunting and other public use; this is the largest of the area's reservoirs and has a large fish population, which attracts eagles and other fish-eating birds.

Red Willow County

Red Willow County is a Republican Valley county with about 2,700 acres of surface water, 7,000 acres of woodlands, and nearly 180,000 acres of grasslands. There are tourist accommodations at Indianola and McCook.

STATE AREAS
Red Willow SRA. Area 5,948 acres, including a 1,628-acre reservoir. Located 12 miles north and 2 miles west of McCook on Highway 83. It consists mostly of High Plains grasslands. Complete camping facilities are present. The associated reservoir (Hugh Butler Lake) extends into Frontier County (see Frontier County).

Furnas County

Furnas County is a Republican Valley county with nearly 5,600 acres of surface water, 8,300 acres of woodlands, and over 175,000 acres of grasslands. There are tourist accommodations at Arapahoe, Beaver City, Cambridge, and Oxford.

FEDERAL AREAS
None

STATE AREAS
Cambridge Diversion Dam. Located 2 miles east of Cambridge. Includes 21 acres of grasslands bordering the Republican River and brushy bottomland.
Bartly Diversion WMA. A small area of grasslands, rolling hills, and scattered trees around a campground. Located 1 mile south and 1.5 miles east of Indianola.

Harlan County

Harlan County is a Republican Valley County with nearly 15,000 acres of surface water (mostly reservoir), nearly 9,000 acres of woodlands, and almost 120,000 acres of grasslands. There are tourist accommodations at Alma, Orleans, and Republican City.

FEDERAL AREAS
Harlan County Dam. Area 17,278 upland acres, 13,338-acre reservoir. This largest reservoir in south-central Nebraska attracts bald eagles, geese (especially Canada geese), and some sandhill cranes during spring and fall and has a population of greater prairie-chickens (on the south side of the reservoir) as well. Near the south end of the dam is an eagle roost. Look for burrowing owls in the prairie dog colony between Republican City and the administration area.

STATE AREAS
South Sacramento Wildlife Area. Area 77 wetland acres, 90 upland acres.
Southeast Sacramento Wildlife Area. Area 140 wetland acres, 45 upland acres.

Counties

Boyd, Holt, Knox, Antelope, Pierce, Garfield, Wheeler, Valley, Greeley, Boone, Madison, Platte, Sherman, Howard, Nance, Polk, Buffalo, Hall, Merrick, Hamilton, York, Kearney, Adams, Clay, Filmore, Franklin, Webster, Nuckolls, and Thayer

The central Platte Valley and nearby Rainwater Basin provide some of the best spring birding opportunities in all of North America; for most of March about 7 million waterfowl and nearly half a million sandhill cranes pour into the region, remaining until late March in the case of the waterfowl and about the second week of April in the case of the sandhill cranes. As the last sandhill cranes are leaving, whooping cranes begin to arrive, as do the earlier shorebirds, continuing the amazing spring spectacle until about the end of April.

Birding in the central Platte Valley during March is a chancy affair in terms of weather; late winter snowstorms may blanket the entire area in a foot of snow, which when melting leaves country roads slippery at best, and driving requires a good deal of care. This is especially true in the Rainwater Basin, an area of clay soils that prevent water from percolating down and thus is rich in temporary wetlands (locally called "lagoons") just at the peak of spring waterfowl populations. This is only true during years when winter snowfalls or spring rains allow the basins to fill; in drier years only the deepest lagoons or those that are kept wet by pumping (Harvard, Massie's, Smith, and the like) can accommodate the hordes of ducks and geese passing through. During such years the stresses caused by bad weather and overcrowding can set off outbreaks of fowl cholera and kill tens of thousands of birds in only a short time. Some of these birds are consumed by wintering bald eagles, hundreds of which occur along ice-free areas of the Platte from late fall until early spring. A good viewing area for these birds is at the J-2 Hydro Plant near Lexington. This area is open to the public on Saturdays and Sundays from 8 A.M. to 2 P.M. with weekday reservations possible for groups.

The best way to watch cranes during the day is observing them field-feeding while remaining quiet and inside a parked car. Opening a door and leaving the car will guarantee that the birds will rapidly depart. Gravel roads on the south side of the Platte River are usually better than those on the north side of I-80. The most rewarding way to watch cranes is from riverside blinds near roosting locations. Such blinds are maintained by the Whooping Crane Trust on Mormon Island at the Audubon Society's Lillian Rowe Sanctuary near Gibbon, and Fort Kearney State Historical Park near Kearney. If it is not possible to arrange a blind viewing, several bridges such as the hike-bike trail bridge near Fort Kearney or the bridge over the middle Platte channel south of Alda provide a less thrilling but still exciting view, both at sunset and sunrise. Information on crane viewing and accommodations can be obtained from the Kearney Visitors Center, the Grand Island Visitors Bureau, or the Adams County Visitors Bureau in Hastings. The Hastings Museum and the Stuhr Museum provide tourist information and sell informative books or pamphlets on local tourist attractions.

An excellent source of both general and specific information on birding in the Platte Valley is available in Gary Lingle's *Birding Crane River: Nebraska's Platte*, which is usually locally available in many stores and the Hastings and Stuhr Museums. It also includes complete county maps and detailed bird-finding advice for seven Platte Valley counties. Other ecological and historical information on the Platte Valley and its natural history exists in *The Platte: Channels in Time* (Johnsgard 1984). The Nebraska Game and Parks Commission in Lincoln can provide free informative materials, including an excellent eight-page "Spring Migration Guide" that focuses on Platte Valley birding.

The Rainwater Basin area is just as attractive as the Platte Valley during early spring, when snow meltwaters accumulate in the clay-rich lowlands and an estimated 7 to 9 million ducks and 2 to 3 million geese pass through. These flocks include 90 percent of the midcontinental greater white-fronted goose population, 50 percent of the midcontinental mallard population, and 30 percent of the entire continent's northern pintail population. More than a million snow geese now also use the area each spring, and some of the shallower wetlands (such as Sandpiper WPA) are of great importance to migrant shorebirds. The Rainwater Basin Wetland Management District comprises about 84 wetlands occupying 28,600 acres (including 21,742 acres federally owned and about 6,900 acres state-owned). Sites such as Harvard, Massie's, and Smith's Lagoons and Mallard Haven are of special value to waterfowl and are prime birding locations in the eastern basin, while Funk Lagoon is of special attraction in the western basin. A bird checklist for the Rainwater Basin, with nearly 200 species including over 100 breeding species, is available from the Kearney office of the U.S. Fish and Wildlife Service.

Visitors to the Platte Valley are urged to dress warmly, drive carefully over the narrow and sometimes slippery roads, and be prepared for rain, snow, and possibly even tornadoes. In recent years tornadoes have struck as early as mid-March and in one case killed tens of thousands of snow geese near York. Sick or dead birds found in the field should never be handled, as they possibly have become infected with diseases such as fowl cholera or botulism. This is especially true in the Rainwater Basin, where these diseases sometimes constitute a special problem.

Information on the Rainwater Basin is available from the U.S. Fish and

Wildlife Service in Kearney or the Rainwater Basin Joint Venture office. The annual Audubon Society's Rivers and Wildlife Celebration, held in Kearney during mid-March of every year, has many expert speakers and field trips.

Boyd County

Boyd County is a Niobrara Valley county with about 2,500 acres of surface water, 1,600 acres of woodlands, and 200,000 acres of grasslands. There are tourist accommodations at Spencer.

FEDERAL AREAS

None

STATE AREAS

Parshall Bridge WMA. Area 230 acres. Located 5 miles south of Butte. Riparian woodlands along the Niobrara River.

Hull Lake WMA. Area 36 acres. Located 3 miles south and 1 mile west of Butte. Hilly uplands, with oaks, conifers, and grasslands around a 3-acre lake.

Holt County

Holt County is a mostly Sandhills county with over 12,000 acres of surface water, almost 69,000 acres of woodlands, and 1.2 million acres of grasslands. There are tourist accommodations at Atkinson and O'Neill.

FEDERAL AREAS

None

STATE AREAS

Atkinson Lake SRA. Area 54 acres. Located at northwest edge of Atkinson. Includes a 14-acre reservoir of the Elkhorn River.

Goose Lake WMA. Area 349 acres. Located 4 miles west of Highway 281 and 2 miles north of Wheeler County boundary (23 miles south and 4 miles east of O'Neill). Mostly lake; also grassy and wooded uplands.

Redbird WMA. Area 433 acres. Located 1 mile south of highway 281 bridge over Niobrara. Mostly bur oak and cedar; bisected by Louse Creek with steep wooded slopes and rolling grasslands.

Spencer Dam WMA. Includes 9 miles of Niobrara River valley. Located 23 miles north of O'Neill on Highway 83.

Knox County

Knox County is bounded by the Niobrara and Missouri Rivers and thus has over 41,000 acres of surface water, 38,000 acres of woodlands, and almost 320,000 acres of grasslands. There are tourist accommodations at Bloomfield, Creighton, Crofton, Niobrara, and Wausa.

FEDERAL AREAS
None

STATE AREAS
Gavin's Point Dam, Lewis and Clark Lake SRA. Area 32,000-acre reservoir and 1,227-acre SRA. See Cedar County.

Bazille Creek WMA. Area 4,500 acres. Bordered for 9 miles by the Missouri River and Lewis and Clark Lake; mixed woods, grasslands, and marshy areas. This area is extensively marshy, as it includes the area where the Missouri River is impounded to form the upper end of Lewis and Clark Lake, and many wetland birds are present.

Niobrara State Park. Area 1,632 acres. This state park is located at the confluence of the Niobrara River and the backwaters of the Missouri River. It is mostly grasslands but also has riparian woodlands. There are more than 12 miles of hiking trails, and a new 2-mile hike-bike trail extends along the park's northern boundary. Woodland birds include whip-poor-wills, and bald eagles and ospreys are seasonally present. There are modern cabins and primitive camping facilities and an interpretive center.

Bohemia Prairie WMA. Area 680 acres. Mainly grasslands with some woods and two ponds.

Greenville WMA. Area 200 acres. Mostly native grasslands and bisected by Middle Verdigre Creek. Deciduous woodlands of the Niobrara Valley within about half a mile of the river itself. Located 10 miles west and 3 miles south of Verdigre.

Antelope County

Antelope County is an Elkhorn Valley county with about 900 acres of surface water, over 21,000 acres of woodlands, and nearly 200,000 acres of grasslands. There are tourist accommodations at Elgin, Neligh, and Orchard.

FEDERAL AREAS
None

STATE AREAS
Ashfall Fossil Beds State Historical Park. Area 360 acres. This extremely important paleontological site preserves the fossils of horses, rhino, camels, and other animals, including crowned cranes killed and interred under a thick layer of volcanic dust that settled here about 10 million years ago. The area is mostly rugged range country, with grassland species most common; however, rock wrens often can be seen near the excavation site.

Grove Lake WMA. Area 1,746 acres, 35-acre reservoir. Mainly mixed hardwoods and grasslands along Verdigre Creek. This is rolling grasslands with scattered trees along East Verdigre Creek. There is also a small reservoir and a trout-rearing facility. Most of the birds are grassland forms, but ospreys and belted kingfishers are also possible.

Hackberry Creek Public Use Area. Area 180 acres. Includes 1 mile of Elkhorn River frontage, several marshy oxbows, mixed woods, and grasslands.

Redwing WMA. Area 320 acres. Includes 1.5 miles of Elkhorn River frontage. Mostly riparian woodlands, with some grasslands and marshes.

Pierce County

Pierce County is a county with about 400 acres of surface water, about 5,000 acres of woodlands, and nearly 125,000 acres of grasslands. There are tourist accommodations at Osmond and Plainview.

FEDERAL AREAS
None

STATE AREAS
Willow Creek SRA. Area 1,600 acres, 700-acre reservoir. A fishing and camping area.

Garfield County

Pierce County is a Sandhills county with about 6,000 acres of surface water (nearly all reservoir), 5,500 acres of woodlands, and over 320,000 acres of grasslands. There are tourist accommodations at Burwell.

FEDERAL AREAS
None

STATE AREAS
Calamus Reservoir SRA. Area 10,312 acres, including a 5,123-acre reservoir. Includes several modern camping facilities, boating and related facilities, and a fish hatchery. No good information on the birds is available. See also Loup County.

Wheeler County

Wheeler County is an eastern Sandhills county with 1,300 acres of surface water, almost 24,000 acres of woodlands, and nearly 330,000 acres of grasslands. There are tourist accommodations at Erickson.

FEDERAL AREAS
None

STATE AREAS
Pibel Lake SRA. Area 42 acres, plus a 45-acre lake. Located 7 miles east and 2 miles south of Erickson. This is a beautiful Sandhills lake with a variety of water birds present during spring and summer.

Valley County

Valley County is a Loup Valley county with nearly 3,000 acres of surface water, 2,300 acres of woodlands, and 200,000 acres of grasslands. There are tourist accommodations at Ord.

FEDERAL AREAS
None

STATE AREAS
Fort Hartsuff State Historical Park. Located about 3 miles north and 1 mile west of Elyria. Mainly of interest for historical reasons, but passerines should be present seasonally.
Davis Creek WMA *and* SRA. Mostly grassy areas surrounding a 1,145-acre reservoir. The WMA includes about 2,000 acres exclusive of the recreation area. Located about 3 miles south of North Loup.
Scotia Canal WMA. Area 180 acres. Located 4.5 miles north of North Loup. Near North Loup River and mostly covered by grassy uplands and mixed woodlands.

Greeley County

Greeley County is an eastern Sandhills and loess hills county with nearly 2,500 acres of surface water (mostly reservoir), almost 3,000 acres of woodlands, and over 200,000 acres of grasslands. There are tourist accommodations near Spalding.

FEDERAL AREAS
None

STATE AREAS
Davis Creek SRA. Area 2,000 acres, 1,145-acre reservoir. See Valley County.

Boone County

Boone County is a mostly loess hills county with about 1,100 acres of surface water, 2,600 acres of woodlands, and over 160,000 acres of grasslands. There are tourist accommodations at Albion.

FEDERAL AREAS
None

STATE AREAS
Beaver Bend WMA. Area 27 acres. Located 1 mile northwest of St. Edward. Located along Beaver Creek, with riparian woodland habitat.

Madison County

Madison County is a loess hills and Elkhorn Valley county with about 800 acres of surface water, about 10,000 acres of woodlands, and over 80,000 acres of grasslands. There are tourist accommodations at Newman Grove, Norfolk, and Tilden.

None

Yellowbanks WMA. Area 680 acres. Includes 1.5 miles frontage of the Elkhorn River. Includes steep riverine bluffs supporting mature hardwood forest and some grassy uplands.
Oak Valley WMA. Area 640 acres. Includes a hardwood bottomland forest bisected by Battle Creek; otherwise grassy uplands.

Platte County

Platte County is a mostly loess hills county with almost 3,000 acres of surface water, 7,400 acres of woodlands, and nearly 90,000 acres of grasslands. There are tourist accommodations at Columbus and Humphrey.

None

George Syas WMA. Area 917 acres. Includes 1.5 miles of Loup River frontage. About half wooded, the rest grasses, crops, and planted shrubs.

Lake Babcock Waterfowl Refuge and Lake Babcock. Area 600 acres. A reservoir on Shell Canal just outside of Columbus.
Lake North. Area 200 acres. A city lake developed for fishing and swimming; probably of limited birding potential.
Looking Glass Creek WMA. Located 11 miles south of Monroe. About half wooded, the rest grasslands, with two small lakes.

Sherman County

Sherman County is a Loup Valley county with over 4,000 acres of surface water (mostly reservoir), 3,000 acres of woodlands, and over 200,000 acres of grasslands. There are tourist accommodations at Loup City.

None

Sherman Reservoir SRA/WMA. Area 3,382 acres, plus a 2,845-acre reservoir. Mostly rolling prairie grasslands, with woody growth along creeks. Includes ten hiking trails.
Bowman Lake SRA. Area 23 acres. A small SRA just outside Loup City.

Howard County

Howard County is a Loup Valley county with nearly 3,000 acres of surface water, over 5,000 acres of woodlands, and about 190,000 acres of grasslands. There are tourist accommodations at Dannebrog and St. Paul.

FEDERAL AREAS
None

STATE AREAS
Harold W. Anderson WMA. Area 272 acres. Located 4 miles south and 2 miles west of St. Paul. Consists of about 12 miles of Loup River frontage, with bottomland timber and a marshy oxbow.
Loup Junction WMA. Area 328 acres. Located 3 miles north and 2 miles east of St. Paul. Bordered on the north by the North Loup and on the south by the Middle Loup; mostly riparian woodlands, with marshes and grassy areas.

Nance County

Nance County is a Loup and Platte Valley county with almost 3,000 acres of surface water, over 5,000 acres of woodlands, and over 190,000 acres of grasslands. There are tourist accommodations at Genoa and Fullerton.

FEDERAL AREAS
None

STATE AREAS
Loup Lands WMA. Area 485 acres.
Prairie Wolf WMA. Area 314 acres. Mainly bottomlands along the Loup River, with some open grasslands and marshes.
Sunny Hollow WMA. Area 160 acres. Mostly grassy uplands, with two marshes and a dugout wetland.

Polk County

Polk County is a Platte Valley county with about 500 acres of surface water, nearly 5,000 acres of woodlands, and 54,000 acres of grasslands. There are tourist accommodations at Osceola.

FEDERAL AREAS
None

STATE AREAS
None

Buffalo County

Buffalo County is a Platte and Loup Valley county with 4,400 acres of surface water, 9,600 acres of woodlands, and nearly 225,000 acres of grasslands. There are tourist accommodations at Elm Creek, Gibbon, and Kearney.

FEDERAL AREAS
None

STATE AREAS

Ravenna Lake SRA. Area 53 acres. Situated along the South Loup River, with a small reservoir.

Blue Hole WMA. Area 530 acres, plus a 30-acre pond and 2 miles of river frontage. Mostly riparian woodlands.

Sandy Channel SRA. Area 133 acres; 11 small lakes and ponds totaling 47 acres.

Union Pacific SRA. Area 26 acres, plus a 15-acre pond.

East Odessa SRA. Area 71 acres, plus a 7-acre pond.

Cottonmill Lake Public Use Area. A hike-bike trail extends 6 miles from this area to the outskirts of Kearney.

Bassway Strip WMA. Area 636 acres, four ponds and 7 miles of river frontage. Includes 90 acres of lakes and sandpits; mostly wooded. In spite of the river frontage, this area is not used by sandhill cranes to any great extent.

War Axe SRA. Area 9 acres, plus a 12-acre pond.

Windmill SRA. Area 168 acres, five ponds.

OTHER AREAS

Lillian Annette Rowe Sanctuary. Area 2,200 acres. This area, the largest Audubon Society refuge in the region, protects prime sandhill and whooping crane habitats near Gibbon and includes nearly 6 miles of river frontage, plus about 260 acres of native prairie. Several riverside blinds are located on the property, and spring sunrise or sunset excursions to the blinds can be arranged between March 1 and April There is also a self-guided hiking/birding trail. The sanctuary headquarters provides information and sells books and other bird-related materials. Summer breeding birds include dickcissel, upland sandpiper, bobolink, and riparian woodland species such as rose-breasted grosbeak and willow flycatcher. Least terns and piping plovers often nest on barren sandbars that are also used by roosting cranes (Lingle 1994). A birding trail begins at the office; no checklists are available. No bird list is yet available for the sanctuary, but a complete list of breeding birds of the central Platte Valley exists, with more than 140 species have been reported. Permission is needed for hiking within refuge grounds.

Grandpa's Steakhouse. Located at the south end of Kearney just north of the Platte River bridge on the east side of the road. It is at the edge of a sandpit lake that attracts tens of thousands of Canada geese from late fall through early spring.

Gibbon Bridge. This bridge, 1.5 miles south of I-80 exit 285, provides a parking area for watching crane roosting flights.

Hall County

Hall County is a Platte Valley county with nearly 2,000 acres of surface water, 3,900 acres of woodlands, and almost 120,000 acres of grasslands. There are tourist accommodations at Alda, Grand Island, and Wood River.

FEDERAL AREAS
None

STATE AREAS
Cornhusker WMA. Area 840 acres. All upland habitats with various planted cover types. The birds include such brush-loving species as Harris' sparrows and American tree sparrows.

Mormon Island SRA. Area 152 acres, plus a 61-acre lake. This area is a popular fishing spot just off I-80 and rarely attracts many waterfowl because of the high disturbance level.

Martin's Reach WMA. Includes 89 acres, with about 0.7 mile of river frontage of the middle channel of the Platte River. Located 1 mile south and 3 miles west of Wood River Exchange.

OTHER AREAS
Hall County Park and Stuhr Museum. This county park offers wooded trails for birding. It is just south of the Stuhr Museum, which sells books of interest to naturalists and has a good exhibit of pioneer artifacts. The surrounding moat sometimes attracts wild ducks.

Mormon Island Whooping Crane Meadows Preserve. Area 2,500 acres. This preserve was the first Platte Valley crane sanctuary to be established and along with the Rowe Sanctuary farther west is the most important in the state. More than 70,000 cranes have been seen on its pristine wet meadows, and nearly 80,000 birds roost along its river shorelines. Nearly 220 bird species have been reported here; the Platte Valley bird list by Faanes and Lingle (1995) is listed in the references. A large crane-observation blind is operated from March 5 to April, and admission is obtained by calling or visiting the Crane Meadows Nature Center. Parking along the narrow road is possible, but requires care because of traffic. Leaving the road without permission is not allowed.

Shoemaker Island Road. A gravel road traverses the length of Shoemaker Island, where many wet meadows attract foraging flocks of cranes. The entire area is privately owned, so birding away from the road requires landowner permission.

Crane Meadows Nature Center. This nature center is located just south of the Alda interstate exit and is where reservations to visit the crane blind on Mormon Island must be made. There is an admission fee. Nearby are about 240 acres of meadows and woodlands, bird feeders, and nearly five miles of hiking trails. The trails are open from 8:30 A.M. to 5:30 P.M. Includes riverine forest, wet meadows, and moist tallgrass prairie. The ecology of this section of the Platte River is described by Johnsgard (1984). No published bird list exists; 212 species had been recorded on the preserve through early 1989. Prairie restoration is under way, and a nature trail with a bridge to an island is available.

Alda Road Bridge. This bridge over the middle Platte channel provides a place where people can watch the sunrise and sunset roosting flights of cranes. It is very near a sandpit lake that may attract up to 40,000 geese, but this lake is on private property and can only be viewed from the highway.

Platte River Road. This paved road going west from Doniphan is a good route for observing field-feeding cranes during the daytime. It continues west to the Kearney area, but the density of crane use varies with location and disturbance. Generally the cranes are best seen from the road nearest the south shore of the Platte River, especially in early morning and late afternoon, among cornfields or the occasional wet meadows that still exist.

Amick Acres. This small subdivision has several sandpit lakes that attract large flocks of Canada geese in early March. Do not stray from the road, as the area is entirely private property.

Nine-mile Bridge. This narrow bridge north of Doniphan provides views of a small crane flock on the downstream side. However, no parking is allowed near the bridge, and so some walking is necessary.

Merrick County

Merrick County is a Platte Valley county with about 600 acres of surface water, over 13,000 acres of woodlands, and more than 113,000 acres of grasslands. There are tourist accommodations at Central City.

FEDERAL AREAS

None

STATE AREAS

None

OTHER AREAS

Bader Memorial Park Natural Area. Area 80 acres. Located 3 miles south of Chapman. A stretch of Platte River woodlands and adjacent native prairie, with trails through all of the local habitat types. American woodcocks occur here, and sandhill cranes sometimes visit during spring. Ducks, geese, marsh birds and shorebirds are abundant during migration. Regal fritillary butterflies are regular in summer.

Hamilton County

Hamilton County is a Platte Valley and Rainwater Basin county with about 1,000 acres of permanent surface water, 2,400 acres of woodlands, and almost 50,000 acres of grasslands. There are tourist accommodations at Aurora.

FEDERAL AREAS

Springer WPA. Area 266 wetland acres, 134 upland acres.

Troesler Basin WPA. Area 123 wetland acres, 37 upland acres.

Nelson WPA. Area 143 wetland acres, 17 upland acres.

STATE AREAS

Pintail WMA. Area 190 wetland acres, 94 upland acres. Includes a shallow seasonal pond and mixed upland and lowland habitats. In wet springs this shallow marsh may attract up to 100,000 snow geese and large numbers of mallards, pintails and other dabbling ducks. In the morning the east side provides best viewing; during the afternoon the west side is better, and the road is closer to the marsh.

Gadwall WMA. Area 90 acres, 70 wetland acres (two dugout wetlands and narrow slough).

Deep Well WMA. Area 78 acres, 35 acres of semipermanent wetlands and 25 acres of permanent wetlands.

York County

York County is a Rainwater Basin county with nearly 3,000 acres of permanent surface water, 2,400 acres of woodlands, and over 50,000 acres of grasslands. There are tourist accommodations at Henderson and York.

FEDERAL AREAS

County Line Marsh WPA. Area 232 wetland acres, 176 upland acres. The county road leads to this marsh, which typically floods the road in spring. Large flocks of dabbling ducks usually gather here in early March.

Waco Basin WPA. Area 159 acres Adjoins Spikerush WMA and has a 15-acre lake stocked with fish.

Sinninger Lagoon WPA. Area 37 wetland acres, 123 upland acres.

STATE AREAS

Spikerush WMA. Area 194 acres. Consists of mixed marsh and upland habitats.

Kirkpatrick Basin WMAs. Area 615 acres. Contains 70 acres of semipermanent wetlands, 175 acres of seasonal wetlands, and the rest upland grasses in the north area; the southern WMA consists of a shallow wetland of 305 acres. This basin is an excellent area in spring for seeing migrating ducks and geese, especially snow geese, and slightly later it attracts a host of shorebirds, including American avocets and long-billed dowitchers. The north area is visible from I-80.

Hidden Marsh WMA. Area 120 acres. Located 2 miles east of Spikerush WMA.

Renquist Basin WMA. Area 107 acres. Consists of mixed uplands and marshlands.

Recharge Lake Natural Resource District Recreation Area. Area 120 acres, 50-acre reservoir. Developed for fishing; there are also hiking trails.

Kearney County

Kearney County is a Platte Valley county with about 200 acres of permanent surface water, 300 acres of woodlands, and over 70,000 acres of grasslands. There are tourist accommodations at Minden.

FEDERAL AREAS

All of the following sites are temporary wetlands of fairly small size but might be attractive to migrant water birds during wet springs.
Bluestem Basin WPA. Area 44 wetland acres, 32 upland acres.
Gleason Lagoon WPA. Area 197 wetland acres, 372 upland acres.
Prairie Dog Marsh WPA. Area 430 wetland acres, 382 upland acres.
Lindau Lagoon WPA. Area 105 wetland acres, 47 upland acres.
Clark Lagoon WPA. Area 227 wetland acres, 222 upland acres.
Youngson Lagoon WPA. Area 113 wetland acres, 70 upland acres.
Frerichs Lagoon WPA. Area 33 wetland acres, 10 upland acres.
Killdeer Basin WPA. Area 36 wetland acres, 2 upland acres.
Jensen Lagoon WPA. Area 187 wetland acres, 278 upland acres.

STATE AREAS

Hike-Bike bridge. This is a very good area for watching sandhill cranes at sunset and sunrise. Sometimes American woodcocks can be seen displaying near the north end of the bridge at sunset. Stop at the Fort Kearney State Historical Park for information and a park permit. The 4-mile trail leads to Bassway Strip WMA along the two northernmost channels of the Platte (see Buffalo County).
Fort Kearney SRA. Area 163 acres. This area has primitive camping facilities and provides nearby parking for the hike-bike bridge.
Fort Kearney State Historical Park. This park has a restored version of Fort Kearney and is also where one can watch field-feeding sandhill cranes. Arrangements for blind visits can be made here (see introduction to region). Park entry permit required.
Northeast Sacramento WMA. Area 30 wetland acres, 10 upland acres.

Adams County

Adams County is a Rainwater Basin county with about 900 acres of permanent surface water, 1,200 acres of woodlands, and over 83,000 acres of grasslands. There are tourist accommodations at Hastings.

FEDERAL AREAS

None

STATE AREAS

Prairie Lake Public Use Area. Area 125 acres, 30-acre lake. Mainly a fishing lake, with limited attractiveness to birds.

Crystal Lake SRA. Area 33 acres. Mostly developed for fishing.

DLD SRA. Area 7 acres. Primitive camping facilities are present.

OTHER AREAS

(See also Lingle 1994.)

Kenesaw Lagoon. A private wetland that attracts many water birds during spring. It is best observed from the south or west side along county roads.

Little Blue River. The wooded riparian zone of this river should be searched for passerines during migration periods.

Hastings Museum and Lake Hastings. The Hastings Museum has a notable exhibit area for a small-town museum, including a diorama with ten whooping cranes. It sells materials of interest to naturalists, has an IMAX theater and planetarium, and provides advice on local attractions. Lake Hastings is a city-owned lake that might seasonally attract some birds and is a short distance north of the museum on Highways 281 and 34. The Adams County Visitors Bureau might also be of assistance.

Air Lake. This is a privately owned seasonal wetland that sometimes attracts good numbers of migrating water birds.

Clay County

Clay County is a Rainwater Basin county with over 4,000 acres of permanent surface water, 900 acres of woodlands, and nearly 76,000 acres of grasslands. There are tourist accommodations at Clay Center and Sutton.

FEDERAL AREAS

These federal areas are WPAs that vary greatly in size and in relative wetland permanence.

Sandpiper WPA. Area 226 wetland acres, 214 upland acres. One of the best sites for seeing migrating shorebirds in the region, especially during late March and April.

Hultine WPA. Area 164 wetland acres, 74 upland acres.

Harvard Marsh WPA. Area 760 wetland acres, 724 upland acres. A deep, permanent marsh that attracts tens of thousands of snow, Canada, and greater white-fronted geese each March. Access from the east is via a narrow, often slippery road but is better from the south, at least to the railroad tracks. Driving beyond is not recommended. There is also a parking area on the north side, but it is located quite far from the nearest water or marshy areas. Later on in spring this area is used by many shorebirds, including several sandpipers and piping plovers, and breeders include northern harriers and short-eared owls. Occasional flocks of sandhill cranes stop, and eagles are regular in early spring. Altogether one of the best birding wetlands in the entire region; up to 500,000 waterfowl have been seen here at the peak of spring migration.

Lange Lagoon WPA. Area 56 wetland acres, 104 upland acres.

Theesen Lagoon WPA. Area 46 wetland acres, 34 upland acres.

Massie's Lagoon WPA. Area 494 wetland acres, 359 upland acres. One of the best of the Clay County lagoons for waterfowl and shorebirds. An observation blind is located close to the parking lot on the south side of the lagoon; this access point is recommended over the others. Water levels in spring are maintained by pumping.

Glenvil Basin WPA. Area 83 wetland acres, 37 upland acres.

Kissinger Basin WPA. Area 342 acres.

Meadowlark WPA. Area 45 wetland acres, 35 upland acres.

Harms WPA. Area 34 wetland acres, 25 upland acres.

Moger WPA. Area 72 wetland acres, 125 upland acres.

Shuck WPA. Area 56 wetland acres, 24 upland acres.

Green Acres WPA. Area 24 wetland acres, 9 upland acres.

Eckhart Lagoon WPA. Area 66 wetland acres, 108 upland acres.

Smith Lagoon WPA. Area 226 wetland acres, 224 upland acres. Another excellent waterfowl site during spring migration.

Greenhead WPA. Area 60 acres. Includes a dugout pond and mainly marshy habitats.

Hansen Lagoon WPA. Area 205 wetland acres, 115 upland acres.

Greenwing WPA. Area 80 acres. Includes marsh, uplands, and scattered thickets.

STATE AREAS

McMurtry Refuge. Area 1,071 acres. No public access and closed to hunting.

Bluewing WMA. Area 160 acres. Includes lowland and seasonal wetland habitat. Located 4 miles west and 0.5 mile south of Edgar.

Bullrush WMA. Area 160 acres. Includes uplands and marshes. Located 3 miles west of Edgar.

Filmore County

Filmore County is a Rainwater Basin county with 1,600 acres of permanent surface water, 2,100 acres of woodlands, and about 55,000 acres of grasslands. There are tourist accommodations at Geneva.

FEDERAL AREAS

County Line Marsh WPA. Area 408 acres.

Real WPA. Area 121 wetland acres, 39 upland acres.

Bluebill WPA *and Marsh Hawk* WPA. Areas 60 acres and 173 acres, respectively. Bluebill includes two marshes separated by higher ground. Marsh Hawk is comprised mostly of seasonal wetlands, with some trees and shrubs.

Wilkins Lagoon WPA. Area 370 wetland acres, 160 upland acres.
Murphy Lagoon WPA. Area 76 wetland acres, 12 upland acres.
Rolland Lagoon WPA. Area 53 acres wetland, 76 upland acres.
Rauscher Lagoon WPA. Area 140 wetland acres, 111 upland acres.
Sandpiper WPA. Area 160 acres. Includes 56 acres of marsh, with plum, willow, cottonwood, and Osage orange present.
Weiss Lagoon WPA. Area 40 upland acres, 120 wetland acres.
Krause Lagoon WPA. Area 303 acres wetland, 224 upland acres.
Mallard Haven WPA. Area 626 wetland acres, 411 upland acres. One of the best marshes for waterfowl during spring, and many wetland birds remain to breed, including northern harriers, great-tailed grackles, and yellow-headed blackbirds. There are several parking lots at access points and an information kiosk at the southeastern corner parking area.

STATE AREAS
None

Franklin County

Franklin County is a Republican Valley county with about 1,500 acres of surface water, over 5,000 acres of woodlands, and over 160,000 acres of grasslands.

FEDERAL AREAS
Ritterbush Marsh WPA. Area 49 wetland acres, 32 upland acres.
Quadhammer Marsh WPA. Area 308 wetland acres, 286 upland acres.
Macon Lakes WPA. Area 498 wetland acres, 466 upland acres.

STATE AREAS
Ash Grove WMA. Area 74 acres. Includes rolling hills, grasses, rock outcrops, and a spring-fed stream.
Limestone Bluffs WMA. Area 479 acres. Includes rolling hills with grasses, rock outcrops, and wooded ravines with a spring-fed stream.

Webster County

Webster County is a Republican Valley county with over 2,600 acres of surface water, nearly 4,000 acres of woodlands, and over 160,000 acres of grasslands. There are tourist accommodations at Red Cloud.

FEDERAL AREAS
None

STATE AREAS
Elm Creek WMA. Area 120 acres. Located 3 miles south of Cowles. Mostly wooded, with a creek and slough at one end.
Indian Creek WMA. Area 114 acres. Located 1 mile south of Red Cloud, containing riparian woods along the Republican River. Wood ducks, ospreys, eagles, woodland woodpeckers, and passerines might be seen.

OTHER AREAS

Willa Cather Prairie. Area 609 acres. Mixed-grass prairie, directly south of Red Cloud, near the Kansas border. Owned by the Nature Conservancy.
C. Bertrand Schulz and Marian Othener Schulz Prairie. Area 640 acres. Mixed-grass prairie 4 miles south and 5 miles east of Red Cloud on the Kansas border. Owned by the Nature Conservancy.

Nuckolls County

Nuckolls County is a Republican Valley county with 1,200 acres of surface water, over 10,000 acres of woodlands, and over 125,000 acres of grasslands. There are tourist accommodations in Nelson and Superior.

FEDERAL AREAS

None

STATE AREAS

Smartweed Marsh WMA. Area 74 wetland acres, 6 upland acres. Located 2 miles south and 2 miles west of Edgar. Mostly grassy lowlands but with some marshy areas.
Smartweed Marsh West WMA. Area 38 acres. Located 1 mile south and 3 miles west of Edgar. Mostly grassy lowlands but with some upland habitats.

Thayer County

Thayer County is a loess plains county with about 1,800 acres of surface water, over 5,000 acres of woodlands, and 113,000 acres of grasslands. There are tourist accommodations at Hebron.

FEDERAL AREAS

None

STATE AREAS

Little Blue WMA. Area 303 acres. Located 3 miles east of Hebron. Mostly flat, wooded bottomland of the Little Blue River, with some grasslands and croplands.
Prairie Marsh WMA. Area 160 acres. Located 2 miles west of Bruning. Consists of seasonal wetlands and adjoining uplands.

The Eastern Region

Counties

Cedar, Dixon, Dakota, Wayne, Thurston, Stanton, Cuming, Burt, Colfax, Dodge, Washington, Butler, Saunders, Douglas, Sarpy, Seward, Lancaster, Cass, Otoe, Saline, Jefferson, Gage, Johnson, Nemaha, Pawnee, and Richardson

The Eastern Region is a land that once was ruthlessly scraped over by glaciers and that later was mantled by tallgrass prairies and riparian deciduous forests with eastern biogeographic affinities. It is bounded to the east by the Missouri River, which is now mostly channeled and much degraded as far as wildlife habitat is concerned. However, some stretches, such as around Ponca Park, provide a faint idea of what the river once was like. The Missouri Valley is still a migratory pathway not only for arctic-breeding waterfowl such as snow geese, which alone now number over 1 million birds using this narrow flyway, but also for myriads of forest-adapted Neotropic migrants, especially warblers and vireos. Remnant stands of mature deciduous forest still exist at Rulo Bluffs Preserve and Indian Cave State Park in Richardson County, Fontenelle Forest and Neale Woods in the Omaha area, DeSoto National Wildlife Refuge near Blair, and Ponca State Park in Dixon County. These are among the best places to see such wonderful birds in early May as they journey north to breeding grounds in the upper Midwest and southern Canada.

The area is the most heavily populated part of the state and thus has the fewest areas of native prairie vegetation, but in these areas grassland species such as greater prairie-chickens still gather at sunrise every spring on traditional display grounds made sacred by decades if not centuries of use. Similarly, other prairie species such as long-billed curlews and upland sandpiper rarely occur here any more, but grassland sparrows still announce their territories from fence posts, and house wrens, gray catbirds, and brown thrashers sing from plum thickets along roadside ditches. Many other eastern or southeastern species occur and presumably breed here, including Henslow's sparrow; Kentucky, northern parula, and prothonotary warblers; and red-shouldered and broad-winged hawks. Furthermore, chuck-will's-widows certainly must nest in the woodlands bordering the southeastern corner of the state (although no nests of this species have been found), and pileated woodpeckers are sometimes also seen here, making it an area of special interest to birders.

Cedar County

Cedar County is a Missouri Valley county with 3,900 acres of surface water, 10,700 acres of woodlands, and almost 134,000 acres of grasslands. There are tourist accommodations at Hartington, Laurel, and Randolph.

FEDERAL AREAS
Gavin's Point Dam. Birding from the dam should offer views of gulls, waterfowl, and other birds, including numerous bald eagles during migration periods. A nature trail and an aquarium are present in the associated Lewis and Clark Recreation Area.

Lake Yankton. Located just below Gavin's Point Dam; partly in South Dakota. A relatively small reservoir below the spillway at Gavin's Point Dam that offers good birding along its wooded shoreline.

STATE AREAS

Chalkrock WMA. Area 130 acres. Located 4 miles south of the Missouri River bridge on Highway 81 and 1.5 miles west of the highway. Consists of 90 upland acres and a 45-acre reservoir.

Wiseman WMA. Area 365 acres. Located 1 mile north and 5 miles east of Wynot. Just south of the Missouri River, this area includes steep wooded bluffs and grassy ridges. The woods are mostly bur oak, cedar, hackberry, and ash.

Dixon County

Dixon County is a Missouri Valley county with over 5,000 acres of surface water, 10,000 acres of woodlands, and over 73,000 acres of grasslands. There are tourist accommodations at Ponca.

FEDERAL AREAS

None

STATE AREAS

Ponca State Park. Area 830 acres. This park is mostly forested with stands of bur oak, walnut, hackberry, and elms; one of the oaks is more than 300 years old. There are 17 miles of hiking trails, and modern cabins are available, as is an undeveloped campground. Whip-poor-wills are common in summer, and bald eagles are present during winter months. Snow geese migrate past the area in spring and fall, and the nearby Missouri River is still unchanneled here, thus resembling its original state. Wildflowers typical of eastern deciduous forests include Dutchman's-breeches, bloodroot, and white fawn lily.

Buckskin Hills WMA. Consists of 340 acres of grasslands and woods around a 75-acre reservoir. Located 2 miles west and 2 miles south of Newcastle.

Dakota County

Dakota County is a Missouri Valley county with 3,400 acres of surface water, 6,800 acres of woodlands, and over 28,000 acres of grasslands. There are tourist accommodations at South Sioux City.

FEDERAL AREAS

None

STATE AREAS

Basswood Ridge WMA. Area 360 acres. Consists of very rugged and heavily wooded uplands, with some Indian petroglyphs near the north end.

Omadi Bend WMA. Area 33 acres. Consists of bottomland forest along an oxbow lake.

Wayne County

Wayne County is a county of glaciated uplands with 180 acres of surface water, 2,100 acres of woodlands, and nearly 46,000 acres of grasslands. There are tourist accommodations at Wayne.

FEDERAL AREAS
None

STATE AREAS
Sioux Strip WMA. Area 25 acres. Located at western edge of Scholes. Consists of upland grasses along an old railroad bed.

Thurston County

Thurston County is a Missouri Valley county with nearly 1,500 acres of surface water, over 22,000 acres of woodlands, and nearly 34,000 acres of grasslands.

FEDERAL AREAS
Missouri River Federal Access Areas. Numerous access points for boats, as posted.

STATE AREAS
None

Stanton County

Stanton County is an Elkhorn Valley county with about 800 acres of surface water, 5,200 acres of woodlands, and 75,000 acres of grasslands.

FEDERAL AREAS
None

STATE AREAS
Red Fox WMA. Area 363 acres. Located 1 mile south of Pilger. Includes a wooded remnant oxbow, a 25-acre sandpit lake, 0.6 mile of Elkhorn River frontage, and 163 acres of grasslands.
Wood Duck WMA. Area 668 acres. Located about 2 miles south and 4 miles west of Stanton. Consists of riparian woodlands bordering the Elkhorn River, with several oxbow lakes and a stream. Many eastern songbirds nest here, and the marshy lakes are used by large numbers of geese, ducks, pelicans, cormorants, and occasional swans. The perimeter roads are often wet and sometimes even flooded during wet springs, so caution is needed when driving.

Cuming County

Cuming County is an Elkhorn Valley county with about 800 acres of surface water, 3,000 acres of woodlands, and over 75,000 acres of grasslands. There are tourist accommodations at Beemer, West Point, and Wisner.

FEDERAL AREAS

None

STATE AREAS

Black Island WMA. Area 240 acres. Located 1 mile north and 4 miles west of Wisner. This area is a mixture of woods and grasslands along 0.75 mile of the Elkhorn River.

OTHER AREAS

Elkhorn River. This river is bounded by deciduous riverine forest along most of its length and should provide for good birding opportunities.

Burt County

Burt County is a Missouri Valley county with 3,200 acres of surface water, 5,500 acres of woodlands, and almost 39,000 acres of grasslands. There are tourist accommodations at Lyons, Oakland, and Tekama.

FEDERAL AREAS

None

STATE AREAS

Decatur Bend WMA. Area 133 acres. Located 3 miles east of Decatur. Accessible only by river.
Pelican Point SRA. Area 36 acres. Located along Missouri River, 3 miles north and 6 miles east of Tekama. Consists of riverine woodlands. Accessible only by river.
Summit Reservoir SRA. Area 535 acres. Located 3 miles west and 1 mile south of Tekama. The 190-acre reservoir is developed for fishing.

Colfax County

Colfax County is a Platte Valley county with about 1,300 acres of surface water, 5,900 acres of woodlands, and 44,000 acres of grasslands. There are tourist accommodations at Schuyler.

FEDERAL AREAS

None

STATE AREAS

Whitetail WMA. Area 216 acres. Located 2 miles south and 1 mile west of Schuyler. Consists of 93 acres of Platte River bottomland forest and 123 acres of islands and river.

Dodge County

Dodge County is an Elkhorn Valley county with 1,800 acres of surface water, over 3,000 acres of woodlands, and almost 40,000 acres of grasslands. There are tourist accommodations at Fremont.

FEDERAL AREAS
None

STATE AREAS
Dead Timber SRA. Area 150 acres, 50-acre lake. A recreational facility developed for fishing.
Powder Horn WMA. Area 289 acres. Consists of riparian woodlands bounding the Elkhorn River plus adjoining grasslands, marshes, and croplands.
Fremont Lakes SRA. Area 670 acres. Includes 20 small sandpit lakes totaling 280 acres, with many recreational facilities.

Washington County

Washington County is a Missouri Valley county with over 3,000 acres of surface water, nearly 15,000 acres of woodlands, and nearly 33,000 acres of grasslands. There are tourist accommodations at Blair.

FEDERAL AREAS
DeSoto National Wildlife Refuge. Area 7,823 acres This important national wildlife refuge is located around an old oxbow of the Missouri River and consists mostly of riverine deciduous forest, an 8-mile oxbow lake, and croplands. It supports an enormous fall population of snow geese (and some Ross' geese), which may reach peak numbers of about 800,000 birds in late October or early November. There is a superb interpretive center, whose large windows face a 788-acre lake, allowing wonderful views of the geese, other waterfowl, and numerous bald eagles in late fall and early spring. There are also outdoor viewing platforms for close viewing. There is a bird checklist of 240 species, including 81 breeders. Peak populations of ducks, mostly mallards, may reach 125,000. There is a 12-mile drive around the refuge and four hiking trails. A summary checklist of the refuge's birds is available upon request. A daily admission fee charged, and there are seasonal driving restrictions.
Boyer Chute National Wildlife Refuge. Located along the Missouri River, this relatively new refuge of about 2,000 acres is a cooperative project involving the U.S. Army Corps of Engineers and several state agencies. The "chute" is a reconstructed side-channel segment of the Missouri River. There are currently several access points, with 2 miles of roads and about 5 miles of trails, including 1 mile of paved trail. The area is being managed from DeSoto National Wildlife Refuge.

Fort Atkinson State Historic Park. Probably mainly of interest for historical reasons, but no doubt migrating passerines use the area to some degree.

Butler County

Butler County is a Platte Valley county with about 700 acres of surface water, 6,000 acres of woodlands, and 84,000 acres of grasslands. There are tourist accommodations at David City.

FEDERAL AREAS

None

STATE AREAS

Redtail WMA. Area 320 acres, 17-acre reservoir. Located 1 mile east of Dwight. Includes grasslands, wooded draws, and a pond.
Timber Point Public Use Area. Area 160 acres, 28-acre lake. Located 1 mile south and 2 miles east of Brainard.

Saunders County

Saunders County is a Platte Valley county with over 4,000 acres of surface water, over 9,000 acres of woodlands, and 103,000 acres of grasslands. There are tourist accommodations at Wahoo.

FEDERAL AREAS

None

STATE AREAS

Jack Sinn WMA. See Lancaster County.
Red Cedar Public Use Area. Area 175 acres, 50-acre lake.
Larkspur WMA. Area 160 acres. Includes 37 acres of bur oak woodlands plus areas of native prairie and seeded grasslands.

OTHER AREAS

Pahuk Natural Area. A privately owned area of riparian woodlands considered sacred by the Pawnee. There is a trail through the woods and adjoining grasslands. Permission to visit is needed from the local landowners.

Douglas County

Douglas County is a Missouri Valley county with 12,800 acres of surface water, over 3,000 acres of woodlands, and 20,000 acres of grasslands. There are tourist accommodations at Omaha.

FEDERAL AREAS

None

STATE AREAS

Two Rivers SRA. SRA area 643 acres, WMA 312 acres. The SRA is developed for recreational purposes. Located just south of the SRA, the WMA consists of timbered riverbottom forest, marshes, and croplands.

OTHER AREAS

Neale Woods Nature Center. Area 554 acres. This privately owned nature center in Omaha includes 9 miles of trails through hardwood forests, prairie uplands, and riverine woodlands. A checklist of 190 species either seen or expected in the area is available. There are 57 likely breeders, including barred owl, whip-poor-wills, ruby-throated hummingbird, eastern wood-pewee, American redstart, scarlet tanager, rose-breasted grosbeak, and indigo bunting. An admission fee is charged.

Standing Bear Lake. Area 135 acres. A flood-control lake that attracts migrant waterfowl seasonally.

Glenn Cunningham Lake. Area 390 acres. A flood-control reservoir that is the largest in the Omaha area and as such is important for migrating waterbirds.

Papio D-4 Lake. Area 30 acres. Located just east of Glenn Cunningham Lake, this reservoir sometimes attracts migrant gulls and waterfowl.

Zorinski Lake. A flood-control reservoir of about 250 acres on a branch of Papillion Creek.

Heron Haven. Located at 118th and Old Maple Road. An Audubon nature center still under development.

Sarpy County

Sarpy County is a Missouri Valley county with 2,500 acres of surface water, nearly 15,000 acres of woodlands, and 21,000 acres of grasslands. There are tourist accommodations at Bellevue, Gretna, and Papillion.

FEDERAL AREAS
None

STATE AREAS

Chalco Hills Recreation Area. Area 1,200 acres; Wehrspan Reservoir 245 acres. This developed area has boating, an arboretum, a nature trail, a natural resource center, a wildlife observation blind, and a hike-bike trail.

Schramm Park Recreation Area. Area 340 acres. This area includes the Ak-Sar-Ben aquarium and some excellent woodlands that teem with warblers during spring migration. There are 5 miles of trails, and an educational center is at the aquarium. Whip-poor-wills can be heard here, and Kentucky warblers may breed. An aquarium features rare fish, such as pallid and lake sturgeons. There is 3-mile loop trail leading from the parking area.

Fontenelle Forest Preserve. Area 1,300 acres. This large area of mature riverine hardwood forest in Omaha includes 17 miles of footpaths as well as a 1-mile-long boardwalk and two recently finished education centers. At least 40 species of deciduous trees occur here; the uplands are dominated by bur oak, shagbark hickory, and basswood. This is about the northern limit for gray squirrels in Nebraska. There is a bird checklist of 246 species that have been reported in the past decade, and more than 100 of these are summering species that potentially breed. Nature trails near the river are particularly rewarding during spring warbler migrations. Summer species and known or potential breeders include American woodcock, red-shouldered hawk (bred in 1995), whip-poor-will, pileated woodpecker (bred in 1999), Acadian flycatcher, Carolina wren, yellow-throated vireo, wood thrush, American redstart, cerulean, prothonotary and Kentucky warblers, brown creeper, and the scarlet and summer tanagers. The yellow-throated warbler has been reported to perhaps nest here, too, a location well to the north of its known breeding range. An observation blind overlooks a marsh, and there are organized bird or nature hikes plus many other programs.

Gifford Point. A 1,300-acre area of riverbottom forest and a 400-acre farm (advance reservations needed for farm). Dedicated to environmental education and located just east of Fontenelle Forest.

Neale Woods Nature Center. See Douglas County.

Seward County

Seward County is a loess plains county with 1,500 acres of surface water, about 6,000 acres of woodlands, and over 110,000 acres of grasslands. There are tourist accommodations at Seward.

FEDERAL AREAS

None

STATE AREAS

Meadowlark NRD Recreation Area. Area 55 acres.

Oak Glen WMA. Area 632 acres. Consists of 260 acres of mature oak woodlands, with some grasslands, including native prairie.

Branched Oak SRA. See Lancaster County.

Bur Oak WMA. Area 143 acres. Located along Highway 84 and comprised of mature bur oak woodlands, with some green ash and native grasslands.

Twin Lakes WMA. Area 1,300 acres. Includes 255- and 50-acre reservoirs plus marshes, wooded bottomlands, upland prairie, grasslands, and small ponds. Although part of the prairie has been re-seeded, other portions are of native prairie. Dickcissels, eastern and western meadowlarks, sedge wrens, eastern bluebirds, and other prairie or forest-edge species occur here. The two lakes do not allow boats with motors, other than electric motors, and no waterfowl hunting is permitted, so undisturbed birding is possible. The area is closed to the public from October 15 to the end of the hunting season for Canada geese.

North Lake Basin WMA. Contains 364 acres of marsh and adjoining uplands. Located 1 mile north of Utica. This area usually attracts large numbers of waterfowl during spring migration.

Freeman Lakes WMA. Includes 42 upland acres and 146 wetland acres. Located northwest of Utica. Similar to North Lake Basin.

Lancaster County

Lancaster County is in an area of glacial till plains with nearly 6,000 acres of surface water, over 7,000 acres of woodlands, and 140,000 acres of grasslands. There are tourist accommodations at Lincoln.

FEDERAL AREAS
None

STATE AREAS
Most of the following state areas are flood-control reservoirs built in the 1960s to control flooding in the Salt Creek valley. Silting-in has affected all of these reservoirs, producing marshlike habitats at the places where creeks feed into the reservoirs, and such areas usually provide the best birding sites. Flooded trees are also usually present, offering perching sites for bald eagles and cormorants.

Pawnee Lake SRA. Area 1,906 acres, 740-acre reservoir. This fairly large reservoir (second to Branched Oak) attracts many migrant waterfowl during spring and also has many prairie species.

Conestoga Lake SRA. Area 486 acres, 230-acre reservoir. Similar to Pawnee Lake in its bird life, this lake is also surrounded by grasslands and some woodlands.

Yankee Hill Lake SRA. Area 728 acres. Includes a 210-acre reservoir with surrounding rolling grasslands and wooded bottomland.

Killdeer WMA. Area 69 acres. Includes a 20-acre reservoir with surrounding marsh, wooded draws, and uplands.

Bluestem Lake SRA. Area 483 acres, 325-acre reservoir. The silted-in upper (northern) end of this reservoir is quite marshlike and is good for finding marsh birds.

Olive Creek SRA. Area 438 acres, 145-acre reservoir. This small lake seems to attract rare waterfowl, especially scoters, which turn up here almost every year.

Teal WMA. Area 66 acres. Includes a 27-acre reservoir, with surrounding wooded bottomlands and rolling uplands.

Stagecoach Lake SRA. Area 412 acres, 120-acre reservoir.

Cottontail Public Use Area. Area 148 acres, 28-acre reservoir.

Wagontrain Lake SRA. Area 750 acres. 315-acre reservoir. Much like the other lakes in the area, with many flooded trees.

Hedgefield Lake WMA. Area 114 acres. Includes a 44-acre reservoir, with surrounding rolling uplands with some wooded vegetation.

Branched Oak Lake SRA. Area 4,406 acres, 1,800-acre reservoir. This is the largest reservoir in the county and attracts many rare birds (gulls, waterfowl, and loons) during fall, winter, and early spring. Vast flocks of snow geese visit in early March, as do Canada and greater white-fronted geese. Eagles are common during the spring when the ice is breaking up, and ospreys may also be seen on migration. Many species of ducks and white pelicans are common during migration. The shorebirds are best during fall, and flooded timber at the northern end of the lake attracts cormorants. Snowy owls and black-billed magpies often turn up here, and the brushy vegetation supports wintering American tree sparrows and Harris' sparrows, among many others.

Wildwood Lake WMA. Area 491 acres. Includes a 107-acre reservoir with surrounding native woodlands, hilly grasslands, and crops. This is a little-visited and an unusually beautiful lake.

Jack Sinn WMA. Area 632 acres. Consists of mostly seasonally wet alkaline lowlands that occur along a creek drainage. Includes some beaver ponds up to approximately 6 feet deep. Some of the best rail and marsh bird habitat in the county occurs here; some of the best saline wetlands in eastern Nebraska are found here. The marsh provides habitat for some saline-dependent species of tiger beetles and halophytic plants. There is an old railroad bed that offers dry walking opportunities in this usually wet environment.

OTHER AREAS

Arbor Lake. Area 63 acres. Consists of a 63-acre saline and semipermanent wetland that seasonally supports great-tailed grackles, migrant ducks and shorebirds, and prairie passerines. This area and the Salt Creek drainage nearby support two rare species, the Salt Creek tiger beetle and saltwort, both limited to highly saline sites.

Lagoon Park. Undeveloped. This little-visited part of Lincoln can only be reached by following 48th Street north to the city landfill office and by stopping there for permission to go on to the marshes and riparian woodlands about a quarter mile beyond. It is well worth the effort; this area is perhaps the best place in Lincoln for seeing large numbers of species. Red-tailed hawks and great horned owls nest here; coyotes and deer are regularly seen; and the marshy areas attract rails, ducks, geese, cormorants, yellow-headed blackbirds, black-crowned night herons, and many other species. Care must be taken in walking through the area, which is partly landfill and sometimes has ground irregularities that one can easily fall into or trip over. Ospreys and shorebirds are sometimes seen along Salt

Creek, which borders the area on the north and west. Visits must be made during the daylight hours that the landfill is open to the public.

Nine-Mile Prairie. One of the few remaining native tallgrass prairies in the county and a classic site of early ecological research by university personnel such as John Weaver. At least 400 plant species have been detected here, and the rare regal fritillary butterfly is common in midsummer. Only a small amount of water exists on the prairie, so the species are mostly upland passerines, such as grasshopper sparrows and dickcissels. Henslow's sparrow is also present.

Oak Lake Park. Area 47 acres. A small park around two lakes that attract gulls, cormorants, geese, ducks, and other waterbirds during migration.

Capitol Beach. The remains of a once-saline lake, the undeveloped east side still supports a good marsh habitat that is now being preserved and has provided records of king rails and other uncommon to rare water-dependent birds. Enter from Sun Valley Boulevard onto Westgate Boulevard, then west on Lake Drive to parking area. Grassy wetland areas are excellent for sparrows.

Pioneer's Park. Area 606 acres. This is the oldest and one of the largest of Lincoln's city parks. It has a few small ponds and some native prairie but is mostly planted to pines and other conifers. These woodlands support great horned owls (year-round), long-eared owls (in winter), red-tailed and sharp-shinned hawks, and many passerines. A nature trail extends out into prairie and brushland and through riparian woodlands along a branch of Salt Creek, where wood ducks are common. A nature center offers views of wintering songbirds (such as Harris' sparrows) that gather near feeders and looks out over a pond that attracts waterfowl. Checklists for the park list 237 bird species and 49 mammal species.

Holmes Lake. Area 555 acres. A city park around a small reservoir that has a resident flock of Canada geese. In some winters ducks such as goldeneyes, scaup, and sometimes hooded mergansers can be seen among the geese.

Wilderness Park. Area 1,455 acres. This long and narrow park follows Salt Creek for about 7 miles and has about 20 miles of hiking and horseback trails. There are good stands of mature bur oak and riparian forest, and several species of owls are common (barred owl, eastern screech-owl, and great horned owl), as well as several woodpeckers including red-bellied. It is probably the best Lincoln location for migrant warblers, and breeding songbirds include tufted titmouse (near its western limits), eastern bluebird, orchard and Baltimore oriole, and Carolina wren (also near its northwestern limits). A bird checklist of 191 species is available from the Lincoln Parks and Recreation Department.

Wyuka Cemetery. Entrance at 37th and Vine, Lincoln. Another excellent location for migrant songbirds such as vireos and warblers, mainly along eastern edge.

Spring Creek Ranch. Area 640 acres. Acquired in 1998 by the National

Audubon Society. Three miles south and 0.5 mile west of Denton. This more than 600-acre and still-undeveloped nature sanctuary is the largest area of unbroken prairie in Lancaster County. It contains some small wetlands including a spring, riparian woodlands, and hilly prairie uplands. No bird list yet exists. Open year-round. Several smaller tallgrass prairies in this southeastern region are managed by Lincoln's Audubon Society, including Dieken Prairie, 1 mile south and 0.75 mile west of Unadilla; Kasl Prairie, 8 miles south and 2.5 miles west of Crete; and Bentzinger Prairie, 9.25 miles south of Syracuse.

Mopac East Trail. This hiking trail, going east from Lincoln, will eventually reach Omaha and is part of the American Discovery Trail, a planned transcontinental trail extending from Delaware to California. The Nebraska segment will extend from Omaha west through Lincoln and along the Platte Valley to Lake McConaughy, then southwest to the Colorado border.

Cass County

Cass County is a Missouri Valley county with about 1,800 acres of surface water, 24,000 acres of woodlands, and almost 56,000 acres of grasslands. There are tourist accommodations at Greenwood.

FEDERAL AREAS

None

STATE AREAS

Eugene T. Mahoney State Park. Area 574 acres. This is a highly developed park, with lodging, cabins, eating facilities, and other popular attractions. It has an excellent population of eastern bluebirds and borders the Platte River, where riparian deciduous forest is well developed. There is a very tall observation tower built in a stand of bur oaks and a 6-mile network of trails.

Platte River State Park. Area 418 acres. A popular park, this area is quite similar to the Eugene T. Mahoney State Park and also to Louisville Lakes State Park. Migrant warblers are abundant, and Kentucky warblers may breed here, as do scarlet and summer tanagers. There are two observation towers. Cabins are available; no camping is allowed.

Louisville Lakes State Park. Area 142 acres. Much like the two previous parks, with both primitive and modern camping and various concessions. The nearby Schramm Park is less crowded and probably offers better birding.

Randall W. Schilling WMA. A 1,500-acre managed waterfowl area, with 25 acres of water and nearby croplands, mainly designed to attract snow geese. Open to the public from April 1 to September 30 and used for controlled-access goose hunting during the fall season.

Rakes Creek WMA. Area 316 upland acres. Located 8 miles east and 1 mile south of Murray.

Otoe County

Otoe County is a Missouri Valley county with 2,500 acres of surface water, over 15,000 acres of woodlands, and nearly 102,000 acres of grasslands. There are tourist accommodations at Nebraska City and Syracuse.

FEDERAL AREAS
None

STATE AREAS
Triple Creek WMA. Area 80 acres. Located 3 miles south and 1 mile west of Palmyra. Contains two intermittent streams and 16 acres of woodlands.
Walnut Creek WMA. Area 41 acres. Located 1 mile south and 3 miles east of Otoe. Includes a 14-acre reservoir and surrounding grasses and shrubs.
Arbor Lodge State Historical Park. Located just west of Nebraska City. There are hundreds of planted but mature trees in the arboretum and a 0.5-mile "tree trail" with identification tags. Worth visiting for historical reasons as well as for its spring migrant birds.

OTHER AREAS
Dieken and Bentzinger Prairies. Dieken Prairie, 1 mile south and 0.75 mile west of Unadilla, and Bentzinger Prairie, 9.25 miles south of Syracuse, are both small but splendid tallgrass prairies managed by the Wachiska (Lincoln) chapter of the National Audubon Society. Dieken Prairie is notable for its stand of prairie fringed orchids and nesting bobolinks.

Saline County

Saline County is in a region of loess and eroded plains with about 1,900 acres of surface water, almost 11,000 acres of woodlands, and about 89,000 acres of grasslands. There are tourist accommodations at Crete, Friend, and Wilber.

FEDERAL AREAS
None

STATE AREAS
Swan Creek WMA. Area 160 acres. Located 9 miles south and 1 mile east of Friend. Consists of a 27-acre lake, marshland, native prairie, woodlands, and croplands.
Walnut Creek Public Use Area. 64 acres. Located 3 miles north and 2 miles east of Crete. Various upland habitats are present.

Jefferson County

Jefferson County is in a region of loess and eroded plains with nearly 2,000 acres of surface water, over 10,000 acres of woodlands, and 148,000 acres of grasslands. There are tourist accommodations at Fairbury.

None

Alexandria Lakes SRA/WMA. Area 778 acres. Located 9 miles west and 6 miles north of Fairbury, or 4 miles east of Alexandria. Consists of a 43-acre lake, marshes, streams, ponds, and woodlands.

Rock Creek Station State Historical Park. Area 393 acres. Located about 2 miles north and 3 miles east of Endicott.

Rock Glen WMA. Includes 706 acres of rolling native uplands and tree-lined drainages. Located 7 miles east and 2 miles south of Fairbury.

Jefferson WMA. Consists of 306 acres of woodlands and grassy uplands surrounding a 77-acre reservoir. Located 3 miles east and 1 mile south of Diller.

Gage County

Gage County is in an area of loess and glacial drift with nearly 2,000 acres of surface water, 17,000 acres of woodlands, and 166,000 acres of grasslands. There are tourist accommodations at Beatrice and Wymore.

Homestead National Monument of America. This monument (160 acres, the size of original Homestead Act allotments) celebrates the 1862 Homestead Act. It includes a 2.5-mile trail passing through riparian woodlands and restored prairie. There is a bird checklist of more than 150 species, a plant list, and an educational exhibit that features pioneer history and related artifacts. Located 3.5 miles west of Beatrice.

Claytonia Public Use Area. Area 115 acres, 40-acre reservoir.

Iron Horse Hike/Bike Trail. Several units, varied areas. There are more than 20 such sites in Gage and Pawnee Counties ranging from 1 to 19 acres.

Rockford Lake SRA. Area 436 acres, 150-acre reservoir.

Wolf Wildcat Public Use Area. Area 160 acres, 42-acre reservoir.

Arrowhead WMA. Area 320 acres. Upland habitats.

Diamond Lake WMA. Area 320 acres, including lakes. Consists of open grasslands and hardwood stands around a 33-acre reservoir. The adjoining Donald Whitney Memorial WMA is much smaller and also includes a reservoir.

Big Indian Public Use Area. Area 77 acres. Includes a lake.

Wildcat Creek Prairie. A 30-acre tallgrass prairie, 6 miles south, 1 mile west, and then 1 mile south of Virginia. Managed by the Wachiska (Lincoln) chapter of the National Audubon Society.

Kasl Prairie. Eight miles south and 2.5 miles west of Crete. Managed by the Wachiska (Lincoln) chapter of the National Audubon Society.

Johnson County

Johnson County is in a region of glacial drift with about 500 acres of surface water, 6,800 acres of woodlands, and over 92,000 acres of grasslands. There are tourist accommodations at Tecumseh.

FEDERAL AREAS
None

STATE AREAS
Osage WMA. Area 778 acres. Mostly comprised of woodlands and intervening grassland habitats plus tree plantings and crops.
Hickory Ridge WMA. Area 250 acres. Includes 60 acres of timber, a small pond, and fairly steep grasslands and creek-bottom habitat.
Twin Oaks WMA. Area 795 acres. Mostly woodlands, with some grasslands and food plots.

Nemaha County

Nemaha County is a Missouri Valley county with nearly 1,800 acres of surface water, over 21,000 acres of woodlands, and 48,000 acres of grasslands. There are tourist accommodations at Auburn and Brownville.

FEDERAL AREAS
None

STATE AREAS
Indian Cave State Park. See Richardson County.
Brownville SRA. Area 220 acres. Located at southeastern edge of Brownville, it provides boating access to the Missouri River.

Pawnee County

Pawnee County is in a region of glacial till with about 1,000 acres of surface water, over 16,000 acres of woodlands, and about 124,000 acres of grasslands, the latter floristically associated with the Flint Hills prairies of Kansas. There are tourist accommodations at Pawnee City and Steinauer.

FEDERAL AREAS
None

STATE AREAS
Iron Horse Trail. Many short sections of railroad bed that are being used as a hike-bike trail. See also Gage County.
Burchard Lake WMA. Area 560 acres, 150-acre reservoir surrounded by native grasslands and some hardwoods. There is a resident flock of greater

prairie-chickens, and two permanent blinds are located on the hilltop that is used as a lek by up to 10 males. Burchard Lake also supports a small population of the rare massasauga rattlesnake.

Bowwood WMA. Area 320 acres. Comprised of wooded areas, croplands, grasslands, and two small ponds.

Pawnee Prairie WMA. Area 1,021 acres. Comprised mostly of native prairie, with some woodlands and a small amount of croplands. Supports a flock of about 20 greater prairie-chickens, which have a lek located near the center of the prairie (about 0.75 mile from the various parking lots). No permanent blinds are present, but temporary blinds are permitted. Massasauga rattlesnakes occur here. Driving on the prairie is not allowed.

Prairie Knoll WMA. Area 120 acres. Includes a small reservoir and a mixture of woodlands, tree plantings, grasslands, and croplands.

Mayberry WMA. Area 195 acres. Consists of grasslands, trees, and a small reservoir. Located 5 miles north of Burchard and 0.5 mile east of Highway 4.

Table Rock WMA. Located just east of Table Rock, on the north side of Highway 4. Includes mainly wooded bottomland along the Nemaha River plus grasslands and croplands.

Richardson County

Richardson County is a Missouri Valley county with over 1,600 acres of surface water, nearly 16,000 acres of woodlands, and 110,000 acres of grasslands. There are tourist accommodations at Falls City.

FEDERAL AREAS

None

STATE AREAS

Indian Cave State Park. Area 3,052 acres. Like Rulo Bluffs (see below), this area has a diverse woodland flora of southeastern affinities and supports such attractive breeding bird species as summer tanagers, Acadian flycatchers, and chuck-will's-widows. The park is 78 percent mature forest and the rest grasslands or developed areas, with both modern and primitive camping facilities and about 36 miles of hiking trails, including a 14-mile "Hardwood Trail" that is moderately difficult owing to the steep loess hillsides. Gray squirrels and southern flying squirrels are common, but the latter are rarely seen. Black rat snakes and timber rattlers are also present.

Verdon Lake SRA. Area 30 acres. Includes a small reservoir.

Kinters Ford WMA. Area 200 acres. Includes riverbottom woodlands, grasslands, and croplands.

Four Mile Creek WMA. Area 160 acres. Mostly upland habitats along a creek bottom.

Iron Horse Trail. Area 210 acres. This hiking trail follows an old railroad right-of-way. See also Gage and Pawnee Counties.

Margrave WMA. Consists of 106 acres along the Nemaha River, including woodlands, croplands, grasses, and marshy areas. Located 3 miles south and 7 miles east of Falls City.

OTHER AREAS

Rulo Bluffs Preserve. Area 424 acres. This highly wooded area, owned by the Nature Conservancy, is located nearly on the Kansas border and has the most southern and eastern hardwood floral affinities of any Nebraska forest, including at least five species of oaks plus many distinctive forest wildflowers. There is no bird checklist yet available, but there should be some southern warbler species in addition to chuck-will's-widow. Trails are few, unmarked, and sometimes steep; it is easy to get lost. Permission to visit must be obtained from the Nature Conservancy Field Office in Omaha.

Index to Selected *Nebraska State Museum Notes*

Most of these titles are available at low cost from the University of Nebraska State Museum in Lincoln.

Mammals

Bats 1983(71):4 pp.
Bison 1969(39):4 pp.
Black-tailed Prairie Dog 1966(31):4 pp.
Fossil Camels 1959(8):4 pp.
Fossil Elephants 1974(52):6 pp.; 1990(77):4 pp.
Fossil Horses 1962(19):4 pp.
Fossil Mammalian Predators 1982(64):4 pp.; 1988(72):4 pp.
Fossil Mammals from Agate Springs 1966(30):6 pp.
Fossil Mammals from Ashfall Park 1992(81):4 pp.
Fossil Mammals 1960(12); 1961(15):4 pp.
Fossil Rhino 1960(11):4 pp.
Mammalian Teeth 1994(89):4 pp. and poster
Nebraska Mammals 1968(35):8 pp.

Birds

Bird Migration 1968(36):4 pp.
Cranes of Nebraska 1996(93):4 pp.
Hummingbirds 1999(103):4 pp. and poster

Nuthatches 1991(79):4 pp. and poster
Trumpeter Swan 1967(32):2 pp.
Waterfowl of Nebraska 1974(50):4 pp.

Amphibians and Reptiles

Amphibians of Nebraska 1999(106):4 pp. and poster
Fossil Plesiosaurs 1965(27):4 pp.; 1965(43):2 pp.
Fossil Tortoises 1977(59):4 pp.
Hognose Snake 1977(57):4 pp.
Snakes of Nebraska 1998(100):4 pp. and poster
Turtles of Nebraska 1996(96):4 pp. and poster

Insects

Butterflies of Nebraska 1993(85):4 pp. and poster
Carrion Beetles 1980(67):4 pp.
Dung Beetles 1980(66):4 pp.
Insect Music 1995(19):4 pp.
Monarch Butterfly 1964(25):2 pp.
Scarab Beetles 1965(42):4 pp.
Sphinx Moths 1991(80):4 pp.
Tiger Beetles 1997(97):4 pp. and poster

Plants

Fossil Flowers 1994(88):6 pp.; 1999(102):4 pp.
Poisonous Plants 1965(41):4 pp.

Index to *Nebraskaland* Notes on Nebraska's Flora and Fauna

(References include year of publication, issue number, and—from 1970 onward—page numbers.)

Mammals

Badger 1953(1); 1966(3); 1978(3):50–51.

Bats 1964(11).

Beaver 1950(4); 1966(11).

Bighorn Sheep 1982(4):50–51

Bison 1965(3); 1976(7):50–51.

Black-footed Ferret 1969(8); 1978(7):50–51; 1996(1) (vanishing species profile).

Black-tailed Jackrabbit 1976(4):50–51.

Black-tailed Prairie Dog 1954(3); 1969(11).

Bobcat 1954(3); 1967(5).

Bushy-tailed Woodrat 1968(2).

Coyote 1956(4).

Crayfish 1967(7)

Eastern Chipmunk 1963(12).

Eastern Cottontail 1951(1); 1967(12); 1980(5):50–51.

Elk 1976(6):50–51.

Fox Squirrel 1956(3); 1971(10):50–51; 1982(7):50–51.

Franklin's Ground Squirrel 1959(4).

Gray Fox 1968(10).

Gray Squirrel 1965(10); 1974(7):42–43.

Least Weasel 1960(8).

Long-tailed Weasel 1958(9); 1975(6):50–51.

Lynx 1964(12).

Mink 1955(1); 1969(1).

Mule Deer 1950(1); 1966(9); 1976(3):50–51.

Muskrat 1952(1); 1967(10).

Northern Grasshopper Mouse 1969(6).

Opossum 1954(1); 1975(5):50–51.

Ord's Kangaroo Rat 1962(5);1987(12):64–65.

Pocket Gopher 1961(11).

Pocket Mouse 1964(2).

Porcupine 1959(1); 1983(11):49–51

Prairie Vole 1970(2):8–9.

Pronghorn 1959(12); 1971(8):48–49.

Raccoon 1955(4); 1974(4):42–43.

Red Fox 1958(2); 1977(1):50–51.

River Otter 1992(11) (vanishing species profile).

Short-tailed Shrew 1960(6).

Southern Flying Squirrel 1967(5); 1994(11) (vanishing species profile).

Spotted Ground Squirrel 1969(9).

Spotted Skunk 1960(4).

Striped Skunk 1952(2); 1970(7):12–13.

Swift (Kit) Fox 1965(12); 1975(12):50–51.

Thirteen-lined Ground Squirrel 1955(3).

Western Harvest Mouse 1965(11).

White-tailed Deer 1958(11); 1974(11):50–51.

White-tailed Jack Rabbit 1958(3).

Woodchuck 1963(1).

Birds

American Avocet 1966(5).

American Bittern 1963(8).

American Coot 1961(10).

American Crow 1962(7).

American Kestrel 1984(9):50–51; 1969(2).

American Robin 1964(6).

American White Pelican 1991(4):50–51.

American Wigeon 1979(1):50–51.

American Woodcock 1985(9):50–51.

Bald Eagle 1960(1); 1993(11) (vanishing species profile).

Barred Owl 1983(12):50–51.

Black-billed Magpie 1961(1).

Blue-winged Teal 1958(8);1974(12):50–51.

Bobwhite Quail 1951(4); 1970(8):12–13; 1979(2):50–51.

Bufflehead 1981(2):50–51.

Burrowing Owl 1969(7).

Canada Goose 1953(4); 1981(4):50–51; 1972(5):52–53.

Canvasback 1954(4); 1976(12):50–51.

Common Goldeneye 1988(8):50–51.

Common (American) Merganser 1960(10); 1979(4) :50–51.

Common Nighthawk 1978(5):50–51.

Common (Wilson's) Snipe 1960(3); 1974(8):42–43.

Cooper's Hawk 1973(3):42–43.

Dark-eyed Junco 1981(12):50–51.

Double-crested Cormorant 1959(6).

Dowitchers 1968(8).

Eastern Screech-owl 1961(2).

Evening Grosbeak 1982(12):50–51.

Franklin's Gull 1963(4).

Gadwall 1962(12); 1974(6):42–43.

Golden Eagle 1964(1).

Gray Partridge 1983(9):65–67.

Great Blue Heron 1982(11):49–51.

Greater Prairie-chicken 1957(4); 1969(10); 1979(12):50–51.

Greater White-fronted Goose 1960(9); 1977(8):50–51.

Great Horned Owl 1957(1).

Green-winged Teal 1965(2); 1976(12):50–51.

Hooded Merganser 1964(10); 1985(3):50–51.

House (English) Sparrow 1966(1).

Interior Least Tern 1997(5) (vanishing species profile).

Killdeer 1961(5).

Lesser Scaup 1958(12); 1986(12):50–51.

Loggerhead Shrike 1962(2).

Long-billed Curlew 1963(3); 1983(6):49–51.

Mallard 1952(4); 1968(12).

Mountain Plover 1994(6) (vanishing species profile).

Mourning Dove 1951(3); 1987(8):66–67.

Northern Cardinal 1965(1).

Northern Harrier (Marsh Hawk) 1959(8).

Northern Pintail 1956(1); 1987(3):50–51.

Northern Shoveler 1961(9); 1982(15):50–51.

Orioles 1986(4):50–51.

Osprey 1962(3).

Pied-billed Grebe 1962(11).

Piping Plover 1995(10) (vanishing species profile).

Red-tailed Hawk 1957(3); 1981(3):50–51.

Redhead 1960(11); 1974(15):42–43.

Ring-necked Duck 1977(2):50–51.

Ring-necked Pheasant 1950(3); 1966(10); 1975(11):50–51.

Ruby-throated Hummingbird 1966(7).

Ruddy Duck 1963(10); 1981(7):50–51.

Sage Grouse 1963(6).

Sandhill Crane 1959(3).

Sharp-tailed Grouse 1958(5); 1974(9):42–43; 1981(11):50–51.

Snow Goose 1958(10); 1963(11); 1975(2):50–51.

Snowy Owl 1980(2):50–51.

Sora Rail 1965(9).

Swainson's Hawk 1968(4).

Trumpeter Swan 1979(11):50–51.

Turkey Vulture 1964(9);1975(8):50–51.

Upland Sandpiper 1981(5):50–51.

Western Bluebird 1967(6).

Western Meadowlark 1960(5);1972(3):50–51.

Whooping Crane 1965(4); 1994(89) (vanishing species profile).

Wild Turkey 1960(12); 1975(1):50–51; 1988(4):50–51.

Wilson's Phalarope 1967(4).

Wood Duck 1972(9):46–47.

Amphibians and Reptiles

Blue Racer 1967(3).

Blue-tailed Skink 1962(10).

Bullfrog 1958(6); 1975(7):50–51.

Bullsnake 1964(7).

Central Plains Spadefoot Toad 1968(1).

Common Water Snake 1970(11):52–53.

Copperhead 1984(4):50–51 :

Eastern Short-horned Lizard 1971(2):52–53.

Garter Snake 1963(7).

Leopard Frog 1966(2).

Ornate Box Turtle 1969(4); 1977(3):50–51.

Prairie Rattlesnake 1961(3):1969(5).

Six-lined Racer 1977(7):42–43.

Snapping Turtle 1964(8).

Spiny Soft-shelled Turtle 1968(3).

Tiger Salamander 1972(6):50–51.

Timber Rattlesnake 1970(1):46–47.

Western Coachwhip 1968(5).

Western Fox Snake 1966(8).

Western Painted Turtle 1965(7).

Black Crappie 1969(3); 1984(8):50–51.
Blacknose Shiner 1992(2) (vanishing species profile).
Blue Catfish 1961(7).
Bluegill 1961(4); 1979(8):50–51.
Brook Trout 1970(10):46–47; 1973(2):42–43; 1980(6):50–51.
Bullhead 1957(2).
Burbot 1962(4); 1972(1):50–51.
Carp 1961(8).
Chain Pickerel 1965(8).
Channel Catfish 1951(2); 1966(4); 1981(9):50–51.
Crappie 1955(2).
Fathead Minnow 1969(12).
Flathead Catfish 1958(4); 1972(4):48–49.
Fresh-water Drum 1961(12).
Gizzard Shad 1960(2).
Grass Pickerel 1963(5).
Johnny Darter 1970(5):8–9.
Largemouth Bass 1953(2); 1967(6); 1979(6):50–51.
Northern Pike 1959(7); 1977(5):50–51.
Orange-spotted Sunfish 1972(12):50–51.
Paddlefish 1959(2); 1982(2):50–51.
Pearl, Northern Redbelly and Finescale Dace 1994(2) (vanishing species profile).
Plains Killifish 1968(7).
Rainbow Trout 1950(2); 1966(6); 1986(7):50–51.
Redear Sunfish 1974(5):42–43.
Redhorse Sucker 1960(7).
Sacramento Perch 1962(1).
Sand Shiner 1974(1):50–51.
Sauger 1964(3); 1987(5):50–51.
Short-nosed Gar 1962(9).
Shovelnose Sturgeon 1963(2).
Smallmouth Bass 1958(7); 1975(3):50–51.
Stickleback 1962(8).
Striped Bass 1980(10):50–51.
Sturgeons 1989(8):50–51; 1993(9) (vanishing species profile).
Tadpole Madtom 1967(2).
Walleye 1954(2); 1966(12).
White Bass 1952(3); 1968(6).
Yellow Perch 1958(1);1983(7):50–51.

Invertebrates

American Burying Beetle 1995(3) (vanishing species profile); 1995(5): 50–51.
Bumblebees 1993(5):58–59.
Dragonflies 1991(7):50–51.
Earthworm 1961(6).
Fragile Heel-splitter (mussel) 1990(8):50–51.
Garden Spider 1996(6):50–51.
Monarch Butterfly 1970(4):8–9.
Moths 1971(11):50–51.
Snails 1971(5):54–55.
Tiger Beetles 1990(7):50–51.

Trees

Black Locust 1990(6):50–51.
Black Walnut 1971(4):52–53.
Bur Oak 1972(8):50–51.
Cottonwood 1986(8):50–51.
Eastern Redbud 1985(7):50–51.
Oaks 1985(12):50–51.
Pawpaw 1982(9):49–51.
Shagbark Hickory 1988(11):50–51.

Shrubs, Forbs, and Grasses

American Bittersweet 1984(12):50–51.
Arrowhead 1973(8):40–41.
Blackberry 1988(3):50–51.
Blowout (Hayden's) Penstemon 1982(6):49–51; 1997(4) (vanishing species profile).
Blowout Plants 1986(3):49–51.
Blue-eyed Grass 1985(5):49–51.
Bush Morning Glory 1988(7):50–51.
Butterfly Weed 1971(6):52–53.
Columbine 1988(5):50–51.
Compass Plant 1983(8):50–51.
Cup-plant and Rosinweed 1987(7):50–51.
Dotted Gayfeather 1972(11):48–49.
Late Goldenrod 1971(7):52–53.
Lead-plant 1987(4):50–51.
Milkweeds 1984(3):49–50.
New England Aster 1972(2):50–51.
Pasque Flower 1983(3):49–51.

Prickly Poppy 1990(3):50–51.
Purple Coneflower 1985(6):50–51.
Sand Cherry 1986(5):50–51.
Smooth Sumac 1972(10):50–51.
Tallgrass Prairie Grasses 1984(7):50–51.
Thick-spike Gayfeather 1989(6) :50–51.
Tufted Evening Primrose 1986(6):50–51.
Western Prairie Fringed Orchid 1993(6) (vanishing species profile).
Yucca 1984(6):49–51.

Glossary

Adaptation. An evolved structural, behavioral, or physiological trait that increases an organism's individual fitness (its individual ability to survive and reproduce).

Advertising behavior. The social behaviors (signals or "displays") of an animal of either sex that may serve to identify and announce its species, sex, relative reproductive capacity, individual social status, and overall vigor. *See also* signals.

Allopatric. Populations of one or more species that are wholly geographically isolated from one another, at least during breeding. "Allospecies" are those allopatric populations that appear to have attained species-level distinction *See also* parapatric, sympatric.

Alluvium (adj., alluvial). Materials transported and deposited by moving water, such as alluvial-based soils. *See also* colluvium, loess, and palustrine.

Annual. A plant that completes its life cycle, from seedling to seed-bearing plant, in a single year and then dies. *See also* biennial, perennial.

Aquifer. A subterranean "reservoir" of saturated sands, gravels, and so forth.

Arboreal. Tree dwelling.

Arthropod. A member of the invertebrate phylum Arthropoda, which includes insects, crustaceans, and other "jointed-legged" animals with chitinous or calcareous exoskeletons. *See also* invertebrate.

Association. A specific type of biotic community having a high degree of floristic uniformity and occurring in similar environments. Associations are usually named for one or more plant species or genera that consistently occur as climax dominants within that community. *See also* climax community, community, and dominant.

Avifauna. The collective bird life of a particular locality or region.

Awn. A long bristle, associated with the seed coverings or leaf tips of some grasses.

Biennial. A plant that completes its life cycle, from seedling to mature plant, in two years and then flowers, sets seeds, and dies. In the first year the plant often assumes a rosette life form of leaves close to the ground. *See also* annual, perennial.

Binomial. Comprised of two terms, such as using the combination of a generic (genus) and specific (species) name to identify an organism, which is the basis for binomial biological nomenclature. *See also* nomenclature.

Biome. A major regional ecosystem, including both the plants and animals. Comparable to an "ecoregion" if defined mainly by landscape geography rather than floristically. *See also* ecoregion, formation.

Boreal forests. Northern forests, especially the coniferous forests of Canada.

Braided stream. A meandering stream having several shallow and interconnected channels.

Brood. As a verb, to cover and apply heat to hatched young (incubation is the same behavior, when applied to eggs); As a noun, a group of young typically tended by a single female or pair.

Browse. To forage on the twigs and leaves of woody plants. *See also* graze.

Bunchgrasses. Grasses that grow in clumps, with intervening areas of bare ground between adjacent clumps; an adaptation to arid conditions. *See also* sod-forming grasses.

Cambrian period. The first major interval of the Paleozoic era, lasting from about 570–590 to 505 million years ago. *See also* Paleozoic era, Precambrian.

Cavernicolous. Living in caves.

Cenozoic era. The geologic era encompassing the past 65 million years, the so-called Age of Mammals. It has traditionally been divided into the (older) Tertiary and (newer) Quaternary periods. Recently the more equally chronologically subdivided Paleogene and Neogene periods have been proposed instead. *See also* Neogene period, Paleogene period, Quaternary period, and Tertiary period.

Chrysalis. The case or cocoon in which a moth or (especially) a butterfly undergoes metamorphosis before emerging as a winged adult. Cocoons that consist entirely of silk are produced by silkworm moths; butterflies typically produce chrysalides that are irregular in shape and often closely resemble their environment.

CITES. Convention on International Trade in Endangered Species, which identifies those species that are in varying degrees of threat of extinction and prevents international trade in them or their body parts.

Climate. The long-term weather conditions of an area, including variations in temperature, precipitation, humidity, wind, and hours of daylight.

Climax community. A stable aggregation of locally interacting species that is self-perpetuating and (theoretically) is no longer undergoing ecological succession. Thus it is unlikely to alter significantly until the climate changes or outside forces are brought to bear. Typically it is characterized or identified by its dominant species. *See also* community, dominant, ecoregion, formation, and succession.

Colluvium (adj., colluvial). Materials accumulated at the base of a slope or cliff; colluvial soils are an example. *See also* alluvium, loess, and palustrine.

Column. The part of an orchid flower that supports the variably fused male and female reproductive structures (stamens, style, and stigma). Also, the lower part of the awn on some grasses.

Community. In ecology, an interacting group of plants, animals, and microorganisms situated in a specific location, tending to recur in different areas but similar habitats, and usually responding similarly to their biotic and physical environments. They may be named for their most characteristic biotic features (such as their dominant plants) or for specific abiotic features (such as rock outcrops or saline wetlands). *See also* ecosystem, habitat.

Congeneric. Belonging to the same genus.

Coniferous. Cone bearing.

Conspecific. Belonging to the same species.

Crepuscular. Active at dawn and dusk. *See also* diurnal, nocturnal.

Cretaceous period. The geologic interval that extended from about 146 million years ago to 65 million years ago, ending the Mesozoic era. *See also* Mesozoic era.

Cursorial. Running.

Deciduous. Descriptive of shedding, especially of trees that drop their leaves after each growing season.

Dicot. A member of a group of plants (dicotyledons) that are characterized by having broad, web-veined leaves; two embryonic leaves within the seed; and flower parts usually grouped in fours or fives and their multiples. Common examples include legumes. *See also* monocot.

Display. An evolved and variably ritualized behavior that communicates information within or between species. *See also* advertising behavior, ritualization, signals, and territoriality.

Diurnal. Active during the daytime. *See also* crepuscular, nocturnal.

Dominant. In ecology, descriptive of plant taxa that exert the strongest ecological effects (control of energy flow) within a community. They may maintain this dominance through sheer biomass, by controlling nutrient or water availability, by maximizing interception of available sunlight, by having antagonistic (or allelopathic) effects on other species (or other individuals of its own species) in the immediate vicinity, or possibly through other controlling influences. *See also* community, keystone species.

Drift. Glacially transported and deposited materials, either unsorted by size (till) or sorted by size (stratified drift). Not to be confused with genetic drift. *See also* glacial erratic, moraine, and till.

Dystrophic. Refers to waters that are brownish because of suspended organic colloids and are poorly provided with nutrients. *See also* eutrophic, oligotrophic.

Ecolocation. The use of high-frequency sounds by most bats, some insectivores, and a few other animals to detect prey or to navigate in the dark.

Ecoregions. Regions having similar climates, geomorphology, and potential natural vegetation groups and composed of clusters of interacting landscapes. *See also* association, biome, formation, and landscape.

Ecosystem. An interacting group of plants, animals, and microorganisms and their physical environment.

Ecotone. An ecologic transition zone that physically connects two quite different biotic communities. Ecotones may be broad or narrow and relatively stable or dynamically changing through time. *See also* seral.

Edge species. A species that is more common in or most characteristic of ecotone communities, such as forest-edge species. "Edge-effect" refers to the fact that edge habitats often exhibit high species diversity, supporting species typical of both adjoining habitat types and sometimes supporting unique edge-adapted species as well.

Endangered. Descriptive of taxa existing in such small numbers as to be in direct danger of extinction without human intervention. Federally endangered species are automatically classified by every state as endangered, but states may independently classify a species as endangered or threatened regardless of its national status. *See also* CITES, threatened.

Endemic. Descriptive of taxa that are both native to and limited to a specific area, habitat, or region. *See also* indigenous.

Environment. The natural surroundings of an organism or community of organisms. *See also* habitat.

Eocene epoch. A major early subdivision of the Cenozoic era, extending from about 58–60 to 34–37 million years ago. *See also* Cenozoic era.

Eolian. Wind-transported materials, such as silt. *See also* loess.

Escarpment (scarp). A steep slope.

Estivation. A period of dormancy during summer. *See also* hibernation.

Eutrophic. Refers to waters that are well provided with plant nutrients. *See also* oligotrophic.

Evolution. Any gradual change. Biological or organic evolution results from changing gene frequencies in successive generations associated with biological adaptations, typically through natural selection. *See also* natural selection.

Extinct. A taxon that no longer exists anywhere. *See also* extirpated.

Extirpated. A taxon that has been eliminated from some part of its range but still exists elsewhere. *See also* extinct.

Fen. A wetland characterized by having a boglike substrate of organic matter (peat or marl) but, unlike typical bogs, having favorable plant nutrition levels (especially calcium availability) and much greater organic productivity. Prairie fens are largely confined to glaciated areas and are rare in Nebraska. *See also* marsh.

Fledging. The initial acquisition of flight by a bird. The period from hatching to fledging is called the "fledging period," which often corresponds to the nestling period or is slightly longer. *See also* nestling.

Fledgling. A newly fledged juvenile bird. *See also* juvenile.

Floret. A small flower, especially of grasses.

Food chain. The sequence of energy transformations in nature, from primary "producers" (green plants), through a series of plant-eaters (herbivores) and meat-eaters (carnivores), collectively called "consumers." Scavengers and parasites are other specialized types of consumers. "Decomposers" such as bacteria recycle organic matter back to its inorganic components.

Forb. An herbaceous plant that is not a grass, rush, or sedge, thus broad-leaved dicots or monocots. *See also* grass, sedge.

Forest. A general term for a community dominated by rather tall trees in which the height of the trees (typically more than 16 feet) is much greater than the average distance between them and the overhead canopy is more or less continuous. *See also* savanna, woodland.

Formation. In ecology, a major type of plant community (or biome, if the animals are included) that extends over broad regions that collectively share similar climates, soils, and biological succession patterns and have similar life forms of dominant plants at their eventually stable or climax vegetational stage. Formations usually include several subcategories of plant "associations," based on their specific climax dominants. Examples of climax plant formations in Nebraska include temperate deciduous forest, coniferous forest, shrubsteppe, and perennial grasslands of varying stature types (e.g., shortgrass, midgrass, and tallgrass). *See also* climax community, community.

Fossorial. Digging.

Gallery forest. Narrow riverine forests that follow waterways out into otherwise nonforested habitats. *See also* riparian, savanna, and woodland.

Genus (adj., generic). A (literally) "general" Latin or Latinized name that is applied to one or more closely related species of plants or animals. If a genus has only a single included species, it is called "monotypic," otherwise it is called "polytypic." The genus is the first (and always

capitalized and italicized) component of a species' two-parted binomial, or scientific, name. *See also* species.

Glacial erratics. Large rocks and boulders that have been glacially transported and randomly deposited over the landscape, having a composition different from that of the bedrock below. *See also* drift, moraine.

Granivorous. Grain eating.

Grass. Herbaceous plants distinguished by having a (usually) hollow stem;
narrow leaves that are parallel-veined and that consist of a stem-
enclosing sheath and a flatter blade; and tiny flowers and seeds borne
on small spikes. *See also* sedge.

Graze. To feed on the leaves of herbs, especially grasses. *See also* browse.

Great Plains. The nonmountainous region of interior North America lying
east of the Rocky Mountains and west of the Central Lowlands of the
Mississippi and lower Missouri drainages.

Guild. A group of species that exploits the same class of environmental
characteristics (such as food types) in a similar way, whether or not
the species are closely related. *See also* niche.

Habitat. The general ecological (biotic and physical) environment in which
a species survives. Thus its natural "address," as opposed to its ecological profession (niche) or its presence in any specific location (biotic community). *See also* community, ecosystem, environment, and
niche.

Halophytic plants. Salt-tolerant plants; often also called "xerophytic"
(drought-tolerant) plants.

Herb. Any plant with no permanent aboveground parts, including grasses,
sedges, rushes, and forbs.

Herpetile (or herp). A member of the amphibian and reptile assemblage.
Herpetology is the study of such animals.

Hibernation. A period of winter dormancy. *See also* estivation.

Holocene (Recent) epoch. The roughly 11,000-year interval extending
from the end of the last (Wisconsinian) glaciation to the present time.
See also Pleistocene epoch.

Home range. The entire area used by an individual, pair, or family over a
specified period. *See also* territory.

Ichthyology. The scientific study of fish.

Imago. Among insects, refers to the adult state.

Incubation. The parental application of body heat to eggs. *See also* brood.

Indigenous. Descriptive of taxa that are native to, but not necessarily
limited to, a particular area or region. *See also* endemic.

Innate. Pertaining to genetically transmitted traits, especially behavioral
ones.

Insectivorous. Having a diet composed mostly of insects and other arthropods.

Instinct. An innate behavioral trait that appears to be more complex than various simpler innate responses such as reflexes and whose performance seemingly depends on specific and transitory internal states as well as rather specific external stimuli. *See also* innate, isolating mechanisms, and signals.

Intergeneric. Pertaining to interactions between genera.

Interspecific. Pertaining to interactions between species.

Intraspecific. Pertaining to interactions within species.

Invertebrates. Animals lacking backbones but often with exoskeletons, such as arthropods. *See also* arthropod, vertebrates.

Isohyet. A line on a map connecting points of equal precipitation.

Isolating mechanisms. Genetically carried (intrinsic) and evolved traits that serve to prevent the exchange of genes between individuals of different species and the production of viable hybrids. *See also* reproductive isolation.

Jurassic period. The second major period of the Mesozoic era, following the Triassic period. It lasted from about 210 to 145 million years ago, or the beginning of the Cretaceous period, and marked the peak of dinosaur abundance and diversity. *See also* Cretaceous period, Triassic period.

Juvenal. The feathers acquired during the nestling period that are carried for a variable time after fledging (initial flight).

Juvenile. The stage in a bird's life during which it carries feathers predominantly of the juvenal plumage and during which it fledges. *See also* juvenal.

Keystone species. A species whose presence in a community has a significant effect on its structure. *See also* dominant.

Lacustrian. Refers to lakes and their biota or substrates, such as lacustrian vegetation or soils. *See also* palustrian.

Landscape. A relatively large area within an ecoregion in which local vegetative communities or subpopulations (fragments or patches) occur repetitively. *See also* ecoregions, patch.

Larva. Immature growth stages, especially of invertebrates having distinctive early body forms (such as grubs and caterpillars) but also used for very early growth stages of fish and of amphibians that have not yet undergone metamorphosis into their adult form.

Latin name. The (usually) two-parted (generic plus specific) name given a species when it is first officially described and by which it thereafter is technically known and properly identified; thus the species' scientific name. *See also* genus, species, and vernacular name.

Legumes. A family of plants (the bean family Fabiaceae or Leguminaceae), many of which harbor nitrogen-converting ("fixing") bacteria in their roots that convert gaseous nitrogen into a molecular form usable by plants.

Life form. A term broadly descriptive of plant categories, such as coniferous or deciduous trees (or forests), broad-leaved shrubs, perennial grasses, and annual forbs.

Litter. Dead, nonwoody plant matter accumulated at the ground surface, intermediate between still-standing herbaceous materials (thatch) and already decomposing matter (duff).

Loam. Soils containing a mixture of particle sizes, including sand, silt, and clay.

Loess (adj., loessal). Silt-sized materials (larger than clays but smaller than sand) that have been transported and deposited by wind. Loess soils show little or no vertical stratification into horizons and are easily eroded. The word is of German origin, meaning "loose," and is pronounced as "luss." *See also* alluvium, till.

Marsh. A wetland type in which the soil is saturated for long periods of time, if not permanently, but in which peat does not accumulate. *See also* fen.

Meristem. An area of plant cell division and growth found at the bases and internodes of grass stems and leaves or in the terminal buds of forbs, shrubs, and trees.

Mesic. A habitat with a moderate level of soil moisture (or other general environmental conditions, such as temperature extremes) between extremes of xeric (dry) and wet (hydric). *See also* xeric.

Mesozoic era. The so-called Age of Dinosaurs, extending from about 225 to 65 million years ago, or to the start of the Cenozoic era. *See* Cenozoic era, Paleozoic era.

Metamorphosis. Bodily transformations during an animal's lifetime. Insects with complete metamorphosis are transformed from wingless larvae (such as caterpillars or grubs) into winged or wingless adults (imagoes) during an intervening immobile, or "rest," period, sometimes occurring within a structure called a cocoon or chrysalis. *See also* imago, larva.

Metapopulations. Spatially distributed populations composed of unstable local subpopulations in discrete habitat patches.

Microtine. A group (subfamily) of rodents that includes voles, lemmings, and muskrats; the word is derived from *Microtus*. Arvicoline is a synonym.

Miocene epoch. The interval within the Tertiary period (and the associated geologic strata deposited during that interval), from the end of the preceding Oligocene epoch (23.3 million years ago) to the start of the Pliocene epoch (5.2 million years ago). *See also* Oligocene epoch, Pliocene epoch.

Mire. A wetland, such as a fen or bog, in which peat accumulates. *See also* fen.

Mixed-grass (or midgrass) prairie. Perennial grasslands that are dominated by grasses of intermediate heights (often 1.5–3 feet tall at maturity) between those typically found in tallgrass and shortgrass prairies, having some species from these two respectively more mesic and xeric types present to varying degrees. Mixed-grass prairies usually occur in areas of limited precipitation, situated geographically between tallgrass and shortgrass prairies. *See also* shortgrass prairie, tallgrass prairie.

Molt. The periodic sequential loss and replacement of bodily parts during a lifetime or a season. Molt in birds involves feather replacement, thus producing a new plumage; in mammals, hair replacement produces a new pelage. Among arthropods, molting refers to the periodic shedding of the exoskeleton during growth and maturation.

Monocot. A member of a group of plants (monocotyledons) characterized by having narrow, parallel-veined leaves, a single embryonic leaf within the seed, and flower parts usually consisting of threes or their multiples. Common examples include lilies, grasses, and sedges. *See also* dicot.

Monogamous. A pair-bonding system characterized by a single male and female remaining together for part or all of a breeding season or sometimes indefinitely.

Monophyletic. Having a single ancestral origin. *See also* polyphyletic.

Moraine. Gently rolling landscapes of glacial drift deposits laid down at a glacier's lateral or terminal margins. *See also* drift, till.

Morph. Any of the phenotypic variants that occur in cases of sexual or nonsexual morphism (e.g., dimorphism, dichromatism). *See also* dimorphism.

Naiad. The aquatic larval stages of dragonflies, mayflies, and some other metamorphosing insects; also the freshwater mussels of the family Unionidae.

Natal plumage. The initial feather covering of a newly hatched bird, often downy in most birds.

Natural selection. In a broad sense, the long-term changes in gene frequencies and associated traits in populations, resulting from differential survival and reproduction of the "fittest" individuals within such interbreeding populations. *See also* sexual selection.

Nearctic Region. The zoogeographic region that includes Greenland and continental North America south to the highlands of Mexico. The lands to the south of this point are called the Neotropical Region.

Neotropical migrants. Migratory birds that winter in the Neotropic Region, namely from Mexico and Central America southward. Some of these are additionally transequatorial migrants, wintering south of the equator.

Nesting success. The percentage of initiated nests that succeed in fledging one or more young. *See also* nestling success.

Nestling. A recently hatched bird still confined to the nest. *See also* fledgling.

Nestling success. The percentage of hatched chicks that produce fledged young; sometimes also called fledging success.

Niche. The behavioral, morphological, and physiological adaptations of a species to its habitat; also sometimes defined from an environmental standpoint, such as the range of ecological conditions under which a species potentially exists (fundamental niche), best survives (preferred niche), or actually survives (realized niche). *See also* habitat.

Nocturnal. Active at night. *See also* crepuscular, diurnal.

Nomenclature. The process of naming objects; binomial biological nomenclature provides a standardized two-part Latin or Latinized name, the generic (general) name or genus, followed by a specific epithet. *See also* binomial, taxonomy.

Nuptial plumage. The definitive breeding plumage of adult birds, typically acquired by a prenuptial molt of variable extent. Now generally termed "alternative plumage" in North America.

Nymph. The aquatic larval stage of some insects, such as dragonflies. *See also* naiad.

Oligocene epoch. The interval within the Tertiary period (and the associated geologic strata deposited during that interval), extending from the end of the Eocene epoch (34–37 million years ago) to the start of the Miocene epoch (23.3 million years ago). *See also* Cenozoic era.

Oligotrophic. Waters poorly supplied with plant nutrients. *See also* dystrophic, eutrophic.

Ornithology. The scientific study of birds.

Pair bond. The establishment of a (usually) monogamous and variably prolonged social bond between two individuals, typically for facilitating reproduction.

Pairing behavior. The epigamic behaviors related to mate choice and associated pair formation (pair-forming signals) and pair-bond maintenance (pair-bonding signals). Sometimes called "courtship," but this term has undesirable human connotations.

Paleozoic era. The geologic era of "ancient life," beginning about 570 million years ago at the start of the Cambrian period and lasting until the beginning of the Mesozoic era, about 225 million years ago. *See also* Mesozoic era, Precambrian period.

Palustrine. Refers to marshes and their edges, such as palustrine vegetation. *See also* lacustrian, riparian.

Parapatric. Populations that come into limited or sometimes extensive contact with one another along some portion of their common borders but do not actually overlap during the breeding season. Parapatric distributions are suggestive of competing, noninterbreeding populations, thus of two separate species. *See also* allopatric, sympatric.

Passerine. Descriptive of members of the avian order Passeriformes; popularly also called "perching birds" (because of their long hind toes) or "songbirds" (because of their complex vocal structures). However, some songbirds (such as crows and ravens) do not sing in the usually understood sense of this word.

Patch. In landscape ecology, a self-contained and rather uniform example of a habitat type that is variable in size but is geographically circumscribed as a single unit and that comprises part of a more diverse landscape. A specific patch may constitute a local site. *See also* landscape.

Pelage. The collective hair and fur coat of a mammal. *See also* plumage.

Perennial. A plant that continues its life cycle from year to year indefinitely. In perennial herbs and grasses, the aboveground parts die back each year at the end of the growing season but are replaced with new growth the following growing season. In shrubs and trees the leaves may fall off, but the aboveground buds persist. *See also* annual.

Pheromones. Odorifous materials produced and released into the air or water by one individual and perceived by another of the same species, producing a behavioral or physiological effect. Allomones are similar materials but have interspecific effects.

Photosynthesis. The biochemical (metabolic) process by which gaseous carbon dioxide and plant-supplied water are converted into simple sugars, using light energy in the presence of chlorophyll. Cellular respiration reverses the process, converting sugars and related molecules back to metabolic water and carbon dioxide, releasing energy that the organism can use.

Phreatophyte. Refers to plants having a high water requirement, often growing only near a steady source of water.

Phylogeny. The evolutionary history or pathway of descent of an organism or group of organisms. Morphological phylogenetics are based on structural similarities, molecular phylogenetics on genetic (often nucleic acid or protein-based) traits. *See also* monophyletic, polyphyletic.

Pistil. The ovary, style, and stigma of a flower; the female component.

Plains. A nontechnical term for a flatland, especially a nonforested flatland. The High Plains of western North America are the arid Great Plains uplands east of the Rocky Mountains that support native shortgrass and shrubsteppe vegetation.

Playa. A shallow, usually temporary, wetland that depends on often unpredictable surface precipitation for recharge.

Pleistocene epoch. The interval extending from about 1.77 million years ago (the end of the Pliocene epoch) to 11,000 years ago (the start of the current Holocene, or Recent, epoch). Popularly referred to as the "Ice Age," this epoch includes part or all of four major glacial periods (and their associated interglacial intervals), of which the Nebraskan glaciation (ca. 2–1.75 million years ago) was the earliest and the Wisconsinian glaciation (ca. 150,000–15,000 years ago) the most recent. The Kansan (ca. 1.4–0.9 million years ago) and Illinoian (ca. 0.55–0.35 million years ago) glaciations were of intermediate age and like the others were named after their southernmost extensions.

Pliocene epoch. The interval (and associated geologic strata deposited during that interval) extending from the end of the preceding Miocene epoch (5.2 million years ago) to the start of the Pleistocene epoch (about 1.8 million years ago).

Plumage. The collective feather coat of a bird. *See also* pelage.

Pollinium. A mass of pollen grains. *See also* anther, viscidium.

Polyphyletic. Descriptive of assemblages of organisms believed to have been derived phyletically from two or more ancestral groups; thus an artificial (but perhaps convenient) evolutionary assemblage. *See also* monophyletic.

Prairie. A native plant community dominated by perennial grasses. Prairies may be broadly classified by the relative stature of their particular dominant grass taxa (tallgrass, midgrass, or shortgrass) or by the characteristic form of these grasses (e.g., bunchgrasses or sodforming grasses). They are described more specifically by their dominant species or genera. Prairie soils are usually rich in organic matter, calcium, and other inorganic nutrients. *See also* shortgrass prairie, steppe.

Precambrian. Refers to that time in the Earth's history occurring before the Cambrian period, or at least 570 million years ago. *See also* Cambrian period, Paleozoic era.

Promiscuity. A mating system in which at least one sex (rarely both) has multiple sexual partners, in the absence of actual pair bonding.

Quaternary period. The last 1.6 million years of the Earth's history, including the Pleistocene and Holocene epochs. *See also* Holocene epoch, Pleistocene epoch.

Refugium. An area where some species can survive during unfavorable environmental conditions.

Reproductive isolation. The impossibility of two populations interbreeding, owing either to preventive environmental barriers (extrinsic isolating factors) or genetic restrictions (inherent isolating mechanisms). *See also* isolating mechanisms.

Rhizome. An underground rootlike stem. *See also* stolen, tiller.

Riparian. Associated with shorelines, especially river or stream shorelines. Gallery forests are a common type of riparian community in grassland habitats. Riverine is a synonym. *See also* gallery forest, palustrine.

Saltatory. Jumping.

Sandsage prairie. Prairies developed over sandy substrates, having an abundance of sand sagebrush and various other sand-adapted species.

Saprophyte. A plant that derives its energy from decaying organic matter rather than producing it through photosynthesis; fungi are typical saprophytes. Vascular plants lacking chlorophyll, such as Indian pipe, are perhaps better called mycotrophic, as they employ mycorrhizae (root-inhabiting fungi) as a food source. Such nonphotosynthetic plants as fungi are collectively termed ecological "decomposers."

Savanna. Originally a Spanish term for describing tropical grasslands that often had scattered trees; now commonly used to describe any grassland community within which scattered trees occur. In typical savannas the distance between the trees is much greater than the average width of the individual tree canopies.

Saxicolous. Dwelling among rocks.

Scansorial. Climbing abilities.

Sedge. Herbaceous, grasslike plants having rounded or angular and solid stems; narrow, parallel-veined leaves that are often arranged in three ranks (as seen from above); and small seeds and flowers borne on spikes. *See also* grass.

Seral. Descriptive of chronologically transitional or "succesional" species or stages in community development. A successional sequence is called a sere. *See also* climax community, succession.

Sexual selection. A type of natural selection in which the evolution and maintenance of traits of one sex result from the social interactions that produce differential individual reproductive success. These include both competitive interactions among members of the same sex (intrasexual, or agonistic, selection) and those occurring between the sexes and involving differential rates of sexual attraction (intersexual, or epigamic, selection). *See also* natural selection.

Shortgrass prairie. A grassland (sometimes called "steppe grassland") dominated by short (well under 3 feet tall at maturity) perennial grasses, such as various species of grama grass and buffalo grass. Perhaps more properly called "steppe," which is a shortgrass ecosystem having much bare soil and arid-adapted woody shrubs variably present. When such shrubs are co-dominant with the grasses, "shrub-steppe" is a useful descriptive term. Shortgrass prairies occur in areas too arid to support midgrass prairie but with more available moisture than those climates typically supporting semidesert, or even more xeric desert, scrub vegetation. *See also* mixed-grass prairie, shrub-steppe, steppe, and tallgrass prairie.

Shrub. A woody plant that is typically less than 12 feet tall at maturity and usually have many aboveground stems. Shrublands are communities dominated by shrubs and having varied degrees of canopy cover. *See also* forest, tree, and woodland.

Shrubsteppe. A semiarid plant community that is variably co-dominated by shrubs—typically various species of sagebrush in western North America—and low-stature grasses (often bunchgrasses) and usually with much bare soil present between the plants. Sandsage prairie is the best Nebraska example.

Signals. Behaviors that have become evolutionarily modified (or "ritu-alized") to transmit information among members of a social group or to other organisms. Such signals are often also called "displays," especially in the case of visual signals, but postures and vocalizations also constitute social signals. *See also* advertising behavior.

Silt. Sedimentary materials of a size intermediate between sand and clay.

Site-fidelity (Site-tenacity). The social attachment of an individual to a specific site (nest site, territory, and so forth) in successive seasons or years.

Sod-forming grasses. Perennial grasses that grow and establish root systems tending to bind the soil substrate in a continuous sod rather than growing in a discontinuous, clumplike, manner. *See also* bunchgrasses.

Songs. The vocalizations of birds that tend to be acoustically complex, prolonged, and often sex-specific and well as individually unique. They frequently are uttered only at particular seasons and may have both intra- and intersexual signal value, including territorial advertisement and sexual attraction. *See also* vocalizations.

Spacial scale. Measurements of abundance or species richness based on a series of units of differing sizes or spacial traits, such as patch and landscape characteristics. *See also* ecoregion, landscape, and patch.

Species. A "kind" of organism, or, more technically, a population whose members share the same isolating mechanism and thus are reproductively isolated from all other populations but are potentially capable of breeding freely among themselves. The term also used as the nomenclatural category below that of the genus and above that of the subspecies (abbreviated *sp.*, plural *spp.*). It occurs as the second (specific) component of a binomial Latin or Latinized name (after the generic name) and is never capitalized in that context. *See also* genus, Latin name.

Stamen. The pollen-bearing structure of a flower, typically an anther on a long filament.

Staminode. A modified, nonfunctional stamen. In some orchids it supports the two functional anthers. Staminodes also occur in penstemons as sterile, hairy-tipped structures.

Steppe. A Russian-based term for native shortgrass communities, especially the vast semiarid areas of grass-covered plains in south-central Asia. Comparable to the popular usage of "plains" in North America. *See also* plains, shortgrass prairie, and shrubsteppe.

Stigma. The part of the pistil that receives pollen.

Stolen. An aboveground horizontal stem that can take root and propagate the plant. *See also* rhizome.

Stomata. The air pores in leaves through which gas exchange occurs and water vapor is lost through transpiration.

Stridulation. Scraping sounds, such as the "songs" of crickets and grasshoppers, made by the rubbing of two parts of the exoskeleton together. *See also* vocalizations.

Subspecies. A geographically defined and recognizable (by morphology or sometimes even behavior) subdivision of a species. Subspecies species (abbreviated *ssp.*, plural *sspp.*) designations follow the species' epithet in a scientific name, comprising the last component of a three-parted, or trinomial, name. The race that shares its name with the species is the first-described, or nominate, subspecies.

Succession. The series of gradual plant and animal changes that occur in biotic communities over time, as relatively temporary (seral) taxa are sequentially replaced by others that are able to persist and reproduce for a more prolonged or even indefinite period (so-called climax species). *See also* climax community, seral.

Superspecies. An assemblage or two or more essentially allopatric populations that are very closely related and might potentially represent biological subspecies; however, the test of reproductive isolation is lacking owing to their allopatric distributions. *See also* allopatric.

Sympatric. Populations that overlap at least in part, especially during the breeding season. By definition, sympatrically breeding but noninterbreeding populations are never considered the same species. *See also* allopatric, parapatric, and reproductive isolation.

Tallgrass prairie. Perennial grasslands that are dominated by tall grasses often at least 6 feet high at maturity, such as various species of bluestems. These species do most of their growing in the warmest part of the summer and thus are often called "warm-season" grasses. These grasslands typically occur in areas that are more mesic than those supporting lesser-stature grasses (midgrass prairies) but are still too arid—or too frequently burned—to support forests. *See also* mixed-grass prairie, shortgrass prairie.

Taxon. A group of organisms comprising a particular category of biological classification or taxonomic group, such as the sparrow genus *Spizella*.

Taxonomy. A system of naming and describing groups of organisms (taxa) and organizing them in ways that reflect perceived phyletic relationships. *See also* nomenclature.

Territoriality. The advertisement and agonistic behaviors associated with territorial establishment and defense. *See also* advertising behavior, display, and signals.

Territory. An area having resources that are defended or controlled by an animal against other of its species (intraspecific territories) or less often against those of other species (interspecific territories). *See also* advertising behavior, home range.

Tertiary period. The first of the two major subdivisions of the Cenozoic era, beginning about 65 million years ago (at the end of the Cretaceous period) and lasting until about 1.6 million years ago (the start of the Quaternary period). *See also* Paleocene period, Neocene period.

Thermal. A rising current of air caused by differential heating rates near the Earth's surface.

Threatened. Descriptive of taxa that have declined and exist in small numbers but are not yet so rare as to be classified as endangered. Somewhat similar to "vulnerable" or "declining," terms used by some conservation agencies to refer to taxa in potential need of conservation measures in order to prevent a threatened or endangered status. States may establish their own criteria for such listings; thus a state might list a species as threatened that is not included in that category nationally. *See also* endangered.

Till. Unsorted glacial drift *See also* drift.

Tiller. A shoot formed at the base of a plant that can propagate it. *See also* rhizome, stolen.

Transpiration. The loss of water vapor by evaporation through the surfaces of leaves, especially their stomata. *See also* stomata.

Tree. A woody plant that is usually well above 12 feet tall at maturity and typically has a single main stem. Trees may be deciduous (usually broad-leaved species in Nebraska) or evergreen (all coniferous in Nebraska). *See also* shrub.

Triassic period. The first major geologic period of the Mesozoic era lasting from about 225 to 210 million years ago, or the start of the Jurassic period. *See also* Jurassic period, Mesozoic era.

Trophic levels. The successive production and consumption levels in a food chain, including primary producers, primary consumers, secondary consumers, and so forth. *See also* food chain.

Ungulate. A hoofed mammal, either even-toed (e.g., deer and antelope) or odd-toed (e.g., horses and rhino).

Vernacular name. The "common" of an organism (usually a species) as opposed to its scientific or Latin name.

Vernal pond (or pool). A temporary wetland produced by melting snow or spring rains. *See also* playa.

Vertebrates. Animals with backbones, including mammals, birds, reptiles, amphibians, and fish. *See also* invertebrates.

Viscidium. The sticky mass of pollen typical of some orchids.

Vocalizations. Utterances, including both songs and calls, generated (in birds) by the syrinx or (in mammals) the larynx and sometimes modulated or resonated by other structures. Nonvocal sound sources include hissing, percussion sounds (such as clapping or foot stamping), and scraping (stridulation) noises. *See also* songs.

Warm-season grasses. Grasses adapted to grow during warmer periods having greater water stress; also called C_4 grasses because their photosynthetic pathway involves an intermediate 4-carbon molecule stage.

Water table. The upper surface of an underground reservoir of water, or aquifer.

Woodland. A partly wooded community in which the height of the trees is usually less (but over 12 feet) than the distances between them, so that the overhead canopy is discontinuous, with more open space than in forests. Tree branches tend to occur lower on the stem and be more spreading than in forest trees and typically consist of a single layer of woody plants. *See also* forest, shrubland.

Xeric. Desertlike or drought-adapted; for example, "xerophilic" vegetation of deserts. *See also* mesic.

Xerothermic period. The period of maximum dryness following the last glaciation, at about 3,000 BC; namely the driest part of the warm postglacial period. The xerothermic period followed a warm but moist "climatic optimum" period at about 6,000 BC.

References

Chapter 1. The Geology and Landforms of Nebraska

Aldrich, J. 1966. Life areas of North America. Poster 102. U.S. Department of Interior, Bureau of Sport Fisheries and Wildlife. (1-page map)

Allen, D. 1967. *The Life of Prairies and Plains*. New York: McGraw-Hill.

Axelrod, D. I. 1985. Rise of the grassland biome, central North America. *Botanical Review* 51:163–201.

Barry, R. 1983. Climatic environment of the Great Plains, past and present. *Transactions of the Nebraska Academy of Sciences* 11:45–55.

Benedict, A. D. 1991. *A Sierra Club Naturalist's Guide: The Southern Rockies*. San Francisco: Sierra Club Books.

Benninghoff, W. S. 1964. The prairie peninsula as a filter barrier to post-glacial plant migration. *Transactions of the Indiana Academy of Sciences* 72:116–24.

Bradbury, J. P. 1980. Late Quaternary vegetation history of the Central Great Plains and its relationship to eolian processes in the Nebraska Sandhills. In *Geologic and paleoecologic studies of the Nebraska Sand Hills*, U.S. Geological Survey Professional Paper 1120-C: 29–36. Washington DC: U.S. Department of the Interior, Geological Survey.

Ducey, J. 1992. Fossil birds of the Nebraska region. *Transactions of the Nebraska Academy of Sciences* 19:83–96.

Graham, A. 1999. *Late Cretaceous and Cenozoic History of North American Vegetation North of Mexico*. New York: Oxford University Press.

Kaul, R. B. 1986. Physical and floristic characteristics of the Great Plains. Pp. 7–10 in *Flora of the Great Plains*, ed. T. M. Barkley. Lawrence: University Press of Kansas.

Kurten, B., and D. C. Anderson. 1983. *Pleistocene Mammals of North America*. New York: Columbia University Press.

Lawson, M. P., K. F. Dewey, and R. E. Neild. 1977. *Climatic Atlas of Nebraska*. Lincoln: University of Nebraska Press.

Lueninghoener, G. C. 1947. *The Post-Kansan Geological History of the Lower Platte Valley*. University of Nebraska Studies No. 2. Lincoln: University of Nebraska.

Lugn, A. L. 1934. Outline of Pleistocene geology of Nebraska. *Bulletin of the University of Nebraska State Museum* B-1-41:1–39.

———. 1935. *The Pleistocene Geology of Nebraska*. Nebraska Geological Survey Bulletin 10, 2nd series.

———. 1939. Classification of the Tertiary system in Nebraska. *Bulletin of the Geological Society of America* 50:245–76.

Mengel, R. M. 1970. The North American Central Plains as an isolating agent in bird speciation. Pp. 280–340 in *Pleistocene and Recent Environments of the Central Great Plains*, ed. W. Dort Jr. and J. K. Jones Jr. Lawrence: University Press of Kansas.

Nebraska Game and Parks Commission. 1994. *The Cellars of Time: Paleontology and Archeology in Nebraska*. Special issue of *Nebraskaland* 72(1).

Palmer, D., ed. 1999. *The Simon and Schuster Encyclopedia of Dinosaurs and Prehistoric Creatures: A Visual Who's Who of Prehistoric Life*. New York: Simon and Shuster. (Excellent review and illustrations of fossil vertebrates)

Sears, P. 1935. Glacial and postglacial vegetation. *Botanical Review* 1:37–51.

Smith, H. T. V. 1965. Dune morphology and chronology in central and western Nebraska. *Journal of Geology* 73:557–78.

Thornberry, W. 1965. *Regional Geomorphology of the United States*. New York: John Wiley and Sons.

Trimble, D. E. 1980. The geologic story of the Great Plains. *Geologic Survey Bulletin 1493*. Washington DC: U.S. Department of the Interior, Geological Survey.

Wayne, W. J., et al. 1991. Quaternary geology of the northern Great Plains. Pp. 441–76 in *Quaternary Nonglacial Geology: Conterminous US. The Geology of North America*, ed. R. B. Morrison. Vol. K-2. Boulder CO: Geological Society of North America.

Wells, P. V. 1970a. Historical factors controlling vegetation patterns and floristic distributions in the Central Plains region of North America. Pp. 211–2l in *Pleistocene and Recent Environments of the Central Great Plains*, ed. W. Dort Jr. and J. K. Jones Jr. Lawrence: University Press of Kansas.

———. 1970b. Postglacial vegetational history of the Great Plains. *Science* 167:1574–82.

———. 1983. Late Quaternary vegetation of the Great Plains. *Transactions of the Nebraska Academy of Sciences* 11:83–89.

Wright, H. E., Jr. 1970. Vegetational history of the Great Plains. Pp. 157–72 in *Pleistocene and Recent Environments of the Central Great Plains*, ed. W. Dort Jr. and J. K. Jones Jr. Lawrence: University Press of Kansas.

Bailey, R. G. 1995. *Description of the Ecoregions of the United States*. 2nd ed. rev. U.S. Department of Agriculture Forest Service Miscellaneous Publication No. 1391. (with separate map)

———. 1996. *Ecosystem Geography*. New York: Springer-Verlag.

Baltensberger, B. M. 1985. *Nebraska: A Geography*. Boulder CO: Westview Press.

Barbour, M. C., and W. D. Billings, eds. 1999. *North American Terrestrial Vegetation*. 2nd ed. Cambridge: Cambridge University Press.

Barkley, T. M., ed. 1977. *Atlas of the Flora of the Great Plains*. Great Plains Flora Association. Ames: Iowa State University Press.

———. 1986. *Flora of the Great Plains*. Lawrence: University Press of Kansas.

Bleed, A., and C. Flowerday. 1989. *An Atlas of the Sand Hills*. Resource Atlas No. 5. Conservation and Survey Division, University of Nebraska–Lincoln.

Bogan, M. A. 1996. Historical changes in the landscape and vertebrate diversity of north central Nebraska. Pp. 105–30 in *Ecology and Conservation of Great Plains Vertebrates*, ed. F. L. Knopf and F. B. Samson. New York: Springer-Verlag.

Bolen, E. G. 1998. *Ecology of North America*. New York: John Wiley and Sons.

Carpenter, J. R. 1940. The grassland biome. *Ecological Monographs* 10: 617–84.

Condra, G. E. 1908. *The Geography of Nebraska*. Lincoln: University Publishing.

Daubenmire, R. 1978. *Plant Geography: With Special Reference to North America*. New York: Academic Press.

Dice, L. R. 1943. *The Biotic Provinces of North America*. Ann Arbor: University of Michigan Press.

Dreezen, V. H. 1973. *Topographic Regions Map, State of Nebraska*. University of Nebraska–Lincoln, Division of Conservation and Survey. (1-page map)

Farrar, J. 1976. Prairie life: Plant succession. *Nebraskaland* 54(5):38–43.

Gaines, J. F. 1946. A survey of the principal ecological studies of the grasslands and adjacent areas of North America. M.S. thesis, University of Nebraska–Lincoln.

Joern, A., and K. Keeler, eds. 1995. *The Changing Prairie: North American Grasslands*. New York: Oxford University Press.

Johnsgard, P. A. 1987. The ornithogeography of the Great Plains states. *Prairie Naturalist* 10:97–112.

———. 1998. Endemicity and regional biodiversity in Nebraska's breeding birds. *Nebraska Bird Review* 66:115–21.

Kantak, G. E., and S. P. Churchill. 1986. A bibliography of the ecological and taxonomic literature of the Nebraska vascular plants. *Transactions of the Nebraska Academy of Sciences* 14:61–78.

Kaul, R. B., and S. B. Rolfsmeier. 1993. *Native Vegetation of Nebraska*. Conservation and Survey Division, University of Nebraska–Lincoln. (map, plus 32-page text supplement)

Kendeigh, S. C. 1974. *Ecology: With Special Reference to Animals and Man*. New York: Prentice Hall.

Knopf, F. L., and F. B. Samson, eds. 1996. *Prairie Conservation: Conserving North America's Most Endangered Ecosystem*. Covelo CA: Island Press.

———, eds. 1997. *Ecology and Conservation of Great Plains Vertebrates*. New York: Springer-Verlag.

Kchler, A. W. 1966. *Potential Natural Vegetation of the Coterminous United States*. American Geographical Society Special Publication 36. New York.

Lawson, M., K. F. Dewey, and R. E. Neild. 1977. *Climatic Atlas of Nebraska*. Lincoln: University of Nebraska Press.

Nebraska Game and Parks Commission. 1996. *Weather and Climate in Nebraska*. Special issue of *Nebraskaland* 74(1).

Omernik, J. M. 1987. Ecoregions of the conterminous United States. *Annals of the Association of American Geographers* 77:118–25. (with map)

Ricketts, T. A., E. Dinerstein, D. Olson, C. Louds, W. Eichbaum, K. DellaSalla, K. Kavanagh, P. Hidao, P. Hurley, K. Carney, R. Abell, and S. Waters. 1999. *Terrestrial Ecosystems of North America: A Conservation Assessment*. Washington DC: Island Press.

Shantz, H. L. 1923. The natural vegetation of the Great Plains region. *Annals of the Association of American Geographers* 13:81–107.

Sims, P. L. 1988. Grasslands. Pp. 265–86 in *North American Terrestrial Vegetation*, ed. M. G. Barbour and W. D. Billings. Cambridge: Cambridge University Press.

Stein, B. A., L. S. Kutner, and J. S. Adams. 1999. *Precious Heritage: The Status of Biodiversity in the United States*. New York: Oxford University Press.

Steinauer, G., and S. Rolfsmeier. 2000. Terrestrial Natural Communities of Nebraska. Manuscript report to Nebraska Game and Parks Commission, Lincoln.

Stone, R. N., and W. T. Bagley. 1961. *The Forest Resource of Nebraska*. Release 4. Fort Collins: U.S. Department of Agriculture Forest Service, Rocky Mountain Forest and Range Experiment Station.

Webb, W. P. 1981. *The Great Plains*. Lincoln: University of Nebraska Press.

Winter, J. M. 1936. An analysis of the flowering plants of Nebraska.

Conservation and Survey Division Bulletin, University of Nebraska
13:1–203.

Chapter 3. The Western Escarpments and Their Coniferous Forests

Bock, C. E. 1970. The ecology and behavior of the Lewis woodpecker (*Asyndesmus lewis*). *University of California Publications in Zoology* 92:1–91.

Cox, M. K., and W. L. Franklin. 1989. Terrestrial vertebrates of Scotts Bluff National Monument. *Southwestern Naturalist* 49:597–613.

Hansen, K. 1992. *Cougar: The American Lion*. Flagstaff AZ: Northland Press.

Johnsgard, P. A. 1995. *This Fragile Land: A Natural History of the Nebraska Sandhills*. Lincoln: University of Nebraska Press.

Kellogg, R. S. 1905. Forest belts of western Kansas and Nebraska. *U.S. Department of Agriculture Forest Service Bulletin* 66:1–44.

Montz, R. L. 1967. Vegetative study of a river bottom in northwest Nebraska. M.S. thesis, Chadron State College, Chadron NE.

Mutel, C. F., and J. C. Emerick. 1984. *From Grassland to Glacier: The Natural History of Colorado*. Boulder CO: Johnson Books.

Nickerson, D. 1993. Western forest. Pp. 72–83 in *Walk in the Woods*. Special issue of *Nebraskaland* 71(1)

Nixon, E. S. 1967. A vegetational study of the Pine Ridge of northwestern Nebraska. *Southwestern Naturalist* 12:134–45.

Oduye, A. 1975. The vascular flora of the Soldier Creek Management Unit in Sioux County, Nebraska. M.S. thesis, Chadron State College, Chadron NE.

Spires, D. E., and R. R. Weedon. 1974. The vegetation and flora of the Pine Ridge of northwestern Nebraska—a progress report. *Proceedings of the Nebraska Academy Sciences* 84:15–16. (Abstract)

Tolstead, W. L. 1947. Woodlands in northwestern Nebraska. *Ecology* 28:80–88.

Turner, R. W. 1974. Mammals of the Black Hills of South Dakota and Wyoming. *University of Kansas Museum of Natural History Miscellaneous Publications* 50:1–78.

Chapter 4. The High Plains and Shortgrass Prairies

Bichel, M. A. 1959. Investigations of a Nebraska and Colorado prairie and their impact on the relict concept. Ph.D. diss., University of Nebraska–Lincoln.

Byers, J. A. 1997. *American Pronghorn: Social Adaptations and the Ghosts of Predators Past*. Chicago: University of Chicago Press.

Cushman, R. C., and S. R. Jones. 1988. *The Shortgrass Prairie*. Boulder CO: Pruett Publishing.

Farrar, J. 1994. Life on the rocks. *Nebraskaland* 72(6):14–29.

Flake, L. D. 1974. Reproduction of four rodent species in a shortgrass prairie of Colorado. *Journal of Mammalogy* 55:213–16.

Henderson, F. R., P. F. Springer, and R. Adrian. 1969. The black-footed ferret in South Dakota. *Technical Bulletin of the South Dakota Department of Game, Fish and Parks* 4:1–37.

Hillman, C. N., and J. C. Sharps. 1978. Return of the swift fox to the northern plains. *Proceedings of the South Dakota Academy of Science* 57:154–62.

Hines, T. 1980. An ecological study of *Vulpes velox* in Nebraska. M.S. thesis, University of Nebraska–Lincoln.

Hitchcock, A. S. 1935. *A Manual of Grasses of the United States*. U.S. Department of Agriculture Miscellaneous Publication No. 200. Washington DC: Government Printing Office.

Huff, D. R. 1988. The reproductive biology of buffalograss *Buchloe dactyloides* (Nutt.) Engel. Ph.D. diss., University of California–Davis.

Klipple, G. E., and D. F. Costello. 1960. Vegetation and cattle responses to different intensities of grazing on short-grass ranges on the central Great Plains. *U.S. Department of Agriculture Technical Bulletin* 1216:1–81.

Maxwell, M. H., and L. N. Brown. 1968. Ecological distribution of rodents on the High Plains of eastern Wyoming. *Southwestern Naturalist* 13:143–58.

Moulton, M. P., J. R. Choate, S. J. Bissell, and R. A. Nicholson. 1981. Associations of small mammals on the central High Plains of eastern Colorado. *Southwestern Naturalist* 26:53–57.

Quinn, J. A., D. P. Mowery, S. M. Emanuele, and R. D. B. Whalley. 1994. The "foliage is the fruit hypothesis": *Buchloe dactyloides* (Poaceae) and the shortgrass prairie of North America. *American Journal of Botany* 81:1545–54.

Tolstead, W. L. 1942. A note on unusual plants in the flora of northwest Nebraska. *American Midland Naturalist* 28:475–81.

Urbatsch, L., and R. Eddy. 1973. A floristic study of Dawes County, Nebraska. *Transactions of the Nebraska Academy of Sciences* 2:190–204.

Young, L. W., II. 1978. Vegetation and flora of the Hudson-Meng bison kill site. *Proceedings of the Nebraska Academy of Sciences* 88:27 (abstract).

Chapter 5. The Loess Hills and the Mixed-Grass Prairies

Branson, F. A. 1952. Native pastures of the dissected loess plains of central Nebraska. Ph.D. diss., University of Nebraska–Lincoln.

Farrar, J. 1997a. Loess hills. *Nebraskaland* 75(7):14–25.

———. 1997b. Red cedar: The good, the bad and the ugly. *Nebraskaland* 5(2):22–31.

Harnett, D. C., A. A. Steuter, and K. R. Hickman. 1997. Comparative ecology of native and introduced ungulates. Pp. 72–100 in *Ecology and Conservation of Great Plains Vertebrates*, ed. F. L. Knopf and F. B. Samson. New York: Springer-Verlag.

Hitchcock, A. S. 1935. *A Manual of Grasses of the United States*. U.S. Department of Agriculture Miscellaneous Publication No. 200. Washington DC: Government Printing Office.

Hopkins, H. H. 1951. Ecology of the native vegetation of the Loess Hills in central Nebraska. *Ecological Monographs* 21:125–47.

———. 1952. Native vegetation in the Loess Hills–Sandhills ecotone in Nebraska. *Transactions of the Kansas Academy of Sciences* 55:266–77.

Hulett, C. K., C. D. Sloan, and G. W. Tomenek. 1968. The vegetation of remnant grasslands in the loessal region of northwestern Kansas and southwestern Nebraska. *Southwestern Naturalist* 13:377–91.

Locklear, J. 1997. Shaggy grass country. *Nebraskaland* 75(5):8–15.

Longfellow, S. 1998. Loess Hills birds. Pp. 95–104 in *The Loess Hills Prairies of Central Nebraska*, ed. H. G. Nagel. Kearney: University of Nebraska–Kearney (Platte Valley Review).

Mutel, C. F. 1989. *Fragile Giants: A Natural History of the Loess Hills*. Iowa City: University of Iowa Press.

Nagel, H. G., ed. 1998. *The Loess Hills Prairies of Central Nebraska*. Kearney: University of Nebraska–Kearney (Platte Valley Review).

Novacek, J. M., D. M. Roosa, and W. P. Pusateri. 1985. The vegetation of the Loess Hills landform along the Missouri River. *Proceedings of the Iowa Academy of Science* 92:199–212.

Pfeiffer, K. E., and D. C. Harnett. 1995. Bison selectivity and grazing response of little bluestem in tallgrass prairie. *Journal of Range Management* 48:26–31.

Rothenberger, S. 1994. Floristic analysis of the C. Bertrand Schulz and Marian Othener Schulz Prairie, a mixed-grass prairie in south-central Nebraska. *Transactions of the Nebraska Academy of Sciences* 21:21–30.

———. 1998. Vegetation of the Loess Hills. Pp. 63–73 in *The Loess Hills Prairies of Central Nebraska*, ed. H. G. Nagel. Kearney: University of Nebraska–Kearney (Platte Valley Review).

Schultz, C. B., and J. C. Frye. 1968. *Loess and Related Aeolian Deposits of the World*. Lincoln: University of Nebraska Press.

Voight, J. W. 1951. Vegetational change on a 25-year subsere in the Loess Hills region of central Nebraska. *Journal of Range Management* 4:254–63.

Weaver, J. E. 1960. Comparison of vegetation of the Kansas-Nebraska drift—loess hills and loess plains. *Ecology* 41:73–88.

Weaver, J. E., and W. E. Bruner. 1948. Prairies and pastures of the dissected loess plains of central Nebraska. *Ecological Monographs* 18:507–49.

Wehrman, K. C. 1961. A study of the transition zone between the Loess Hills and the Sand Hills in central Nebraska. M.S. thesis, University of Nebraska–Lincoln.

Chapter 6. The Sandhills and the Sandsage Prairies

Barnes, P. W. 1986. Variation in the big bluestem (*Andropogon gerardii*)–sand bluestem (*Andropogon hallii*) complex along a local dune/meadow gradient in the Nebraska Sandhills. *American Journal of Botany* 73:172–84.

Baumann, W. L. 1982. Microhabitat use in three species of rodents on a Nebraska bleed, Sandhills prairie. M.S. thesis, University of Nebraska–Lincoln.

Bleed, A., and C. Flowerday, eds. 1989. *An Atlas of the Sand Hills*. Resource Atlas No. 5. Conservation and Survey Division, University of Nebraska–Lincoln.

Daley, R. H. 1972. The native sand sage vegetation of eastern Colorado. M.S. thesis, Colorado State University.

Farrar, J. 1993a. Sandsage prairies: History of the land. *Nebraskaland* 71(7):22–29.

———. 1993b. Sandsage prairies: The Cinderella sandhills. *Nebraskaland* 71(7):30–41.

Flessner, T. R. 1988. Propagation, establishment and ecological characteristics of *Penstemon haydeni*. M.S. thesis, University of Nebraska–Lincoln.

Henzlik, R. E. 1965. Biogeographic extensions into a coniferous forest plantation in the Nebraska Sandhills. *American Midland Naturalist* 74:87–94.

Joern, A. 1982. Distributions, densities, and relative abundances of grasshoppers (Orthoptera, Acrididae) in a Nebraska Sandhills prairie. *Prairie Naturalist* 14:37–45.

Johnsgard, P. A. l995. *This Fragile Land: A Natural History of the Nebraska Sandhills*. Lincoln: University of Nebraska Press.

Keech, C. F., and R. Bentall. 1971. *Dunes on the Plains*. Resource Report 4. Conservation and Survey Division, University of Nebraska–Lincoln.

Keeler, K., A. T. Harrison, and L. S. Vescio. 1980. The flora and Sandhills prairie communities of Arapaho Prairie, Arthur County, Nebraska. *Prairie Naturalist* 12:65–78.

Pool, R. J. 1914. A study of the vegetation of the Sandhills of Nebraska. *Minnesota Botanical Studies* (University of Minnesota, Minneapolis) 4(3):189–312.

Ramaley, F. 1939. Sand-hill vegetation of northeastern Colorado. *Ecological Monographs* 9:1–51.

Reed, K. M., and J. R. Choate. 1986. Natural history of the Plains pocket mouse in agriculturally disturbed sandsage prairie. *Prairie Naturalist* 18:79–80.

Riley, C. V. 1892. The yucca moth and yucca pollination. *Report of the Missouri Botanic Garden* 3:99–158.

Stubbendieck, J., T. R. Flessner, C. H. Butterfield, and A. A. Steuter. 1993. Establishment and survival of the endangered blowout penstemon. *Great Plains Research* 3:3–19.

Twedt, C. M. 1974. Characteristics of sharp-tailed grouse display grounds in the Nebraska Sandhills. Ph.D. diss., University of Nebraska–Lincoln.

Weeden, R. R., J. Stubbendieck, and M. Beel. 1982. Blowout penstemon. *Nebraskaland* 60(6):49–51.

Chapter 7. The Niobrara Valley and Its Transitional Forest

Brogie, M. A., and M. J. Mossman. 1983. Spring and summer birds of the Niobrara Valley Preserve area, Nebraska. *Nebraska Bird Review* 51:44–51.

Catling, P. M. 1983. Pollination in North American *Spiranthes* (Orchidaceae). *Canadian Journal of Botany* 61:1080–93.

Churchill, S. P., C. C. Freeman, and G. E. Kantak. 1988. The vascular flora of the Niobrara Valley Preserve and adjacent areas in Nebraska. *Transactions of the Nebraska Academy of Sciences* 16:1–15.

Cingel, N. A. van der. 1995. *An Atlas of Orchid Pollination: European Orchids*. Rotterdam: A. A. Balkema.

Ducey, J. 1989. Birds of the Niobrara River valley, Nebraska. *Transactions of the Nebraska Academy of Sciences* 17:37–60.

Harrison, A. T. 1980. The Niobrara Valley Preserve: Its biogeographic importance and description of its plant communities. A working report to the Nature Conservancy, Minneapolis.

Hearty, P. J. 1978. The biogeography and geomorphology of the Niobrara River valley near Valentine, Nebraska. M.S. thesis, University of Nebraska–Omaha.

Johnsgard, P. A. 1995. *This Fragile Land: A Natural History of the Nebraska Sandhills*. Lincoln: University of Nebraska Press.

Kantak, G. E. 1995. Terrestrial plant communities of the middle Niobrara Valley, Nebraska. *Southwestern Naturalist* 40:129–38.

Kantak, G. E., and S. P. Churchill. 1993. The Niobrara Valley Preserve: Inventory of a biogeographical crossroads. *Transactions of the Nebraska Academy of Sciences* 20:1–12.

Kaul, R. B., G. E. Kantak, and S. P. Churchill. 1988. The Niobrara River valley, a postglacial migration corridor and refugium of forest plants and animals in the grasslands of central North America. *Botanical Review* 54:44–81.

Morgan, G. R. 1962. The geographical distribution and environment of white birches along the Niobrara River in north-central Nebraska. M.A. thesis, University of Nebraska–Lincoln.

Nilsson, L. A. 1992. Orchid pollination biology. *Trends in Ecology and Evolution* 7–8:255–59.

Rising, J., and N. J. Flood. 1998. Baltimore Oriole (*Icterus galbula*). In *The Birds of North America*, No. 384, ed. A. Poole and F. Gill. Philadelphia and Washington DC: Academy of Natural Sciences and American Ornithologists' Union.

Sipes, S. D., and V. J. Tepedino. 1995. Reproductive biology of the rare orchid *Spiranthes diluvialis*: Breeding system, pollination and implications for conservation. *Conservation Biology* 9:929–38.

Steinauer, G. 1993. The Niobrara Valley forests. Pp. 84–85 in *Walk in the Woods*. Special issue of *Nebraskaland* 71(1).

Tolstead, W. L. 1942. Vegetation of the northern part of Cherry County, Nebraska. *Ecological Monographs* 12:255–92.

Winner, C. 1995. Rare beauty (Ute ladies' tresses). *Wyoming Wildlife* 59(8)24–29.

Chapter 8. *The Platte River Valley and its Riparian Forests*

Abrams, M. D. 1985. Fire history of oak gallery forest in a northeast Kansas tallgrass prairie. *American Midland Naturalist* 114:188–91.

Becker, D. A. 1980. Floristic analysis of a natural area on the Lower Platte River floodplain. *Transactions of the Nebraska Academy of Sciences* 8:15–30.

Bray, T., and B. L. Wilson. 1992. Status of *Plantathera praeclara* Sheviak and Bowles (western prairie orchid) in the Platte River valley in Nebraska from Hamilton to Garden Counties. *Transactions of the Nebraska Academy of Sciences* 19:57–62.

Colt, C. J. 1996. Breeding bird use of riparian forests along the central Platte River: A spatial analysis. M.S. thesis, University of Nebraska–Lincoln.

Currier, P. J. 1982. The floodplain vegetation of the Platte River: Phytosociology, forest development and seedling establishment. Ph.D. diss., Iowa State University.

———. 1993. Riparian forest. Pp. 37–48 in *Walk in the Woods*. Special issue of *Nebraskaland* 71(1).

Faanes, C. A. 1984. Wooded islands in a sea of prairie. *American Birds* 38:3–6.

Johnsgard, P. A. 1984. *The Platte: Channels in Time*. Lincoln: University of Nebraska Press.

———. 1995. *This Fragile Land: A Natural History of the Nebraska Sandhills*. Lincoln: University of Nebraska Press.

Johnson, W. C. 1961. Woodland expansion in the Platte River, Nebraska: Patterns and causes. *Ecological Monographs* 64:45–84.

Johnson, W. C., and S. E. Boettcher. In press. The presettlement Platte: Wooded or prairie river? *Great Plains Research*.

Krapu, G., ed. 1981. *The Platte River Ecology Study*. Special Research Report. Jamestown ND: Northern Prairie Wildlife Research Station, U.S. Fish and Wildlife Service.

Nagel, H. G., K. Geisler, J. Cochran, J. Fallesen, B. Hadenfelt, J. Mathews, J. Nickel, S. Stec, and A. Walters. 1980. Platte River island succession. *Transactions of the Nebraska Academy of Sciences* 8:77–90.

Rand, P. J. 1973. *The Woody Phreatophyte Communities of the Republican River Valley in Nebraska*. Final Report, Research Contract, U.S. Bureau of Reclamation. Botany Department, University of Nebraska–Lincoln.

Chapter 9. The Eastern Glaciated Plains and the Tallgrass Prairies

Boettcher, J. F. 1981. Native tallgrass prairie remnants of eastern Nebraska: Floristics and effects of management, topography, size and season of evaluation. M.S. thesis, University of Nebraska–Omaha.

Boettcher, J. F., T. B. Bragg, and D. M. Sutherland. 1993. Floristic diversity in ten tallgrass prairie remnants of eastern Nebraska. *Transactions of the Nebraska Academy of Sciences* 20:33–34. (217 spp., including 30 nonnative spp.)

Cole, T. P. 1975. A comparative study of two grassland communities. M.S. thesis, University of Nebraska–Omaha.

Davies, R. W. 1986. The pollination ecology of *Cypripedium acaule*. *Rhodora* 88:445–50.

Dressler, R. L. 1981. *The Orchids: Natural History and Classification*. Cambridge: Harvard University Press.

Farney, D. 1980. The tallgrass prairie: Can it be saved? *National Geographic Magazine* (Jan.), pp. 37–61.

Fichter, E. 1954. An ecological study of invertebrates of grassland and deciduous shrub savanna in eastern Nebraska. *American Midland Naturalist* 51:3221–39.

Hitchcock, A. S. 1935. *A Manual of Grasses of the United States*. U.S. Department of Agriculture Miscellaneous Publication No. 200. Washington DC: Government Printing Office.

Kaul, R. B., and S. B. Rolfsmeier. 1987. The characteristics and phytogeographic affinities of the flora of Nine-mile Prairie, a western tallgrass prairie in Nebraska. *Transactions of the Nebraska Academy of Sciences* 15:23–35.

Klier, K., M. J. Leoschke, and J. F. Wendel. 1991. Hybridization and introgression in white and yellow ladyslipper orchids (*Cypripedium candidum* and *C. pubescens*). *Journal of Heredity* 82:305–18.

Knapp, A. K. 1985. Effect of fire and drought on the ecophysiology of *Andropogon gerardii* and *Panicum virgatum* in a tallgrass prairie. *Ecology* 66:1309–20.

Luer, C. A. 1975. *The Native Orchids of the United States and Canada.* New York: New York Botanical Garden.

Madson, J. 1993. *Tallgrass Prairie.* Helena MT: Falcon Press.

———. 1995. *Where the Sky Began: Land of the Tallgrass Prairie.* Ames: Iowa State University Press.

Mentzer, L. W. 1951. Studies of plant succession in true prairie. *Ecological Monographs* 21:255–67.

Nagel, H. G., and O. A. Kolstad. 1987. Comparison of plant species composition of Mormon Island Crane Meadows and Lillian Annette Rowe Sanctuary in central Nebraska. *Transactions of the Nebraska Academy of Sciences* 15:37–48.

Nicholson, R. A., and G. K. Hulett. 1969. Remnant grassland vegetation in the Great Plains of North America. *Journal of Ecology* 77:599–612.

Nicholson, R. A., and M. G. Marcotte. 1979. Vegetation of the Willa Cather Memorial Prairie. *Journal of Range Management* 32:104–8.

Reichman, O. J. 1977. *Konza Prairie: A Tallgrass Natural History.* Lawrence: University Press of Kansas.

Risser, P. G., E. C. Berry, H. D. Blocker, S. W. May, W. J. Parton, and J. A. Weins. 1981. *The True Prairie Ecosystem.* Stroudsbury PA: Hutchinson Ross Publishing.

Samson, F., and F. Knopf. 1994. Prairie conservation in North America. *BioScience* 44:418–21.

Sheviak, C. J., and M. L. Bowles. 1986. The prairie fringed orchids: A pollinator isolated pair. *Rhodora* 88:267–90.

Smith, A. 1996. *Big bluestem: Journey into the Tallgrass Prairie.* Tulsa OK: Council Oak Books.

Steiger, T. L. 1930. Structure of prairie vegetation. *Ecology* 11:170–211.

Steinauer, G. 1998. In search of prairie orchids. *Nebraskaland* 76(2):9.

Vinton, M. A., and D. C. Hartnett. 1992. Effects of bison grazing on *Andropogon gerardii* and *Panicum virgatum* in burned and unburned tallgrass prairie. *Oecologia* 90:374–82.

Weaver, J. E. 1954. *North American Prairie.* Lincoln NE: Johnson Publishing.

———. 1958. Native grassland of southwestern Iowa. *Ecology* 39:733–50.

————. 1965. *Native Vegetation of Nebraska.* Lincoln: University of Nebraska Press.

————. 1968. *Prairie Plants and Their Environment: A Fifty-Year Study in the Midwest.* Lincoln: University of Nebraska Press.

Weaver, J. E., and F. W. Albertson. 1956. *Grasslands of the Great Plains: Their Nature and Use.* Lincoln: Johnson Publishing.

Weaver, J. E., and T. J. Fitzpatrick. 1954. The prairie. *Ecological Monographs* 4:109–295. Reprint, 1980, Aurora NE: Prairie-Plains Resources Institute.

Zimmerman, J. L. 1993. *The Birds of Konza: The Avian Ecology of the Tallgrass Prairie.* Lawrence: University Press of Kansas.

Chapter 10. The Missouri Valley and Its Deciduous Forests

Aikman, J. M. 1929. Distribution and structure of the forests of eastern Nebraska. *Botanical Seminar, University of Nebraska Botanical Survey,* New Series 5:1–75.

Beightol, D. A., and T. B. Bragg. 1993. Woody vegetation of a disjunct bur oak (*Quercus macrocarpa*) forest in east-central Nebraska. *Transactions of the Nebraska Academy of Sciences* 20:41–46.

Braun, E. L. 1967. *Deciduous Forests of Eastern North America.* New York: Hafner Publishing.

Costello, D. F. 1931. Comparative study of river bluff succession on the Iowa and Nebraska sides of the Missouri River. *Botanical Gazette* 91:295–307.

Currier, P. 1993. Riparian forest. Pp. 37–48 in *Walk in the Woods.* Special issue of *Nebraskaland* 71(1).

Davies. R. W. 1986. The pollination ecology of *Cypripedium acaule. Rhodora* 88:445–50.

Hanson, H. C. 1918. The invasion of a Missouri River alluvial flood plain. *American Midland Naturalist* 5:196–201.

Holch, A. E. 1932. Forest vegetation in southeastern Nebraska. *Journal of Forestry* 30:72–74.

Lawrey, J. D. 1973. The Missouri River floodplain plant communities from Yankton, South Dakota, to Rulo, Nebraska: Their successional relationships and effects of river bank stabilization. M.A. thesis, University of South Dakota.

Peattie, D. C. 1950. *A Natural History of Trees of Eastern and Central North America.* Boston: Houghton Mifflin.

Reynolds, H. C. 1942. Flora of Richardson County, Nebraska. M.S. thesis, University of Nebraska–Lincoln.

Rothenberger, S. 1993. Eastern deciduous forest. Pp. 20–34 in *Walk in the Woods.* Special issue of *Nebraskaland* 71(1).

Rozmajzl, M. K. 1988. Presettlement savanna in eastern Nebraska. M.A. thesis, University of Nebraska–Omaha.

Weaver, J. 1960. Flood plain vegetation of the central Missouri Valley and contacts of woodland with prairie. *Ecological Monographs* 30:37–64.

Chapter 11. *Rivers, Lakes, Marshes, and Other Wetlands*

Bentall, R. 1989. Streams. Pp. 93–134 in *An Atlas of the Sand Hills*, ed. A. Bleed and C. Flowerday. Resource Atlas No. 5. Conservation and Survey Division, University of Nebraska–Lincoln.

Bleed, A. 1989. Groundwater. Pp. 67–92 in *An Atlas of the Sand Hills*, ed. A. Bleed and C. Flowerday. Resource Atlas No. 5. Conservation and Survey Division, University of Nebraska–Lincoln.

Bleed, A., and M. Ginsberg. 1989. Lakes and wetlands. Pp. 115–22 in *An Atlas of the Sand Hills*, ed. A. Bleed and C. Flowerday. Resource Atlas No. 5. Conservation and Survey Division, University of Nebraska–Lincoln.

Bouc, K. 1983. The Cedar, pp. 50–57, The Blues, pp. 82–89, and The Missouri, pp. 90–101, in *Nebraska Rivers*, ed. J. Farrar. Special issue of *Nebraskaland* 61(1).

Cunningham, D. 1983. The Calamus, pp. 22–27, and The Platte, pp. 28–39, in *Nebraska Rivers*, ed. J. Farrar. Special issue of *Nebraskaland* 61(1).

Currier, P. J., G. R. Lingle, and J. G. VanDerwalker. 1985. *Migratory Bird Habitat on the Platte and North Platte Rivers in Nebraska*. Grand Island NE: Whooping Crane Trust.

Delich, C. 1984. The Nemahas. Pp. 114–19 in *Nebraska Rivers*, ed. J. Farrar. Special issue of *Nebraskaland* 61(1).

Ducey, J. 1987. Biological features of saline wetlands in Lancaster County, Nebraska. *Transactions of the Nebraska Academy of Sciences* 16:5–14.

Farrar, J. 1982. The Rainwater Basin: Nebraska's vanishing wetlands. *Nebraskaland* 60(3):18–41.

———. 1983. The Dismal, pp. 8–15, The Snake, pp. 40–49, and The Niobrara, pp. 102–33, in *Nebraska Rivers*, ed. J. Farrar. Special issue of *Nebraskaland* 61(1).

———. 1991. Nebraska salt marshes: Last of the least. *Nebraskaland* 69(6):18–41.

———. 1996. Nebraska's Rainwater Basin. *Nebraskaland* 74(2):18–35.

———, ed. 1983. *Nebraska Rivers*. Special issue of *Nebraskaland* 61(1).

Flowerday, C. A., ed. 1993. *Flat Water: A History of Nebraska and Its Water*. Resource Report No. 12. Conservation and Survey Division, University of Nebraska–Lincoln.

Floyd, S. 1995. Pollination structure, dynamics and genetics of *Guara neomexicana* ssp. *coloradensis* (Onagraceae), a rare semelparous perennial. M.A. thesis, University of Colorado.

Forsberg, M. 1996. Wet meadows of the Platte. *Nebraskaland* 74(4):36–47.

Grier, R. 1984. The Elkhorn, pp. 66–71, and The Pine Ridge streams, pp. 120–27, in *Nebraska Rivers*, ed. J. Farrar. Special issue of *Nebraskaland* 61(1).

Hiskey, R. M. 1981. The trophic dynamics of an alkaline-saline Nebraska Sandhills lake. Ph.D. diss., University of Nebraska–Lincoln.

Hoffman, R. 1984. The Frenchman, pp. 16–21, and The Republican, pp. 58–65, in *Nebraska Rivers*, ed. J. Farrar. Special issue of *Nebraskaland* 61(1).

Jenkins, A., ed. 1993. *The Platte River: An Atlas of the Big Bend Region.* Kearney: University of Nebraska–Kearney.

Johnsgard, P. A. 1984. *The Platte: Channels in Time.* Lincoln: University of Nebraska Press.

Kuzila, M. S., D. C. Rundquist, and J. A. Green. 1991. Methods for estimating wetland loss: The Rainbasin Region of Nebraska 1927–1991. *Journal of Soil and Water Conservation* 46:441–45.

LaGrange, T. 1997. *Guide to Nebraska Wetlands and Their Conservation Needs.* Lincoln: Nebraska Game and Parks Commission.

Mahoney, D. L. 1977. Species richness and diversity of aquatic vascular plants in Nebraska with special reference to water chemistry parameters. M.S. thesis, University of Nebraska–Lincoln.

McCarraher, D. B. 1977. *Nebraska's Sandhills Lakes.* Lincoln: Nebraska Game and Parks Commission.

McMurtry, M. S., R. Craig, and G. Schildmann. 1972. *Nebraska Wetland Survey.* Lincoln: Nebraska Game and Parks Commission.

Morrison, J. L. 1935. The development and structure of the vegetation on the sandbars and islands of the Lower Platte River. M.S. thesis, University of Nebraska–Lincoln.

Moulton, M. P. 1972. The small playa lakes of Nebraska: Their ecology, fisheries and biological potential. Pp. 15–23, in *Playa Lakes Symposium Transactions.* Lubbock: International Center for Arid and Semi-Arid Land Studies, Texas Tech University.

Nielsen, E. L. 1953. Revegetation of alkali flood plains adjoining the North Platte River, Garden County, Nebraska. *American Midland Naturalist* 49:915–19.

Rolfsmeier, S. B. 1993. The saline wetland-meadow vegetation and flora of the North Platte River valley in the Nebraska Panhandle. *Transactions of the Nebraska Academy of Sciences* 20:12–24.

Rundquist, D. C. 1983. *Wetland Inventories of Nebraska's Sandhills.* Resource Report No. 9. Conservation and Survey Division, University of Nebraska–Lincoln.

Seinauer, J. 1998. The Loupe: Lifeblood of central Nebraska. *Nebraska-land* 76(5):24–33.

Shirk, C. J. 1924. An ecological study of the vegetation of an inland saline area. M.S. thesis, University of Nebraska–Lincoln.

Sidle, J. G., E. D. Miller, and P. J. Currier. 1989. Changing habitats in the Platte Valley of Nebraska. *Prairie Naturalist* 21:91–104.

Steinauer, G. 1992. Sandhills fens. *Nebraskaland* 70(6):16–32.

———. 1994. Alkaline wetlands of the North Platte River valley. *Nebraskaland* 72(5):18–43.

Steinauer, G., S. B. Rolfsmeier, and J. P. Hardy. 1996. Inventory and floristics of Sandhills fens in Cherry County, Nebraska. *Transactions of the Nebraska Academy of Sciences* 23:9–21.

Ungar, I. A., W. Hogan, and M. McClelland. 1969. Plant communities of saline soils at Lincoln, Nebraska. *American Midland Naturalist* 82:564–77.

van der Valk, A., ed. *Northern Prairie Wetlands*. Ames: Iowa State University Press.

Weller, M. W. 1987. *Freshwater Marshes: Ecology and Wildlife Management*. Minneapolis: University of Minnesota Press.

Chapter 12. *Vanishing Habitats and Changing Times*

Askins, R. A. 1993. Population trends in grassland, shrubland and forest birds in eastern North America. *Current Ornithology* 11:11–34.

Dobkin, D. S. 1994. *Conservation and Management of Neotropical Migrant Landbirds in the Northern Rockies and Great Plains*. Moscow: University of Idaho Press.

Farney, D. 1980. The tallgrass prairie: Can it be saved? *National Geographic Magazine* (Jan.), pp. 37–61.

Farrar, J. 1998. Wildlife and the land. *Nebraskaland* 76(7):12–23.

Finch, D. M. 1992. *Threatened, Endangered, and Vulnerable Species of Terrestrial Vertebrates in the Rocky Mountain Region*. General Technical Report RM-215. Fort Collins CO: U.S. Department of Agriculture Forest Service, Rocky Mountain Forest and Range Experiment Station.

Goriup, P., ed. 1988. *Ecology and Conservation of Grassland Birds*. ICBP Technical Publication No. 7. International Council for Bird Preservation, Cambridge.

Hagen, J. M., III, and D. W. Johnston, eds. 1992. *Ecology and Conservation of Neotropical Migrant Landbirds*. Washington DC: Smithsonian Institution Press.

Herkert, J. R. 1994. The effect of habitat fragmentation on midwestern grassland bird communities. *Ecological Applications* 4:461–71.

Houtcooper, W. C., D. J. Ode, J. A. Peterson, and G. M. Vandel III. 1985.

Rare animals and plants of South Dakota. *Prairie Naturalist* 17:143–65.

Johnsgard, P. A. 1999a. The geese from beyond the north wind. Pp. 148–57, in *Earth, Water and Sky: A Naturalist's Stories and Sketches*. Austin: University of Texas Press.

————. 1999b. Where have all the curlews gone? Pp. 139–46 in *Earth, Water and Sky: A Naturalist's Stories and Sketches*. Austin: University of Texas Press.

————. 2001. *Prairie Birds: Fragile Splendor in the Great Plains*. Lawrence: University Press of Kansas.

Knopf, F. L. 1996. Prairie legacies—birds. Pp. 13–48 in *Prairie Conservation: Conserving North America's Most Endangered Ecosystem*, ed. F. B. Samson and F. L. Knopf. Covelo CA: Island Press.

Knopf, F. L., and F. B. Samson, eds. 1997. *Ecology and Conservation of Great Plains Vertebrates*. New York: Springer-Verlag.

Lemaire, R. J. 1968. A list of plants rare in Nebraska which should be protected. Lincoln: Nebraska Game and Parks Commission.

Manning, R. 1998. *Grasslands: The Biology, Politics and Promise of the American Prairie*. New York: Penguin Books.

Martin. T. E., and D. M. Finch, eds. 1995. *Ecology and Management of Neotropical Migratory Birds: A Synthesis and Review of Critical Issues*. New York: Oxford University Press.

Miller, B., R. Reading, and S. Forrest. 1996. *Prairie Night: Black-footed Ferrets and the Recovery of Endangered Species*. Washington DC: Smithsonian Institution Press.

Nicholson, R. A., and G. K. Hulett. 1969. Remnant grassland vegetation in the Great Plains of North America. *Journal of Ecology* 77:599–612.

Price, J., S. Droege, and A. Price. 1995. *The Summer Atlas of North American Birds*. New York: Academic Press.

Rodenhouse, N. L., L. B. Best, R. J. O'Conner, and E. K. Bollinger. 1993. Effects of temperate agriculture on Neotropical migrant landbirds. Pp. 280–95 in *Status and Management of Neotropical Migratory Birds*, ed. D. M. Finch and P. W. Stangel. General Technical Report RM-229. Fort Collins CO: U.S. Department of Agriculture Forest Service, Rocky Mountain Forest and Range Experiment Station.

Ryder, R. A. 1980. Effects of grazing on bird habitats. Pp. 51–66, in *Workshop Proceedings: Management of Western Forests and Grasslands for Nongame Birds*, ed. R. M. DeGraff and N. G. Tilghman. *General Technical Report INT-86*. Ogden UT: U.S. Department of Agriculture Forest Service.

Samson, F., and F. Knopf. 1994. Prairie conservation in North America. *BioScience* 44:418–21.

————, eds. 1996. *Prairie Conservation: Conserving North America's Most Endangered Ecosystem*. Covelo CA: Island Press.

Skinner, R. M. 1975. Grassland use patterns and prairie bird populations. Pp. 171–80 in *Prairie: A Multiple View*, ed. M. K. Wali. Grand Forks: University North Dakota Press.

Stein, B. A., L. S. Kutner, and J. S. Adams, eds. 2000. *Precious Heritage: The Status of Biodiversity in the United States*. New York: Oxford University Press.

Thompson, F. R., III, ed. 1995. *Management of Midwestern landscapes for the Conservation of Neotropical Migratory Birds*. General Technical Report NC-187. Detroit MI: U.S. Department of Agriculture, North Central Forest Experiment Station.

Thompson, F. R., III, S. J. Lewis, J. Green, and D. Ewert. 1993. Status of Neotropical migrant landbirds in the Midwest: Identifying species of management concern. Pp. 145–55 in *Status and Management of Neotropical Migratory Birds*, ed. D. M. Finch and P. W. Stangel. General Technical Report RM-229. Fort Collins CO: U.S. Department of Agriculture Forest Service, Rocky Mountain Forest and Range Experiment Station.

Torres, J., S. Bissell, G. Craig, W. Graul, and D. Langlois. 1978. Wildlife in danger, the status of Colorado's threatened or endangered birds, fish and mammals. *Colorado Outdoors* (supplement) 27(4).

U.S. Department of the Interior. 1989. *Endangered and Threatened Wildlife and Plants*. 50 CFR 17.11–17.12.

Wingfield, G., and G. Steinauer. 1992. Nebraska's vanishing species. *Nebraskaland* 70(10):18–35.

Wyoming Game and Fish Department. 1989. *Endangered and Nongame Bird and Mammal Investigations*. Wyoming Game and Fish Department Annual Completion Report.

Zimmerman, J. 1996. Avian community responses to fire, grazing, and drought in the tallgrass prairie. Pp. 167–80 in *Ecology and Conservation of Great Plains Vertebrates*, ed. F. L. Knopf and F. B. Samson. New York: Springer-Verlag.

References for Appendixes

Mammals

Armstrong, D. M. 1972. Distribution of mammals in Colorado. *University of Kansas Museum of Natural History Monographs* 3:1–425.

Barbour, R. W., and W. H. Davis. 1969. *Bats of America*. Lexington: University Press of Kentucky.

Baumann, W. L. 1982. Microhabitat use in three species of rodents on a Nebraska Sandhills prairie. M.S. thesis, University of Nebraska–Lincoln.

Bee, J. W., G. E. Glass, R. S. Hoffmann, and R. R. Patterson. 1981. *Mammals of Kansas*. University of Kansas Museum of Natural History Public Education Series 7.

Benedict, R. A., H. H. Genoways, and P. W. Freeman. 2000. Shifting distributional patterns of mammals in Nebraska. *Transactions of the Nebraska Academy of Sciences* 26:55–84.

Benedict, R. A., P. W. Freeman, and H. H. Genoways. 1996. Prairie legacies—Mammals. Pp. 149–66 in *Prairie Conservation: Conserving North America's Most Endangered Ecosystem,* ed. F. B. Samson and F. L. Fritz. . Covelo CA: Island Press.

Bouc, K. 1998. White-tailed deer and mule deer. *Nebraskaland* 76(7):56–63.

Burt, W. H., and R. P. Grossenheider. 1976. *A Field Guide to the Mammals.* 3rd ed. Boston: Houghton Mifflin.

Campbell, T. M., III, and T. W. Clark. 1981. Colony characteristics and vertebrate associates of white-tailed and black-tailed prairie dogs in Wyoming. *American Midland Naturalist* 105:269–75.

Chapman, J. A., and C. A. Feldhammer, eds. 1982. *Wild Mammals of North America.* Baltimore: Johns Hopkins University Press.

Clark, T. W., ed. 1986. The black-footed ferret. *Great Basin Naturalist Memoirs* No. 8. Provo UT: Brigham Young University.

Clark, T. W., and M. R. Stromberg. 1987. *Mammals in Wyoming.* University of Kansas Museum of Natural History Publication Series 10.

Cochrum, E. L. 1952. *Mammals of Kansas.* University Kansas Museum of Natural History Publication Series 7.

Downhower, J. F., and E. R. Hall. 1966. The pocket gopher in Kansas. *University of Kansas Museum of Natural History Miscellaneous Publications* 44:1–32.

Eisenberg, J. E. 1963. The behavior of heteromyid rodents. *University California Publications in Zoology* 69:1–114.

Epperson, C. 1978. The biology of the bobcat in Nebraska. M.S. thesis, University of Nebraska–Lincoln.

Farrar, J. 1992. Musquash . . . grazer of the marsh. *Nebraskaland* 70(5): 14–23.

———. 1996. Porky in the pigweeds. *Nebraskaland* 74(7):32–35.

———. 1999. "Egg-sucking" weasels. *Nebraskaland* 77(6):30–35.

Finley, R. B., Jr. 1958. The wood rats of Kansas: Distribution and ecology. *University of Kansas Museum of Natural History Publications* 10:213–552.

Forbes, R. B. 1964. Some aspects of the biology of the silky pocket mouse, *Perognathus flavus. American Midland Naturalist* 72:438–42.

Forsberg, M. 1997. Fox squirrels: Acrobats of the trees. *Nebraskaland* 75(2):8–15.

Forsyth, A. 1999. *Mammals of North America: Temperate and Arctic Regions.* Buffalo NY: Firefly Books.

Freeman, P. 1989. Mammals. Pp. 181–88 in *An Atlas of the Sand Hills,* ed. A. Bleed and C. Flowerday. Resource Atlas No. 5. Conservation and Survey Division, University of Nebraska–Lincoln.

Freeman, P., K. N. Geluso, and J. S. Altenbach. 1997. Nebraska's flying mammals. *Nebraskaland* 75(6):38–45.

Freeman, P., and R. A. Benedict. 1993. Flat water mammals. *Nebraskaland* 71(6):24–35.

Garner, H. W. 1974. Population dynamics, reproduction and activities of the kangaroo rat, *Dipodomys ordii,* in western Texas. *Graduate Studies, Texas Tech University* 7:1–28.

Grier, R. 1988. The smallest and rarest. (Swift fox) *Nebraskaland* 66(4): 18–25.

———. 1995. Nebraska's elk. A plan for the future. *Nebraskaland* 73(7):34–39.

———. 1998a. Bighorns. *Nebraskaland* 76(2):24–27.

———. 1998b. Elk and bighorn sheep. *Nebraskaland* 76(7):68–71.

———. 1998c. Pronghorn. *Nebraskaland* 76(7):64–67.

———. 1999. A new century of promise for bighorns. *Nebraskaland* 77(9):18–23.

Haberman, C. G., and E. D. Fleharty. 1972. Natural history notes on Franklin's ground squirrel in Boone County, Nebraska. *Transactions of the Kansas Academy of Sciences* 74:76–80.

Hall, E. R. 1981. *The Mammals of North America.* 2 vols. New York: John Wiley and Sons.

Henderson, F. R. 1960. Beaver in Kansas. *University of Kansas Museum of Natural History Publications* 26:1–85.

Jameson, E. W., Jr. 1947. Natural history of the prairie vole (mammalian genus *Microtus*). *University of Kansas Museum of Natural History Publications* 1:125–51.

Jochum, E., R. Case, and R. A. Lock. 1982. The durable red fox. *Nebraskaland* 60(13):20–25.

Jones, J. K., Jr. 1964. Distribution and taxonomy of mammals of Nebraska. *University of Kansas Publications of the Museum of Natural History* 16:1–356.

Jones, J. K., Jr., D. M. Armstrong, R. S. Hoffmann, and C. Jones. 1983. *Mammals of the Northern Great Plains*. Lincoln: University of Nebraska Press.

Jones, J. K., Jr., D. M. Armstrong, and J. R. Choate. 1985. *Guide to Mammals of the Plains States*. Lincoln: University Nebraska Press.

Jones, J. K., and J. R. Choate. 1980. Annotated checklist of mammals of Nebraska. *Prairie Naturalist* 12:43–53.

Kaufman, D. W., and E. D. Fleharty. 1974. Habitat selection by nine species of rodents in north-central Kansas. *Southwestern Naturalist* 18:443–51.

King, J. A. 1955. Social behavior, social organization, and population dynamics of a black-tailed prairie dog town in the Black Hills of South Dakota. *Contributions from the Laboratory of Vertebrate Biology, University of Michigan* 67:1–123.

———. 1968. Biology of *Peromyscus. American Society of Mammalogists, Special Publication* 2:1–593.

Mahan, C. J. 1980. Winter food habits of Nebraska bobcats (*Felix rufus*). *Prairie Naturalist* 12:59–63.

Meserve, P. L. 1969. Some aspects of the biology of the prairie vole, *Microtus ochrogaster*, in western Nebraska. M.S. thesis, University of Nebraska–Lincoln.

Moulton, M. P., J. R. Choate, S. J. Bissell, and R. A. Nicholson. 1981. Associations of small mammals on the central High Plains of eastern Colorado. *Southwestern Naturalist* 26:53–57.

Murie, J., and C. D. Michener, eds. 1984. *The Biology of Ground-Dwelling Squirrels*. Lincoln: University of Nebraska Press.

Nebraska Game and Parks Commission. 1997. The deer of Nebraska. *Nebraskaland* 75(3):18–35.

Novak, M., J. A. Baker, M. E. Obbard, and B. Malloch, eds. 1987. *Wild Furbearer Management and Conservation in North America*. Toronto: Ontario Ministry of Natural Resources.

O'Neal, G. T., J. T. Flinders, and W. P. Clary. 1987. Behavioral ecology of the Nevada kit fox (*Vulpes macrotis nevadensis*) on a managed desert rangeland. *Current Mammalogy* 1:443–81.

Packard, R. L. 1956. The tree squirrels of Kansas. *University of Kansas Museum of Natural History Miscellaneous Publications* 11:1–67.

Parker, S. P., ed. 1990. *Grzimek's Encyclopedia of Mammals.* 5 vols. New York: McGraw-Hill.

Phillips, G. L. 1966. Ecology of the big brown bat (Chiroptera: Vespertilionidae) in northeastern Kansas. *American Midland Naturalist* 75:168–98.

Quimby, D. C. 1951. The life history of the jumping mouse, *Zapus hudsonicus. Ecological Monographs* 21:61–95.

Roze, U. 1989. *The North American Porcupine.* Washington DC: Smithsonian Institution Press.

Sather, J. H. 1958. Biology of the muskrat in Nebraska. *Wildlife Monographs* 2:1–35.

Schwartz, C. W., and E. R. Schwartz. 1981. *The Wild Mammals of Missouri.* 2nd ed. Columbia and Jefferson City: University Missouri Press and Missouri Department of Conservation.

Scott-Brown, J. M., S. Herrero, and J. Reynolds. 1987. Swift fox. Pp. 423–41 in *Wild Furbearer Management and Conservation in North America*, ed. M. Novak, J. A. Baker, M. E. Obbard, and B. Malloch. Toronto: Ontario Ministry of Natural Resources.

Seal, U., E. T. Thorne, M. Bogan, and S. Anderson, eds. 1989. *Conservation Biology and the Black-footed Ferret.* New Haven CT: Yale University Press.

Shober, W. 1984. *The Lives of Bats.* London: Croom Helm.

Smith, R. E. 1967. Natural history of the prairie dog in Kansas. *University of Kansas Museum of Natural History Miscellaneous Publications* 49:1–39.

Steinauer, G. 1999. Buffalo: The native grazer. *Nebraskaland* 77(6):10–19.

Tamarin, R. H., ed. 1985. Biology of New World *Microtus. American Society of Mammalogists Special Publication* 8:1–893.

Toweill, D. E., and J. E. Tabor. 1982. River otter. Pp. 688–703 in *Wild Mammals of North America*, ed. J. A. Chapman, and C. A. Feldhamer Jr. Baltimore: Johns Hopkins University Press.

Wallmo, O. C., ed. 1981. *Mule and Black-tailed Deer of North America.* Lincoln: University of Nebraska Press.

Wilson, D. E., and S. Ruff. 1999. *The Smithsonian Book of North American Mammals.* Washington DC: Smithsonian Institution Press.

Birds

Andrews, R., and R. Richter. 1992. *Colorado Birds: A Reference to Their Distribution and Habitat.* Denver: Denver Museum of Natural History.

Bicak, T. K. 1977. Some eco-ethological aspects of a breeding population of long-billed curlews (*Numenius americanus*) in Nebraska. M.A. thesis, University of Nebraska–Omaha.

Bock, C. E. 1970. The ecology and behavior of the Lewis' woodpecker. *University of California Publications in Zoology* 92:1–100.

Bomberger, M. F. 1982. Aspects of the breeding biology of Wilson's phalarope in western Nebraska. M.S. thesis, University Nebraska–Lincoln.

Brogie, M. A., and M. J. Mossman. 1983. Spring and summer birds of the Niobrara Valley Preserve area, Nebraska. *Nebraska Bird Review* 51:44–51.

Brown, C. R., and M. B. Brown. 1998. *Swallow Summer*. Lincoln: University of Nebraska Press.

Brown, C. R., M. B. Brown, P. A. Johnsgard, J. Kren, and W. C. Scharf. 1996. Birds of the Cedar Point Biological Station area, Keith and Garden Counties, Nebraska. *Transactions of the Nebraska Academy of Sciences* 29:91–108.

Clawson, S. D. 1980. Comparative ecology of the northern oriole (*Icterus galbula*) and the orchard oriole (*Icterus spurius*) in western Nebraska. M.S. thesis, University of Nebraska–Lincoln.

Cunningham, D. 1998. Greater prairie chicken and sharp-tailed grouse. *Nebraskaland* 76(7):38–41.

Desmond, M., J. A. Savidge, and R. Ekstein. 1997. Prairie partners. *Nebraskaland* 75(9):16–25.

Dickinson, M., ed. 1999. *Field Guide to the Birds of North America*. 3rd ed. Washington DC: National Geographic Society.

Ducey, J. 1988. *Nebraska Birds: Breeding Status and Distribution*. Omaha NE: Simmons-Boardman Books.

———. 1989. Birds of the Niobrara River valley. *Transactions of the Nebraska Academy of Sciences* 17:37–60.

Faanes, C. E., and G. R. Lingle. 1995. Breeding birds of the Platte Valley of Nebraska. Northern Prairie Wildlife Research Center Home Page, Jamestown ND. URL: *http://www.npwrc.usgs.gov/resources/distr/birds/platte/platte(version 16JUL97)*

Farrar, J. 1984. Trumpeters. *Nebraskaland* 62(2):22–29.

———. 1991. Marsh birds. *Nebraskaland* 69:(4):8–21.

———. 1997. Cormorants. *Nebraskaland* 75(4):18–27.

———. 1998. Ring-necked pheasant. *Nebraskaland* 76(7):26–33.

Forsberg, M. 1996. The enigmatic magpie. *Nebraskaland* 74(3):8–17.

———. 1998a. Bobwhite quail. *Nebraskaland* 76(7):34–37.

———. 1998b. The solitary pied-billed grebe. *Nebraskaland* 76(6):17–23.

Grier, R. 1998. Wild turkey. *Nebraskaland* 76(7):72–77.

Gubanyi, J. 1995. Silent masters of the night. *Nebraskaland* 73(8):16–21.

Haig, S. 1992. Piping plover (*Charadrius melodus*). In *The Birds of North*

America, No. 2, ed. A. Poole and F. Gill. Philadelphia: The Birds of North America.

Johnsgard, P. A. 1974. Waterfowl of Nebraska. *Museum Notes* (University of Nebraska State Museum) 50.

———. 1979. *Birds of the Great Plains: Breeding Species and Their Distribution.* Lincoln: University of Nebraska Press.

———. 1991. *Crane Music: A Natural History of American Cranes.* Washington DC: Smithsonian Institution Press.

———. 1996. The cranes of Nebraska. *Museum Notes* (University of Nebraska State Museum) 93.

———. 1998. Endemicity and regional biodiversity in Nebraska's breeding birds. *Nebraska Bird Review* 66:115–21.

———. 1999a. Buzz-wings: The hummingbirds of Nebraska. *Museum Notes* (University of Nebraska State Museum) 103.

———. 1999b. The ultraviolet birds of Nebraska. *Nebraska Bird Review* 67:103–5.

———. 2000. *The Birds of Nebraska.* Lincoln NE: Published by author.

Knopf, F. L. 1996a. Mountain plover (*Charadrius montanus*). In *The Birds of North America*, No. 211, ed. A. Poole and F. Gill. Philadelphia: The Birds of North America.

———. 1996b. Prairie legacies—birds. Pp. 13–48 in *Prairie Conservation: Conserving North America's Most Endangered Ecosystem*, ed. F. B. Samson and F. L. Knopf. Covelo CA: Island Press.

Labedz, T. 1989. Birds. Pp. 161–80, in *An Atlas of the Sand Hills*, ed. A. Bleed and C. Flowerday. Resource Atlas No. 5. Conservation and Survey Division, University of Nebraska–Lincoln.

Mollhoff, W. J. 2001. *The Nebraska Breeding Bird Atlas.* Lincoln: Nebraska Game and Parks Commission.

Nebraska Game and Parks Commission. 1984. Prairie grouse of Nebraska. *Nebraskaland* 62(9):18–35.

———. 1985. *Birds of Nebraska.* Special issue of *Nebraskaland* 63(1): 1–146.

———. 1990. The bobwhite quail in Nebraska. *Nebraskaland* 69(8): 18–35.

———. 1996. The wild turkey in Nebraska. *Nebraskaland* 74(8):18–35.

Pettingill, O. S., Jr. 1981. *A Guide to Bird-finding West of the Mississippi.* 2nd ed. New York: Oxford University Press.

Rosche, R. C. 1982. *Birds of Northwestern Nebraska and Southwestern South Dakota.* Chadron NE: Published by author.

Sharpe, R. W., W. R. Silcock, and J. G. Jorgensen. 2001. *Birds of Nebraska: Their Distribution and Temporal Occurrence.* Lincoln: University of Nebraska Press.

Short, L. L., Jr. 1965. Hybridization in the flickers (*Colaptes*) of North

America. *Bulletin of the American Museum of Natural History* 129:311–428.

Sibley, C. G., and L. L. Short Jr. 1964. Hybridization in the orioles of the Great Plains. *Condor* 66:130–50.

Sibley, C. G., and D. A. West. 1959. Hybridization in the rufous-sided towhees of the Great Plains. *Auk* 76:326–38.

Sisson, L. 1976. *The Sharp-tailed Grouse in Nebraska*. Lincoln: Nebraska Game and Parks Commission.

Thompson, B. C., J. A. Jackson, J. Burger, L. A. Hill, E. M. Kirsh, and J. L. Atwood. 1997. Least Tern (*Sterna albifrons*). In *The Birds of North America*, No. 290, ed. A. Poole and F. Gill. Philadelphia: The Birds of North America.

Vickery, P. D., and J. R. Herkert, eds. 1999. *Ecology and Conservation of Grassland Birds of the Western Hemisphere*. Cooper Ornithological Society, Studies in Avian Biology No. 19.

Wingfield, G. 1982. Tough times for terns. *Nebraskaland* 60(5):37–42.

Amphibians and Reptiles

Ballinger, R. E., J. E. Lynch, and P. H. Cole. 1979. Distribution and natural history of amphibians and reptiles in western Nebraska, with ecological notes on the herptiles of Arapaho Prairie. *Prairie Naturalist* 22:65–74.

Baxter, G. T., and M. D. Stone. 1980. *Amphibians and Reptiles of Wyoming*. Bulletin No. 16. Cheyenne: Wyoming Game and Fish Department.

Benedict, R. 1996. Snappers, soft-shells, and stinkpots: The turtles of Nebraska. *Museum Notes* (University of Nebraska State Museum) 96 (4 pp. and color poster).

Collins, J. T. 1993. *Amphibians and Reptiles in Kansas*. University of Kansas Museum of Natural History Public Education Series No. 13.

Conant, R. 1998. *A Field Guide to the Reptiles and Amphibians of Eastern and Central North America*. 3rd ed., expanded. Boston: Houghton Mifflin.

Droge, D. L. 1980. Seasonal patterns of aggression in the lizards *Sceloperus undulatus* and *Holbrookia maculata*. M.S. thesis, University of Nebraska–Lincoln.

Ernst, C. H., J. E. Lovich, and R. W. Barbour. 1994. *Turtles of the United States and Canada*. Washington DC: Smithsonian Institution Press.

Farrar, J. 1998. Box turtles: Life in the fast lane. *Nebraskaland* 76(5):24–33.

Freeman, P. 1989. Amphibians and reptiles. Pp. 157–160. in *An Atlas of the Sand Hills*, ed. A. Bleed and C. Flowerday. Resource Atlas No. 5. Conservation and Survey Division, University of Nebraska–Lincoln.

Hammerson, G. A. 1982. *Amphibians and Reptiles in Colorado*. Denver: Colorado Division of Fish and Wildlife.

Holycross, A. 1995. Serpents of the Sandhills. *Nebraskaland* 73(6):28–35.

Hudson, G. E. 1958. *The Amphibians and Reptiles of Nebraska*. Bulletin 22. Conservation and Survey Division, University of Nebraska–Lincoln.

Iverson, J. B., and G. R. Smith. 1993. Reproductive ecology of the painted turtle (*Chrysemys picta*) in the Nebraska Sandhills. *Copeia* 1993:1–21.

Jones, S. M., and R. E. Ballinger. 1987. Comparative life histories of *Holbrookia maculata* and *Sceloperus undulatus* in western Nebraska. *Ecology* 68:1828–36.

Lynch, J. D. 1985. Annotated checklist of the amphibians and reptiles of Nebraska. *Transactions of the Nebraska Academy of Sciences* 13:33–57.

Native Fishes

Cross, F. B. 1975. Handbook of fishes of Kansas. *University of Kansas Museum of Natural History Publications* 45:1–357.

Cross, F. B., and J. T. Collins. 1975. *Fishes in Kansas*. University of Kansas Museum of Natural History Public Education Series No. 3.

Hubbs, C. L., and G. P. Cooper. 1936. Minnows of Michigan. *Cranbrook Institute of Science Bulletin* 8:1–84.

Johnson, R. E. 1942. The distribution of Nebraska fishes. Ph.D. diss., University of Michigan.

Jones, D. 1963. *A History of Nebraska's Fishery Resources*. Lincoln: Nebraska Game and Parks Commission.

Lynch, J. D., and B. R. Roh. 1996. An ichthyological survey of the forks of the Platte River in western Nebraska. *Transactions of the Nebraska Academy of Sciences* 23:65–84.

Madsen, T. I. 1985. The status and distribution of the uncommon fishes of Nebraska. M.S. thesis, University of Nebraska, Omaha. 97 pp.

Morris, J., L. Morris, and L. Witt. 1972. *The Fishes of Nebraska*. Lincoln: Nebraska Game and Parks Commission.

Nebraska Game and Parks Commission. 1987. *The Fish Book*. Special issue of *Nebraskaland* 65(1).1–132.

Priegel, G. R., and T. L. Wirt. 1977. The lake sturgeon: Its life history, ecology and management. *Wisconsin Department of Natural Resources Publication* 4-3600(77):1–20.

Stasiak, R. H. 1972. The morphology and life history of the finescale dace, *Pfrille neogaea*, in Itasca State Park, Minnesota. Ph.D. diss., University of Minnesota.

Tomelleri, J. R., and M. E. Eberle. 1990. *Fishes of the Central United States*. Lawrence: University Press of Kansas.

Wallace, C. R. 1969. Acoustic, agonistic and reproductive behavior of three species of bullheads. Ph.D. diss., University of Nebraska–Lincoln.

Zuerlein, G. 1998. Bringing back the pallid sturgeon. *Nebraskaland* 76(2):24–27.

Insects

Borrer, D. J., and R. E. White. 1970. *A Field Guide to the Insects.* Boston: Houghton Mifflin.

Milne, L., and M. Milne. 1981. *National Audubon Society Field Guide to North American Insects and Spiders.* New York: A. Knopf.

Beetles

Carter, M. R. 1989. The biology and ecology of the tiger beetles (Coleoptera Cicindelidae) of Nebraska. *Transactions of the Nebraska Academy of Sciences* 17:1–18.

Hoffman, R. 1997. In search of the American burying beetle. *Nebraskaland* 76(3):10–21.

Leonard, J. G., and R. T. Bell. 1999. *Northeastern Tiger Beetles: A Field Guide to Tiger Beetles of New England and Eastern Canada.* Boca Raton FL: St. Lucie Press.

Lomolino, M. V., J. C. Creighton, G. D. Schnell, and D. L. Certain. 1995. Ecology and conservation of the endangered American burying beetle. *Conservation Biology* 9:605–14.

Ratcliffe, B. C. 1991a. Scarab beetles. *Nebraskaland* 66(4):30–36.

———. 1991b. *The Scarab Beetles of Nebraska.* Bulletin of the University of Nebraska State Museum 12.

———. 1996. *The Carrion Beetles (Coleoptera: Silphidae) of Nebraska.* Bulletin of the University of Nebraska State Museum 13.

Scott, M. P. 1989. Brood guarding and the evolution of male parental care in burying beetles. *Behavioral Ecology and Sociobiology* 26:31–39.

Scott, M. P., and J. F. A. Traniello. 1989. Guardians of the underworld. *Natural History* (June 1989):32–37.

Spomer, S. M., and L. G. Higley. 1993. Population status and distribution of the Salt Creek tiger beetle, *Cicindela nevadica lincolniana* Casey (Coleoptera: Cicindelidae). *Journal of the Kansas Entomological Society* 66(4):392–98.

Grasshoppers, Mantids, and Walking-Sticks

Alexander, G. 1941. Keys for the identification of Colorado Orthoptera. *University of Colorado Studies* (Boulder), Ser. D 1(3):129–64.

Capinera, J. L., and T. S. Sechrist. 1981. *Grasshoppers (Acrididae) of Colorado: Identification, Biology and Management.* Colorado State University Experiment Station Bulletin 5845.

Hagen, A. F. 1970. *An Annotated List of Grasshoppers (Orthoptera: Acrididae) from the Eleven Panhandle Counties of Nebraska.* University of Nebraska Agricultural Experiment Station Research Bulletin 238.

Hagen, A. F., and G. C. Rabe. 1991. *Distribution Maps of Grasshopper Species in Nebraska, Based on Three Studies.* University of Nebraska, Agricultural Experiment Station Report 16(1).

Hauke, H. A. 1949. An annotated list of the Orthoptera of Nebraska. Pt. 1. The Blattidae, Mantidae and Phasmidae. *Bulletin of the University of Nebraska State Museum* 3(5):63–75.

———. 1953. An annotated list of the Orthoptera of Nebraska. Pt. 2. The Tettigidae and Acrididae. *Bulletin of the University of Nebraska State Museum* 3(9):1–79.

Joern, A. 1982. Distributions, densities, and relative abundances of grasshoppers (Orthoptera, Acrididae) in a Nebraska Sandhills prairie. *Prairie Naturalist* 14:37–45.

Knutson, S., G. F. Smith, and M. H. Blust. 1983. *Grasshoppers, Identifying Species of Economic Importance.* Kansas State University Cooperative Extension Series S-21.

McDaniel, B. 1987. *Grasshoppers of South Dakota.* South Dakota State University Agricultural Experiment Station Bulletin TB 89.

Newton, R. C., and A. B. Gurney. 1956. *Distribution Maps of Range Grasshoppers in the United States.* U.S. Department of Agriculture Cooperative Economic Insect Report 6(43).

Otte, D. 1981. 1984. *The North American Grasshoppers.* 2 vols. Cambridge: Harvard University Press.

Pfadt, R. E. 1986. *Keys to the Wyoming Grasshoppers: Acrididae and Tettigidae.* University of Wyoming Agricultural Experiment Station Circular 210.

———. 1994. *Field Guide to Common Western Grasshoppers.* 2nd ed. Wyoming Agricultural Experiment Station Bulletin 912.

Shotwell, R. L. 1941. *Life Histories and Habits of Some Grasshoppers of Economic Importance on the Great Plains.* U.S. Department of Agriculture Technical Bulletin 774.

Butterflies

Dankert, N., H. Nagel, and T. Nightengale. 1993. *Butterfly Distribution Maps—Nebraska.* Kearney: Department of Biology, University of Nebraska–Kearney.

Ely, C., M. D. Schwilling, and M. E. Rolfs. 1986. *An Annotated List of *the Butterflies of Kansas*. Fort Hays Studies, 3rd Series (Science), No. 7. Fort Hays KS.

Farrar, J. 1993. The making of a monarch. *Nebraskaland* 71(6):34–41.

Ferris, C. D., and F. M. Brown. 1980. *Butterflies of the Rocky Mountain States*. Norman: University of Oklahoma Press.

Heitzman, J. R., and J. E. Heitzman. 1987. *Butterflies and Moths of Missouri*. Rev. ed. Jefferson City: Missouri Department of Conservation.

Johnson, K. 1973. The butterflies of Nebraska. *Journal of Research on the Lepidoptera* 11(1):1–64.

Opler, P. A. 1992. *A Field Guide to Eastern Butterflies*. Boston: Houghton Mifflin.

Pyle, R. M. 1981. *The Audubon Society Field Guide to North American Butterflies*. New York: A. A. Knopf.

Scott, J. 1986. *Butterflies of North America*. Stanford CA: Stanford University Press.

Tilden, J. W., and A. C. Smith. 1986. *A Field Guide to Western Butterflies*. Boston: Houghton Mifflin.

Moths

Cody, J. 1996. *Wings of Paradise: The Giant Saturniid Moths*. Chapel Hill: University of North Carolina Press.

Covell, C. V. 1984. *A Field Guide to the Moths of Eastern North America*. Boston: Houghton Mifflin.

D'Abrera, B. 1986. *Sphingidae Mundi: Hawk Moths of the World*. Faringdon UK: E. W. Classey.

Holland, W. H. 1934. *The Moth Book*. Garden City NY: Doubleday Doran.

Jordison, J. 1996. Jewels of the night. *Nebraskaland* 74(4):8–19.

Messenger, C. 1997. The sphinx moths of Nebraska. *Transactions of the Nebraska Academy of Sciences* 24:89–141.

Ratcliffe, B. 1993. Nebraska's giant silkmoths. *Nebraskaland* 71(3):22–29.

Selman, C. L. 1975. *A Pictorial Key to the Hawkmoths (Lepidoptera: Sphingidae) of Eastern North America (except Florida)*. Columbus: Ohio Biological Survey, Biological Notes No. 9.

Tuskes, P. M., J. P. Tuttle, and M. M. Collins. 1986. *The Wild Silk Moths of North America*. Ithaca NY: Cornell University Press.

Dragonflies and Damselflies

Beckemeyer, R. J., and D. Huggans. 1997. Checklist of Kansas dragonflies. *Kansas School Naturalist* 43(2):1–16. (See also checklist of Nebraska Odonata at Roy Beckemeyer's dragonfly website: *http://www.south-wind.net/˜roy6/odonata.html*)

Corbett, P. S. 1963. *A Biology of Dragonflies*. Chicago: Quadrangle Books.

———. 1999. *Dragonflies: Behavior and Biology of Odonata*. Ithaca NY: Cornell University Press.

Dunkle, S. W. 2000. *Dragonflies through Binoculars*. New York: Oxford University Press.

Keech, C. F. 1934. The Odonata of Nebraska. M.S. thesis, University of Nebraska–Lincoln.

Molnar, D. R., and R. J. Lavigne. 1994. *The Odonata of Wyoming (Dragonflies and Damselflies)*. University of Wyoming Agricultural Experiment Station Science Monograph 37R.

Needham, J. G., and H. B. Heyward. 1929. *A Handbook of the Dragonflies of North America*. Springfield IL: Charles C. Thomas.

Needham, J. G., and M. J. Westfall. 1955. *A Manual of the Dragonflies of North America (Anisoptera): Including the Greater Antilles and the Provinces of the Mexican Border*. Berkeley: University California Press.

Walker, E. M., and P. S. Corbet. 1953–73. *The Odonata of Canada and Alaska*. 3 vols. Toronto: University Toronto Press.

Westfall, M. J., Jr., and M. L. May. 1996. *Damselflies of North America*. Gainesville FL: Scientific Publishers.

Mussels

Freeman, P., and K. Perkins. 1994. Water nymphs of the Platte. *Nebraskaland* 72(3):15–25.

Hoke, E. 1994. A survey and analysis of the unionid mollusks of the Elkhorn River basin, Nebraska. *Transactions of the Nebraska Academy of Sciences* 21:31–50.

———. 1995. A survey and analysis of the unionid mollusks of the Platte rivers of Nebraska and their minor tributaries. *Transactions of the Nebraska Academy of Sciences* 22:49–72.

———. 1996. The unionid mollusks of the Big and Little Nemaha River basin of southeastern Nebraska and northeastern Kansas. *Transactions of the Nebraska Academy of Sciences* 23:37–57.

Oesch, R. D. 1984. *Missouri Naiades: A Guide to the Mussels of Missouri*. Jefferson City: Missouri Department of Conservation.

Other Invertebrates

Forsberg, M. 1997. Cicadas: The ushers of summer. *Nebraskaland* 75(6):16–25.

Golick, D., and M. Ellis. 2000. Bumble boosters. University of Nebraska Cooperative Extension EC 00-1564-S. (picture key to 19 species of Nebraska bumblebees)

Gordon, C. C., L. D. Flake, and K. F. Higgins. 1990. Aquatic invertebrates in the Rainwater Basin area of Nebraska. *Prairie Naturalist* 22:191–200.

Pflieger, W. L. 1996. *The Crayfishes of Missouri.* Jefferson City: Missouri Department of Conservation.

Regional Floras and General Guides

Barkley, T. M., ed. 1977. *Atlas of the Flora of the Great Plains.* Ames: Iowa State University.

———. 1986. *Flora of the Great Plains.* Lawrence: University Press of Kansas.

Johnson, J. R., and G. E. Larson. 1999. *Grassland Plants of South Dakota and the Northern Great Plains.* Brookings: South Dakota State University.

Larson, G. E., and J. R. Johnson. 1999. *Plants of the Black Hills and Bear Lodge Mountains.* Brookings: South Dakota State University.

Trees

Brockman, C. F. 1986. *Trees of North America.* 2nd ed. New York: Golden Press.

Elias, T. S. 1986. *Field Guide to North American Trees.* Danbury CT: Grolier Book Clubs.

Kuhns, M., and D. P. Mooter. 1992. *Trees of Nebraska.* Circular EC 92-1774X. Nebraska Cooperative Extension Service, University of Nebraska–Lincoln.

Nebraska Game and Parks Commission. 1994. *Walk in the Woods.* Special issue of *Nebraskaland* 71(1).

Peattie, D. C. 1964. *A Natural History of Trees of Eastern and Central North America.* 2nd ed. Boston: Houghton Mifflin.

———. 1980. *A Natural History of Western Trees.* Lincoln: University of Nebraska Press.

Petrides, G. A. 1958. *A Field Guide to Trees and Shrubs.* Boston: Houghton Mifflin.

———. 1988. *A Field Guide to Eastern Trees.* Boston: Houghton Mifflin.

Pool, R. 1961. *Handbook of Nebraska Trees.* Rev. ed. Bulletin No. 32. Conservation and Survey Division, University of Nebraska–Lincoln.

Preston, R. J. 1976. *North American Trees.* Ames: Iowa State University Press.

Readers Digest. 1998. *North American Wildlife: Trees and Non-flowering Plants.* Pleasantville NY: Readers Digest.

Richardson, D. M., ed. 1998. *Ecology and Biogeography of Pinus.* Cambridge: Cambridge University Press.

Schmidt, T., and T. D. Wardle. 1998. *The Forest Resources of Nebraska*. Publication NC-332. St. Paul MN: U.S. Department of Agriculture Forest Service, North Central Research Station.

Stephens, H. A. 1969. *Trees, Shrubs and Woody Vines of Kansas*. Lawrence: University Press of Kansas.

Shrubs and Woody Vines

Petrides, G. A. 1958. *A Field Guide to Trees and Shrubs*. Boston: Houghton Mifflin.

Stephens, H. A. 1969. *Trees, Shrubs and Woody Vines of Kansas*. Lawrence: University Press of Kansas.

————. 1973. *Woody Plants of the North-Central Plains*. Lawrence: University Press of Kansas.

Stubbendieck, J., J. T. Nichols, and C. H. Butterfield. 1989. *Nebraska Range and Pasture Forbs and Shrubs*. Circular 89-118. Nebraska Cooperative Extension Service, University of Nebraska–Lincoln.

Forbs

Bare, J. 1979. *Wildflowers and Weeds of Kansas*. Lawrence: Regents (University) Press of Kansas.

Barkley, T. M. 1983. *Field Guide to the Common Weeds of Kansas*. Lawrence: Regents (University) Press of Kansas.

Blackwell, W. H. 1990. *Poisonous and Medicinal Plants*. Englewood Cliffs NJ: Prentice Hall.

Cingel, N. A. van der. 1995. *An Atlas of Orchid Pollination: European Orchids*. Rotterdam: A. A. Balkema.

Denison, E. 1983. *Missouri Wildflowers*. Jefferson City: Missouri Department of Conservation.

Dressler, R. L. 1981. *The Orchids: Natural History and Classification*. Cambridge: Harvard University Press.

Emboden. W. 1979. *Narcotic Plants*. New York: Macmillan.

Farrar, J. 1990. *Field Guide to Wildflowers of Nebraska and the Great Plains*. Lincoln: Nebraska Game and Parks Commission.

Freeman, C. C., and E. K. Shofield. 1991. *Roadside Wildflowers of the Southern Great Plains*. Lawrence: University Press of Kansas.

Ladd, D. 1995. *Tallgrass Prairie Wildflowers*. Helena MT: Falcon Press.

Lommasson, R. 1973. *Nebraska Wild Flowers*. Lincoln: University of Nebraska Press.

Luer, C. A. 1975. *The Native Orchids of the United States and Canada*. New York: New York Botanical Garden.

Nebraska Game and Parks Commission. 1990. *A Wildflower Year*. Special issue of *Nebraskaland* 68(1).

Nilsson, L. A. 1992. Orchid pollination biology. *Trends in Ecology and Evolution* 7–8:255–59.

Owensby, C. E. 1980. *Kansas Prairie Wildflowers*. Ames: Iowa State University Press.

Pool, R. J., and E. G. Maxwell. 1952. *Wild Flowers in Nebraska*. Circular 5-168. Nebraska Cooperative Extension Service, University of Nebraska–Lincoln.

Runkel, S. T., and D. M. Roosa. 1989. *Wildflowers of the Tallgrass Prairie*. Ames: Iowa State University Press.

Stubbendieck, J., G. Y. Frissoe, and M. R. Bolick. 1995. *Weeds of Nebraska and the Great Plains*. 2nd ed. Lincoln: Nebraska Department of Agriculture.

Stubbendieck, J., J. T. Nichols, and C. H. Butterfield. 1989. *Nebraska Range and Pasture Forbs and Shrubs*. Circular 89-118. Nebraska Cooperative Extension Service, University of Nebraska–Lincoln.

Stubbendieck, J., S. L. Hatch, and C. H. Butterfield. 1997. *North American Range Plants*. 5th ed. Lincoln: University of Nebraska Press.

Vance, F. R., J. R. Jowsey, and J. S. McLean. 1984. *Wildflowers of the Northern Great Plains*. Minneapolis: University of Minnesota Press.

van der Pijl, L., and C. H. Dodson. 1966. *Orchid Flowers: Their Pollination and Evolution*. Coral Gables FL: University of Miami Press.

Grasses and Sedges

Brown, L. 1979. *Grasses: An Identification Guide*. Boston: Houghton Mifflin.

Frolik, A. L., and F. D. Keim. 1942. *Common Native Grasses of Nebraska*. Circular 59. Nebraska Cooperative Extension Service, University of Nebraska–Lincoln.

Hallsten, G. P., Q. D. Skinner, and A. A. Beetle. 1987. *Grasses in Wyoming*. 3rd ed. Research Journal 202. University of Wyoming, Laramie.

Hitchcock, A. S. 1935. *Manual of the Grasses of the United States*. U.S. Department of Agriculture Publication No. 200. Reprint. 1971, New York: Dover.

Jensen, P. N. 1965. *A Key to Native and Introduced Grasses in Nebraska*. Lincoln NE: U.S. Department of Agriculture, Soil Conservation Service.

Knobel, E. 1980. *Field Guide to the Grasses, Sedges and Rushes of the United States*. New York: Dover.

Stubbendieck, J., J. T. Nichols, and K. K. Roberts. 1985. *Nebraska Range and Pasture Grasses*. Circular 85-170. Nebraska Cooperative Extension Service, University of Nebraska–Lincoln.

Sutherland, D. M. 1975. A vegetative key to Nebraska grasses. Pp. 283–

316 in *Prairie, a Multiple View*, ed. M. K. Wali. Grand Forks: University of North Dakota Press.

———. 1984. Vegetative key to grasses in the Sand Hills region of Nebraska. *Transactions of the Nebraska Academy of Sciences* 12:23–60.

U.S. Department of Agriculture. 1948. *Grass: The Yearbook of Agriculture*, 1948. Washington DC: U.S. Government Printing Office.

Shoreline and Aquatic Plants

Larson, G. 1993. *Aquatic and Wetland Vascular Plants of the Northern Great Plains*. Fort Collins: Rocky Mountain Forest and Range Experiment Station.

Lindstrom, L. E. 1968. The aquatic and marsh plants of the Great Plains of central North America. Ph.D. diss., Kansas State University.

Muenscher, W. C. 1944. *Aquatic Plants of the United States*. Ithaca NY: Comstock Publishing.

Oberholser, H. C., and W. L. McAtee. 1920. Waterfowl and their food plants in the Sandhill region of Nebraska. *U.S. Department of Agriculture Bulletin* 794:1–79.

Whitley, J. R., B. Bassett, J. D. Dillar, and R. A. Haefer. 1990. *Water Plants for Missouri Ponds*. Jefferson City: Missouri Department of Conservation.

A Guide to Nebraska's Natural Areas

Boyle, W. J., and R. H. Bauer. 1994. Birdfinding in forty National Forests and Grasslands. *Birding* (supplement) 26(2):1–186.

Ducey, J. 1989. Birds of the Niobrara River valley, Nebraska. *Transactions of the Nebraska Academy of Sciences* 17:37–60.

Johnsgard, P. A. 1984. *The Platte: Channels in Time*. Lincoln: University of Nebraska Press.

Jones, J. O. 1990. *Where the Birds Are: A Guide to All 50 States and Canada*. New York: William Morrow.

Knue, J. 1997. *Nebraskaland Magazine Wildlife Viewing Guide*. Special issue of *Nebraskaland* 75(1).

Lingle, G. R. l994. *Birding Crane River: Nebraska's Platte*. Grand Island NE: Harrier Publishing.

Nebraska Game and Parks Commission. 1972. *The Nebraska Fish and Wildlife Plan*. Vol. 1. *Nebraska Wildlife Resources Inventory*. Lincoln: Nebraska Game and Parks Commission.

———. 1982. *Discover Nebraska Travel Guide*. Special issue of *Nebraskaland* 60(2).

———. 1996. Nebraska parklands. *Nebraskaland* 74(3):18–35.

Pettingill, O. S., Jr. 1981. *A Guide to Bird-finding West of the Mississippi.*
2nd ed. New York: Oxford University Press.
Rosche, R. C. 1990. Birding pristine Nebraska. *Winging It* 2(6):1–6.
———. 1994. Birding in western Nebraska. *Birding* 26:179–89, 416–23.

General Index

This index provides references to all species selected for profiles and vignettes, to illustrations of these and other plants and animals (in *italics*), Nebraska landmarks and other important locations, natural community types or ecological regions, geological time periods or strata, cited authorities, and other selected topics. The appendixes are not indexed, nor are those species that are only briefly mentioned in the main text.

McKelvie Division, Nebraska National
    Forest, 77
meadowlark. *See* western and eastern
    meadowlarks
Merriam, C. H., 16
Mesozoic geological era, 4, 5, 6, 11
Middle Loup River, 164
Miocene geological epoch, 4, 11, 27, 115
Missouri River (and valley), 9, 10, 12, 13,
    20, 23, 40, 67, 93, 95, 110, 123, 124,
    125, 142–55, 163, 165, 169, 174, 183
Missouri Valley Riparian Forest (region),
    145
Mixed Coniferous Woodland (community),
    21, 99
Mixed-grass Prairie (community), 8, 10, 16,
    21, 61–66
Mixed-grass Prairies (sub-ecoregion), 17
Mormon Trail, 23
mountain lion, 39–40
Mountain Mahogany Shrubland
    (community), 21, 32
mountain plover, 57–58, 57
mule deer, 30, 39, 186
Murphy, Robert Cushman, 188
muskrat, 167–68

Neale Woods, 142, 191
Nebraska National Forest, 77
Nebraska Sandhills and Border Prairies
    (sub-ecoregion), 17, 78
Nebraska Sandhills (region), 77–92
Nemaha River, 165
Newton, Isaac, 13
Nine-Mile Prairie, 128, 136, 191
Niobrara geological formation, 63
Niobrara River (and valley), 9, 11, 12–13,
    18, 20, 31, 32, 33, 80, 93–109, 123,
    125, 144, 161–62, 164, 171, 173
Niobrara Valley Preserve, 94, 97, 145, 189
Niobrara Valley Riparian Forest (region),
    94
Northeastern Upland Forest (community),
    21, 98, 144
northern flicker, 120–21
northern oriole. *See* Baltimore and Bullock's
    orioles
northern redbelly dace, 170–71, 172
Northern Springbranch Canyon Forest
    (community), 21, 98
North Loup River, 164
North Platte River (and valley), 9, 28, 58,
    106, 115, 164

Northwestern Canyonbottom Deciduous
    Woodland (community), 21, 31
Northwestern Glaciated Plains (ecoregion),
    17, 18, 62
Northwestern Great Plains (ecoregion), 17
Northwestern Mixed-grass Prairie
    (community), 21, 48

Oak Woodland (community), 21
Ogallala aquifer, 10, 12, 79, 80
Ogallala Formation, 11, 115
Oglala National Grassland, 44, 47
Oligocene geological epoch, 4, 11, 115, 191
Omernik, J. M., 17, 94, 111, 143
opossum, 148–49
Ordovician geological period, 4
Ord's kangaroo rat, 30, 80, 88–90, 88, 95
ornate box turtle, 30, 80, 88–90, 89, 95

Paleozoic geological era, 3, 4, 5, 11
pallid sturgeon, 169–70, 170
Panorama Point, 9
paper birch, 98, 99–100
Pawnee National Grassland (Colorado), 60
Pawnee Prairie (Wildlife Management
    Area), 136
Peorian loess, 12
peregrine falcon, 36, 125, 135, 183–85, 184
Permian geological period, 4, 5
Pierre Hills, 44
Pierre shale, 95
Pine Ridge (region), 9, 12, 24, 27–42
Pine Ridge Sandy-slope Prairie
    (community), 21, 33
Pioneers Park Nature Center, 191
piping plover, 111, 124–25, 124
Platte River (and valley), 7, 9, 12, 13, 15,
    18, 20, 23, 33, 79, 80, 93, 97, 106,
    110–26, 144, 161, 165, 174, 183, 186,
    189
Platte River Valley Riparian Forest (region),
    111
Pleistocene geological epoch, 4, 11, 12, 148
Pliocene geological epoch, 4, 11
Ponca State Park, 65, 142, 145
ponderosa pine, 20, 24, 28, 34–36, 99
Ponderosa Pine Escarpments (sub-
    ecoregion), 17, 29
prairie-chicken. *See* greater prairie-chicken
prairie dog. *See* black-tailed prairie dog
prairie falcon, 30, 34, 36–37, 37, 58, 59, 97
prairie wolf. *See* gray wolf
pronghorn, 8, 45, 49, 51–53, 52, 82

County Index

This index includes county names mentioned in the text proper as well as the individual county descriptions provided in Appendix 2 (shown in italics). The locations of all Nebraska counties are mapped on pages 18–19 and 164–65.

Adams, *309–10*
Antelope, *300–301*
Arthur, *287–88*

Banner, 9, *278*
Blaine, *287*
Boone, *302*
Box Butte, 58, 92, *275–76*
Boyd, 13, *299*
Brown, 31, 65, 99, 100, 171, *284*
Buffalo, 143, *304–5*
Burt, 144, 163, *317*
Butler, *319*

Cass, *325–26*
Cedar, *314–15*
Chase, 82, 83, *292–93*
Cherry, 31, 39, 75, 90, 92, 98, 99, 100, 139, 163, 171, 173, *282–84*
Cheyenne, *280*
Clay, *310–11*
Colfax, *317*
Cuming, 143, *317*
Custer, 171, *288*

Dakota, 144, *315–16*
Dawes, 30, 44, 46, 47, 173, *274–75*
Dawson, 75, *292*
Deuel, 30, *281*
Dixon, 145, 146, *314*, *315*
Dodge, *318*
Douglas, 142, 143, *319–20*
Dundy, 82, *295*

Filmore, 136, *311–12*
Franklin, *312*
Frontier, *293–94*
Furnas, 39, *296*

Gage, 136, *327–28*
Garden, 92, *280–81*
Garfield, *301*
Gosper, 39, *294*
Grant, *286*
Greeley, *302*

Hall, 139, *306–7*
Hamilton, *307–8*
Harlan, *296*
Hayes, 82, *293*
Hitchcock, *295*
Holt, *299*
Hooker, *286*
Howard, 141, *304*

Jefferson, 13, 163, *326–27*
Johnson, 142, *328*

Kearney, *309*
Keith, 33, 65, 79, 171, *299–300*
Keya Paha, 31, 171, 173, *284*
Kimball, 9, 31, 46, 57, 58, 177, *278*
Knox, 146, *299–300*

Lancaster, 136, 139, 164, 175, 177, 178, *322–25*
Lincoln, 75, 82, 171, *290–93*
Logan, 18, *288*
Loup, *287*

Madison, *302–3*
McPherson, *288*
Merrick, 188, *307*
Morrill, 92, *279*

Nance, *304*
Nemaha, 136, *328*